CW01376252

PlanePlotter

A User Guide
for the Aviation Enthusiast

by Lionel K Anderson MSc

PlanePlotter
A User Guide
for the Aviation Enthusiast

First Edition 2010 by Las Atalayas Publishing

Author and Editor Lionel K Anderson MSc
Copyright © Lionel K Anderson 2010

The moral rights of the author have been asserted. All rights reserved. No part of this publication may be reproduced, stored in a retrieval system or transmitted in any form or by any means, electronic, mechanical or otherwise without the written permission of the Publisher.

ISBN 978-1-4461-3063-6

Design and Typesetting by kenandglen.com

Contents

1 Introduction	1
2 6.67μsecs	5
3 Radios and Virtual Radars	19
4 Software Installation	23
5 First Time Use & Setup Wizard	31
6 PlanePlotter Menus	41
7 PlanePlotter Toolbar	73
8 Charts, Maps & Overlays	83
9 Sharing	107
10 Logs & Databases	111
11 Multilateration	119
12 Google Earth, Google Maps & Other Add-Ons	133
13 Understanding Alerts	141
14 Tips and Tricks	147
15 User Support	153
Appendix A - Radar Frequency Bands	155
Appendix B - NTP Timekeeping Utility	157
Appendix C - PlaneGadgets Radar	163
Appendix D - ACARS Messages	168
Appendix E - ICAO/IATA Airport codes	173
Bibliograpy	207
Index	208

Glossary

ACARS	Aircraft Communications Addressing and Reporting System
ACAS	Airborne Collision Avoidance System
ADS-B	Automatic Dependent Surveillance - Broadcast
ADS-C	Automatic Dependent Surveillance - Contract (also known as ADS or ADS-A)
AIS	Aeronautical Information Services
AIS-SART	AIS- Search and Rescue Transponder
APW	Area Proximity Warning
ASAS	Airborne Separation Assistance System
ASDE	Airport Surface Detection Equipment
ASM	Airspace Management
ATC	Air Traffic Control
ATFM	Air Traffic Flow Management
ATM	Air Traffic Management
ATCRBS	Air Traffic Control Radar Beacon System
AVI	Audio Video Interleave
BMP	Bitmap graphic format file extension
BSB	BSB a proprietary chart format
CAP	Controller Access Parameters
CDTI	Cockpit Display of Traffic Information
CLB	Calibration file extension
CNS	Communications, Navigation And Surveillance
COAA	Centro de Observação Astronómica no Algarve
CPDLC	Controller-Pilot Data Link Communications
CSTDMA	Carrier Sense Time Division Multiple Access
DDE	Dynamic Data Exchange
DERA	Defence Evaluation & Research Agency
DF	Direction Finding
DGPS	Differential Global Positioning System
DLIC	Data Link Initiation Capability
DoS	Denial of Service
DPSK	Differential Phase-Shift Keying
DSC	Digital Selective Calling
ELT	Emergency Locator Transmitter
ETA	Estimated Time of Arrival
EW	Electronic Warfare
FANS	Future Air Navigation Systems
FCS	Frame Check Sequence
FIR	Finite Impulse Response
FIS-B	Flight Information Service - Broadcast
FMS	Flight Management System
FTP	File Transfer Protocol
GMSK	Gaussian Minimum Shift Keying
GNSS	Global Navigation Satellite System
GPS	Global Positioning System
GPW	Ground Proximity Warning
GPX	GPS Exchange Format
GS	Ground Station
HDLC	High-Level Data Link Control

HF	High Frequency
HFDL	High Frequenct Data Link
HMI	Human-Machine Interface
HTTP	HyperText Transfer Protocol
I/O	Input/Output
IAF	Initial Approach Fix
ICAO	International Civil Aviation Organization
IFF	Identification Friend or Foe
IFR	Instrument Flight Rules
JPEG	Joint Photographic Experts Group
JPG	JPEG graphic format file extension
KML	Keyhole Markup Language
LAN	Local Area Network
MDS	Multistatic Dependent Surveillance
MLAT	Multilateration
MSAW	Minimum Safe Altitude Warning
MTCD	Medium Term Conflict Detection
MU	Master User
NIMA	National Imagery and Mapping Agency
NTP	Network Sharing Protocol
OSM	Open Street Map
PC	Personal Computer
PGR	PlaneGadgets Radar
PSR	Primary Surveillance Radar
Radar	Radio Aid to Detection And Ranging
RCS	Radar Cross Section
RPT	Regular Passenger Transport
RRE	Royal Radar Establishment
RSRE	Royal Signals and Radar Establishment
SAR	Search and Rescue
SMC	Surface Movement Control
SQL	Structured Query Language
SSR	Secondary Surveillance Radar
STCA	Short Term Conflict Alert
TCAS	Traffic Alert and Collision Avoidance System
TCP/IP	Transmission Control Protocol/Internet Protocol
TDMA	Time Division Multiple Access
TDOA	Time Difference of Arrival
TIS-B	Traffic Information Service - Broadcast
TRE	Telecommunications Research Establishment
UDP/IP	User Datagram Protocol/Internet Protocol
UHF	Ultra High Frequency
USB	Universal Serial Bus
UTC	Coordinated Universal Time
VEC	A graphic file extension associated with Daylon Vectre Vector Graphic
VFR	Visual Flight Rules
VHF	Very High Frequency
VMAP	Vector Smart Map
VMC	Visual Meteorological Conditions
WAM	Wide Area Multilateration
WAN	Wide Area Network
XML	Extensible Markup Language

About the Author

Lionel K Anderson developed an early interest in aviation and aircraft during his schooldays living on the Isle of Wight. This encouraged him to build model aeroplanes and experiment with radio control designing and building a simple system using thermionic valves and electro mechanical escapements.

Those were the days when great things were happening in aviation and he recalls seeing the Fairey Delta 2, flown by Peter Twiss, overhead in 1956 when a new World Air Speed Record of 1132mph was made, a significant achievement considering the old record had only been set the previous year by an American F100 Super Sabre, and was so much slower.

As a 14 year old Army Cadet he was fortunate enough to visit his local Territorial Army unit, a Battery of a Heavy Anti Aircraft Regiment. There he was introduced to the 3Mk7 Field Control Radar and the 4Mk7 Tactical Control Radar which led to a life long interest in the science and engineering of Radar. Leaving school he joined the British Army as an Apprentice Radar Mechanic pursuing a career in the maintenance of aircraft tracking and surveillance radars followed by the development of radar controlled guided missiles. In 1963 he was seconded to Elliott Bros (London) to learn about the Elliott 803B Digital Computer that was then used to create a working facsimile of the US Army's Field Artillery Digital Automatic Computer (FADAC) that then led to the deevlopment of the British Army's Field Artillery Computer Equipment (FACE). In 1965 and 1966, during the Indonesian Confrontation, he was deployed to Borneo as part of a team responsible for the maintenance and upkeep of a number of Green Archer mortar locating radars

On leaving the Army he then naturally progressed into the world of electronic design and computing going on to gain Masters Degree and becoming a professional electronics engineer involved in the design and development of a number of sophisticated engineering systems.

Now retired he continues his many interests and activities in electronics and computing publishing a number of books on various allied subjects. He still misses the 'pop, sizzle and crack' of the sounds, smells and excitement of working in the high voltage high megawatt power world experience in some early military radar systems.

Acknowledgements

I'd first like to thank Kevin Daws for his excellent photograph of the Red Airs and Vulcan formation flying at the Farnborough Air Show in July 2010 that was used as the basis of the books cover.

I'd also like to thank Ian Kirby, Lars Magnusson, Mateusz Lukasek, Trevor Marshall and Lee Shand for permission to include their excellent photographs which have been used to help illustrate this book.

Many thanks to Curt Deegan for his development and support of the PlanePlotter to Google Map, FindFlight and ZoneMe add-ons .

My sincere thanks to Faith Carroll for he assistance in correcting my grammar and the typographical errors that crept in occasionally.

I must also add my thanks to Hamish McTorsk and Mike Cogan of the Radarspotters Forum for their expertise on all matters appertaining to Virtual Radar.

Last but by no means least to Bev Ewen-Smith at COAA for creating and maintaining the PlanePlotter software and for his never ending help and support to all its users.

Disclaimer

The use of radio receivers and scanners to monitor radio frequencies is not permitted in every country of the world. Similarly some countries and authorities prohibit the use of information received. Thus readers are advised to check with their licensing authorities prior to doing so as the author cannot be held responsible for any legal consequences.

Sir Robert Watson-Watt

Chapter 1
Introduction

PlanePlotter is a unique software program that enables a user to have a live radar type display of aircraft not only in their own local area but also regions and airways around the world. The PlanePlotter software program receives radio signals and decoded messages received with digital data transmitted from a variety of sources including ACARS, ADS-B and Mode-S systems by using VHF and UHF radio receivers scanners or from a number of readily available *Virtual Radar boxes*. PlanePlotter can also display position reports from aircraft over an almost global range using an HF receiver and decoding HFDL traffic.

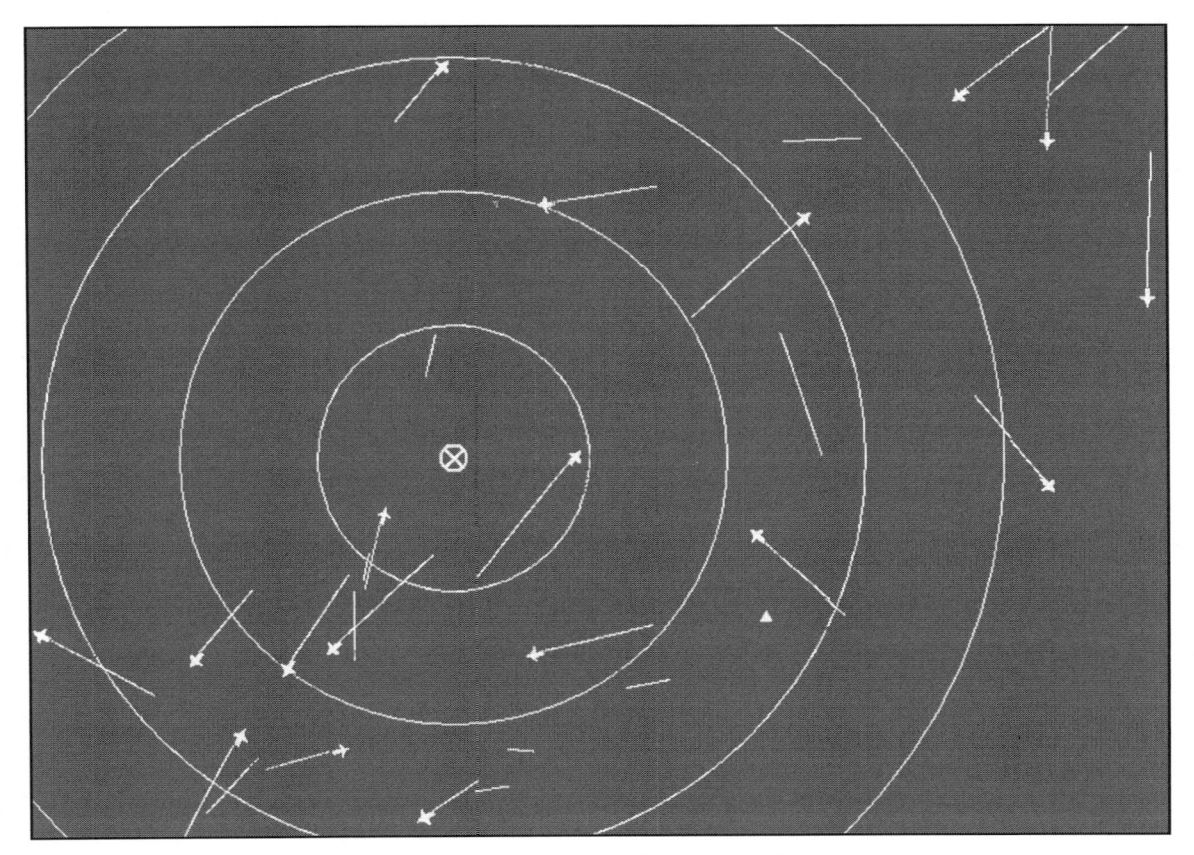

Figure 1 - PlanePlotter simple radar type display

As well as radar type displays, PlanePlotter can show aircraft position and information on a variety of charts created from graphic-image files, satellite-image downloads, downloaded OpenStreetMaps or custom-created aeronautical-type charts. As well as those graphical displays, PlanePlotter can also display messages received and decoded from live aircraft transmissions in a number of tabular formats. The software archives all the digital data that it receives and decodes it to log files that can be analysed at a later time.

PlanePlotter can decode HF SelCal messages displaying the called codes PlanePlotter has a unique Direction Finding (DF) feature that allows the software to detect the direction of any aircraft making voice transmissions providing a suitable antenna array and switch has been installed.

As well as displaying the position of aircraft received using local receivers/scanner or virtual radar boxes PlanePlotter can display data collected and shared by other PlanePlotter users by connecting to the Internet, thus a user in the centre of Europe can see quasi-live aircraft movements from all around the globe.

Additional third party software enabling aircraft positions and movements to be displayed using standard Web pages by accessing the PlanePlotter data files can be easily implemented providing visual interfaces using Google Maps and Google Earth

Decoded data can include the following information

- Aircrafts Registration Number
- Aircrafts Call Sign
- Aircrafts ICAO Number
- Aircrafts Latitude
- Aircrafts Longitude
- Aircrafts Altitude
- Aircrafts Heading
- Aircrafts Speed
- Aircrafts Type
- Aircrafts Route
- Aircrafts Squawk

ACARS

Aircraft Communications Addressing and Reporting System (ACARS) is a digital data-link system for the transmission of short and relatively simple messages between aircraft and ground stations and designed to utilise existing ground station and aircraft radio equipment, and enhance air-ground-air communications.

Introduction

ACARS messages can range from simple arrival/departure reports to lengthy aircraft computer downlinks of navigation, engine, and performance data as well as weather observations, flight plans, navigation positions, aircraft and engine performance data, arrival/departure/delay reports, equipment malfunction reports, crew reports and connecting gate lists. ACARS messages are either transmitted from an aircraft to a company ground station, called a downlink message, or transmitted from ground stations up to aircraft, called uplink messages.

More information on ACARS can be found in the book *ACARS - A Users Guide for the Aviation Enthusiast ISBN 978-1-4457-8847-0* readily available from the online Amazon bookshop.

ADS-B

Automatic Dependant Surveillance - Broadcast (ADS-B) is a system where aircraft determine their own position using global navigation satellite systems and then periodically broadcast position and other information to ground stations and other aircraft.

Mode-S

Mode S Beacon Radar System is a combined beacon radar and ground-air-ground data link system that was designed to replace the aging Air Traffic Control Radar Beacon Systems (ATCRBS). Mode S provides more accurate positional information and minimizes interference. This is accomplished by discrete interrogation of each Mode S transponder-equipped aircraft and improved processing of aircraft replies. In addition, Mode S provides the medium for a digital data link which can be used to exchange information between aircraft and various air traffic control functions and weather databases.

HFDL

HFDL is a High Frequency data link protocol as defined in ARINC Speciation 635-3. The HFDL service is operated by Aeronautical Radio Incorporated (ARINC) as a global service through a worldwide network of HF stations. HFDL is also referred to as HF ACARS.

SelCal

SelCal is a selective-calling radio system that can alert an aircraft's crew that a ground radio station wishes to communicate with the aircraft. SelCal uses a ground-based encoder and radio transmitter to broadcast an audio signal that is picked up by a decoder and radio receiver on an aircraft.

Figure 2 - Royal Artillery 3Mk7 Fire Control Radar

The 3MK7 FC was a mobile anti-aircraft fire control radar station from an original design of the mid late1940s and remained in service until the 1970s. The radar could both search and track aircraft. The radar was deployed in Heavy Anti Aircraft batteries to feed information via a No 11 Predictor to 3.7" guns. The radar operated in the X Band frequency range of 3,000MHz (10cm) and had a transmit output power of 200kW peak. The 3Mk7 was able to detect targets at up to 36,000 yards.

Chapter 2
6.67μsecs

In the early days of aviation, the only way that people on the ground could detect and locate an aircraft in the sky was by using their eyes and ears. In fact, prior to World War II, sound ranging (locating) was used extensively as a form of anti-aircraft defence.

In the 1930s the War Office established the Telecommunications Research Establishment (TRE) to investigate the use of high-power radio waves and their effects. This work was carried by Robert Watson-Watt and Arnold Wilkins, who discovered that transmitted radio signals would be reradiated back towards the transmitter from objects, particularly metallic ones, and that the signals struck in their propagation through the ether. So important was this discovery that it was essential to keep the work secret, and the TRE was moved from Worth Matravers in Dorset to Malvern in Worcestershire and later renamed as the Army Radar Establishment. Later the name was changed to the Radar Research Establishment (RRE) then renamed again in 1957 as the Royal Radar Establishment. It became the Royal Signals and Radar Establishment (RSRE) and in 1997 renamed again this time to Defence Evaluation & Research Agency (DERA). In 2000 it was privatised by the Labour Government into the private sector company QinetiQ.

The work carried out by Watson-Watt led to the design of a practical radar system as shown in Figure 3 below. A pulse of high-energy electromagnetic waves propagated from an antenna on striking an object such as an aircraft causes a small proportion of that energy to be reradiated in many directions. A basic radar system uses the same antenna system to

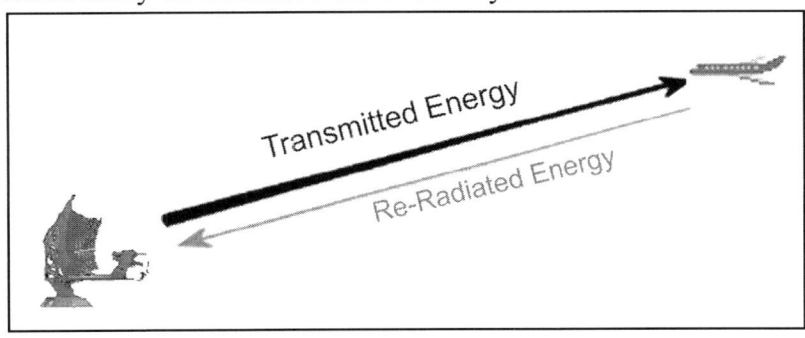

Figure 3 - The Basic Radar Principle

both transmit the high-energy signal and receive the low-energy signal reradiated form the aircraft or other reflecting surface. Figure 4 shows a very basic radar where a switch called a "duplexer" switches the connection to the antenna from the transmitter to the receiver. The receiver signal after processing is then shown on a display of some type.

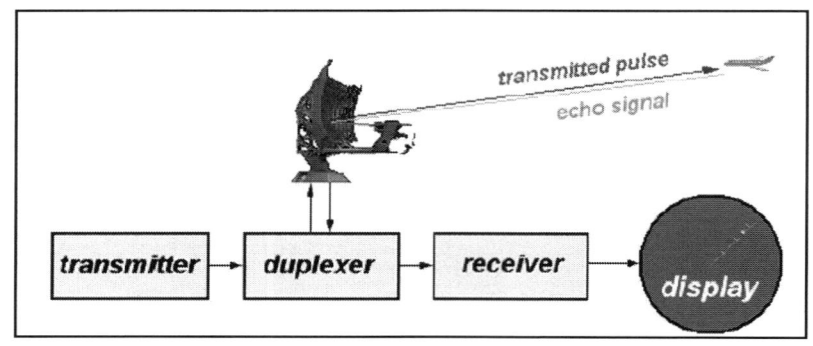

Figure 4 - A Basic Radar System

A radar system not only detects an aircraft but also calculates how far away from the transmitter/receiver it is. Thus radar stands for **RA**dio **D**etection **A**nd **R**anging. The range is calculated by knowing that electromagnetic waves travel through air at approximately the speed of light: 300,000 kilometres per second or 186,000 miles per second. By measuring the time between transmitting a pulse of energy and receiving a reradiated portion of it back, the distance to an aircraft can be calculated.

Figure 5 shows how these pulses may be displayed on a cathode ray oscilloscope (CRO). High-energy pulses are transmitted repetitively and reradiated pulses received a short time later.

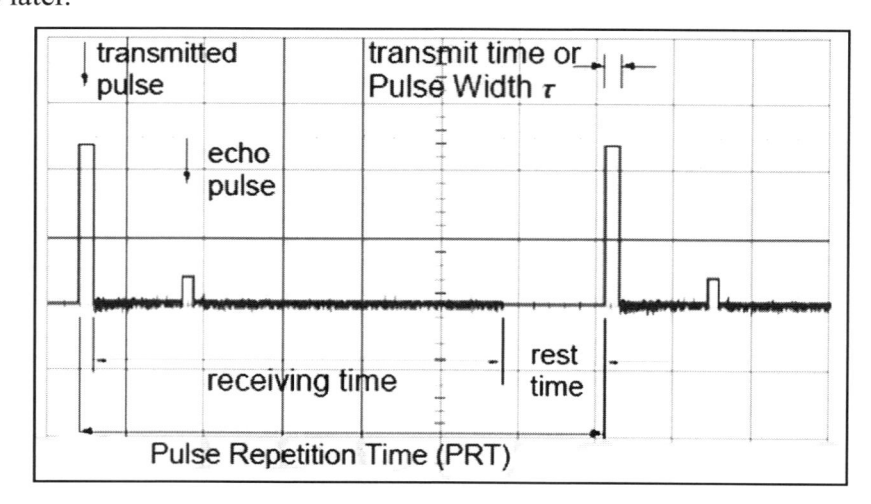

Figure 5 - Radar Pulse Timing

Knowing the speed of light we can determine that for an aircraft 1,000 yards away, the time taken between transmission and reception of the pulse is 6.1µsecs. *(As an Apprentice Radar Mechanic, it is a number I learnt and have never forgotten.)* For an aircraft 20,000 yards distance the time would be 122µsecs. In metric units the time is 6.67µsec per 1000 metres.

A quick word about pulse width. The receiver does not listen during the transmitting pulse, because it needs to be disconnected from the transmitter during transmission to avoid damage to the front end of the receiver. In that case, the echo pulse comes from a very close aircraft and aircraft at a range equivalent to the pulse width from the radar will not be detected. A typical value of 1µs for the pulse width of short range radar corresponds to a minimum range of about 150m; however, radars with a longer pulse width suffer a relatively large minimum range. Some typical pulse widths for different type of systems are shown in Table 1.

System	Pulse Width	Minimum Range
Airport Surface Movement Radar	100ns	25m
Anti Aircraft Fire Control Radar	250ns	65m
ATC Surveillance Radar	1.5 µs	250m
Air Defense Tactical Radar	800µs	120km

Table 1 - Transmitter Pulse Width vis Minimum Range

The maximum range at which an aircraft can be detected by a radar is a function of several parameters including the power of the transmitter, the sensitivity of the receiver, the gain of the antenna and the radar cross section of the aircraft. The theoretical maximum range can be calculated using the Radar Range Equation.

$$R_{max} = \sqrt[4]{\frac{P_{tx} \cdot G^2 \cdot \lambda^2 \cdot \sigma_t}{(4\pi)^3 \cdot P_{MDS} \cdot L_s}}$$

Where

R_{MAX} = the maximum range
P_{tx} = the radiated power of the transmitter
P_{MDS} = the minimum discernable signal
G = the gain of the antenna
λ = the transmitters wavelength
σ = radar cross section
L_s = the sum of the losses

For example with
P_{tx} = 1Mw, G = 1,200, = 10cm, MDS = $5*10^{-15}$, = 20 & Ls = 118.89 the theoretical maximum range to detect a jet airliner would be 125km.

Before moving on a brief word about the Radar Cross Section (RCS) of an aircraft. The RCS, or reflectivity to radar, of an aircraft obviously depends on a number of factors including the airplane's physical geometry and exterior features, the direction of the illuminating radar, the radar transmitters frequency and the type of material in the reflecting surface. As seen in the example on calculating the maximum range a value of 20 was used and is typical for a jet airliner. Table 2 shows that if the aircraft had been a jumbo jet then the maximum detectable range would have increased to over 157km.

Type of Aircraft	RCS [m2]	RCS [dB]
Jumbo Jet	100	20
Jet Airliner	20 … 40	13 … 16
Large Fighter	6	7.8
Helicopter	3	4.7
Executive Jet	2	3
Small Aeroplanes	1	0
Stealth Jet	0.1 … 0.01	-10 …-20

Table 2 - Radar Cross Section

A radar can detect just one or several aircraft at any given time depending upon the type of the radar and more importantly on the type of antenna. A Fire Control Radar whose task is to track an aircraft and pass on information as to its range, altitude and azimuth would a very narrow-shaped beam as shown in Figure 7. Only Aircraft A which is in the beam will be tracked and the radar will not be aware of aircrafts B & C.

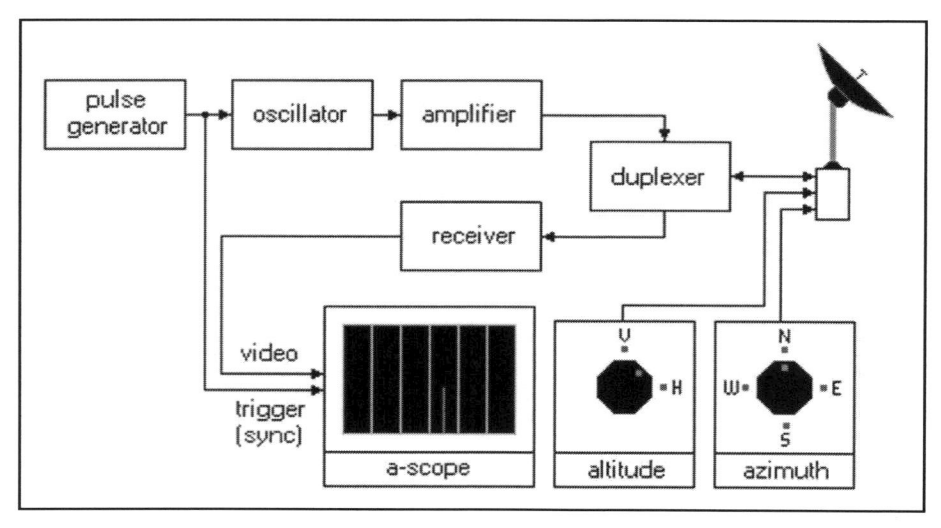

Figure 6 - Basic Pulse Radar system with steerable parabolic dish typical of the Fire Control Radar shown in Figure 2

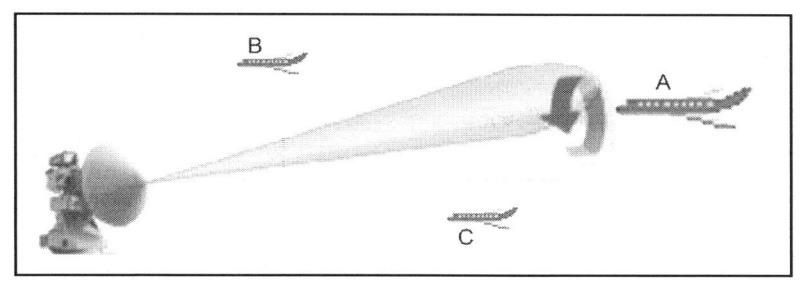

Figure 7 - Conical Beam Pattern from a parabolic dish antenna

A surveillance radar could use a Cosecant Squared Pattern antenna as shown below in Figure 8. This beam pattern will detect aircraft from 50m up to 6,000m altitude and display them simultaneously. In this case all three aircraft A, B & C will be detected.

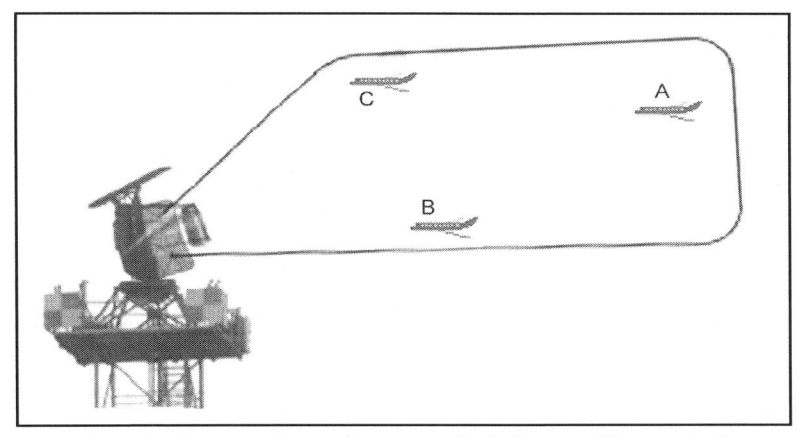

Figure 8 - Cosecant Squared Pattern typical of a surveillance radar8

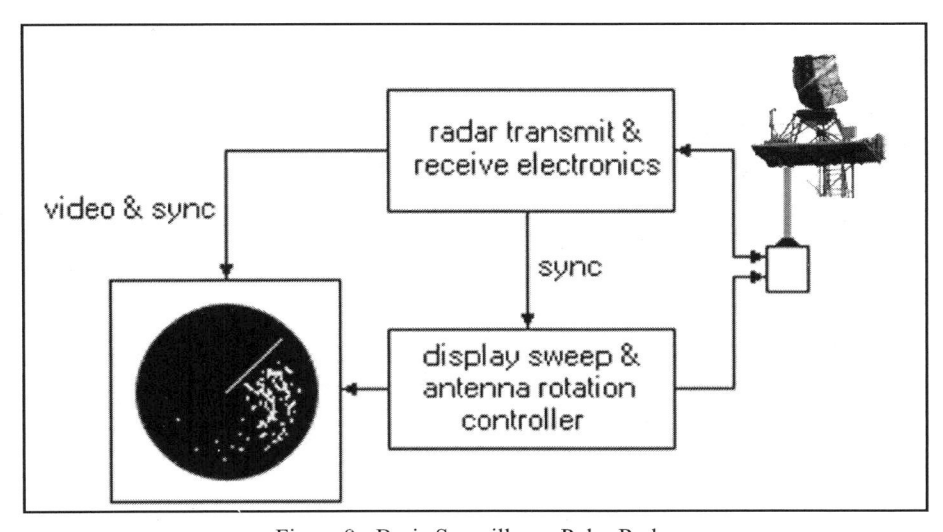

Figure 9 - Basic Surveillance Pulse Radar

IFF

In the early wartime days of radar, the detection of several aircraft at a time led to problems in identifying whether the aircraft were from our own air forces or from the enemies. Robert Watson-Watt was the inventor of the **I**dentification **F**riend or **F**oe system (IFF) and oversaw its development for use by Britain's Royal Air Force during WWII. The basis for the system was a ground-based transmitter, the interrogator that broadcast an additional radio signal to the aircraft and a transponder on the aircraft that would receive and reply to the signal. Interrogations within IFF used very specific types of signals, or modes. Aircraft participating in IFF would be equipped to respond to recognize and respond correctly in these modes. If a plane did not respond correctly to the IFF interrogation, the system determined that the target was an enemy aircraft.

As well as seeing hostile aircraft, it soon became apparent that radar was a good tool to see friendly aircraft and hence be able to control and direct them. If an aircraft is fitted with a transponder it can return the normal radar echo and also send back an encoded response signal to the interrogated signal. Figure 10 shows a simple IFF system where the IFF interrogator is part of the ground-station radar, but with a separate antenna and the aircraft is fitted with a transponder to reply to the interrogation.

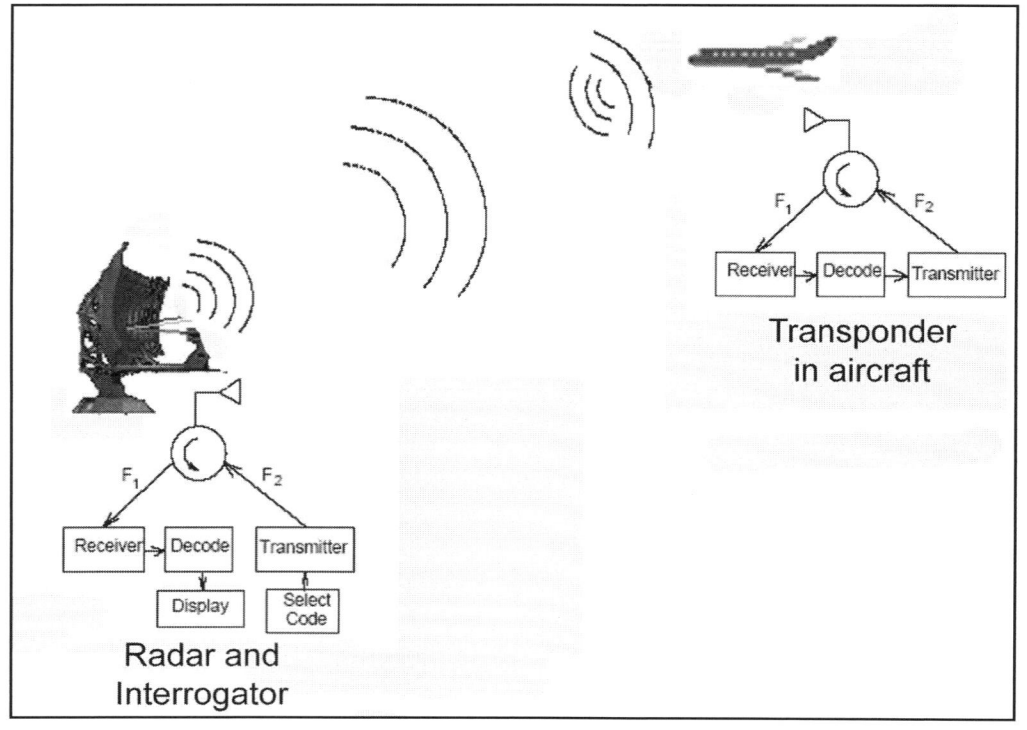

Figure 10 - Simple IFF System

At the end of the Second Word War, the Berlin Airlift in 1949 and the Korean War in 1951 provided a need for a much improved IFF system The IFF Mk10 was adopted by NATO and the limitations of the basic reply coding led to the development of the Selective Identification Feature (SIF). Incidentally, the terms IFF used by air defence systems, SIF by military Air Traffic Control (ATC) and Secondary Surveillance Radar (SSR) used by the civil ATC are all synonymous.

The IFF Mk10 also heralded the move to the 1030Mhz interrogation and 1090Mhz response frequencies that are still used today and was followed by the development of the Air Traffic Control Radar Beacon System (ATCRBS), a secondary surveillance radar system providing more precise position reporting of planes. This was then superseded in the 1960s Airborne Collision Avoidance System (ACAS) though the widespread international adoption of SSR for ATC lead to all aircraft flying in controlled airspace being required to be fitted with an Airborne Collision Avoidance System (ACAS).

All of those systems use the same basic technology, that of the interrogator transmitting a pair of pulses which are recognised by the transponder and replied to it appropriately. Figure 11 shows the six basic IFF modes. Each pair of pulses is 0.8μsecs wide and the separation between pulses varies from 3 to 25μsecs.

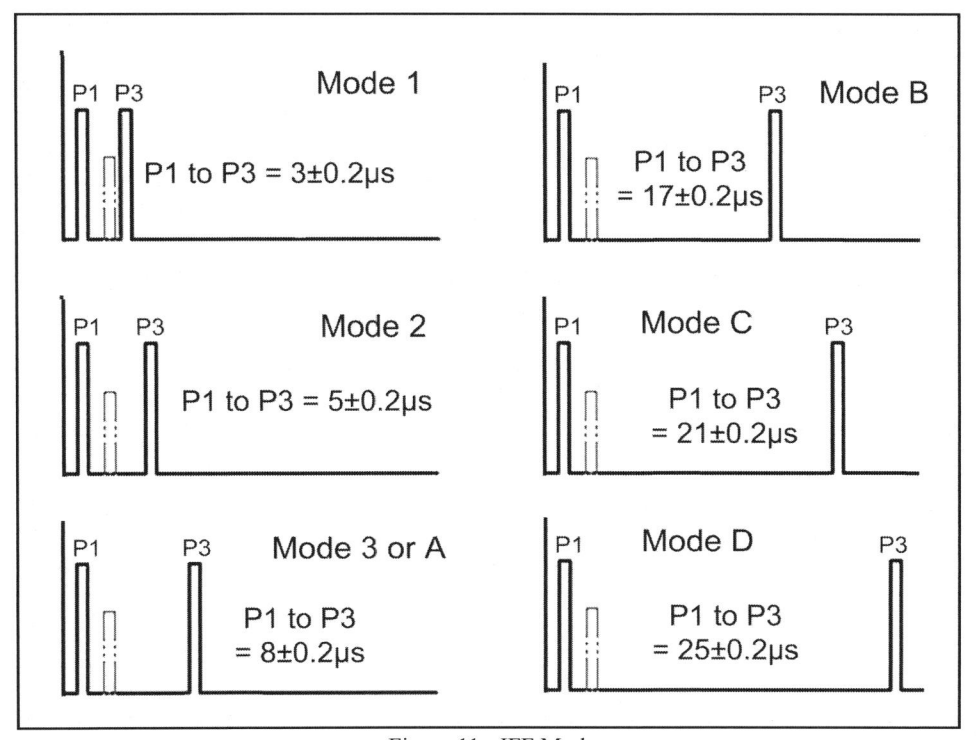

Figure 11 - IFF Modes

Mode 1 is a military mode used to indicate the role, mission, or type of aircraft (hence several aircraft may give the same Mode 1 reply value). This mode is not in common use in a normal peacetime environment.

Mode 2 is usually used to indicate an individual aircraft airframe by using one of 4,096 codes that are set in the aircraft usually before take off.

Mode 3, now often referred to as Mode A, is used by both civil and military aircraft. It provides 4,096 IDs and is the most commonly used mode.

Mode C is use to extract that other essential information required by ATC, namely the aircraft's altitude as derived from the aircraft's pressure altimeter or its radar altimeter.

Modes B and D have never been used for civil ATC purposes, although they were originally defined in the specifications; thus, present day civil SSR systems are normally referred to as SSR Mode A/C.

The observant reader will notice that in Figure 11 there is a second and smaller pulse immediately after P1. This is 9db down on the main P1 and P3 pulses and is to prevent side lobes triggering the transponder.

Mode-S

In the 1980s even with IFF, ATCRBS and ACAS there were a significant number of mid-air collisions thus an improvement to those systems had to be found and was found with the development of the Mode S technology.

The basic ATCRBS system relied on pulses of microwave energy as its mean of communication with an aircraft. Mode S rather than transmitting two pulses still uses the 1030Mhz interrogation frequency but modulates the carrier wit a Differential Phase Shift Keying (DPSK) providing greater efficiency without interfering with other Mode A/C interrogations. When the interrogation is received by the aircraft's transponder, it will verify the request and integrity of the signal replying on the 1090Mhz frequency with a pulse-position modulation of the carrier wave.

A Mode-S system can generate one of four different types of interrogation. The first is where the narrow P4 pulse is transmitted and does not invoke a reply from an aircraft fitted with a Mode S transponder, although it will be detected solely. The second is an "ATCRBS all call" where the interrogation consists of P1, P3 and a 0.8 µs P4 pulse. The P2 side lobe-suppression signal is transmitted as normal. All ATCRBS transponders reply with the 4096 identification code for Mode A interrogations and altitude data for Mode C, whilst Mode-S transponders do not reply at all. The third type of interrogation is called an

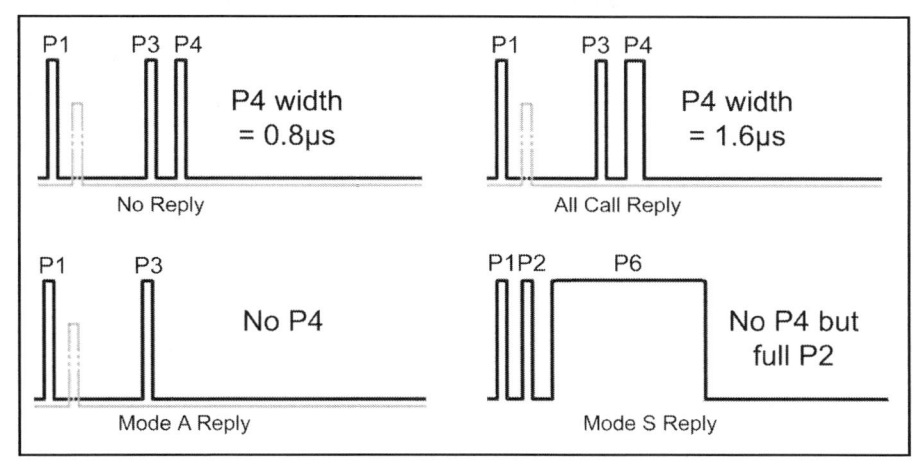

Figure 12 - Mode S Interrogation Pulses

"ATCRBS/Mode-S all call reply" This interrogation is similar to the last except that the P4 pulse width is now 1.6 μsec. Standard ATCRBS transponders reply with the 4096 code or altitude data as per the ATCRBS all call, whilst Mode S transponders reply with a special code containing the identity of the aircraft by its discrete address. The fourth type is a "Mode S discrete interrogation." Here the interrogation is transmitted to a specific Mode-S transponder-equipped aircraft. The interrogation consists of P1, P2 and P6 pulses. In this case the P2 pulse is transmitted from a directional antenna and is the same amplitude as the P1 and P3 pulses effectively suppressing any ATCRBS transponders from replying. The P6 pulse is not actually a pulse as such but a block of DPSK data containing either a 56 bit or 112 bit message. This modulated block of data produces a spread spectrum signal having greater immunity to interference.

Having received a valid Mode-S discrete interrogation, the aircraft's transponder will return a reply 128μs after reception that is transmitted on 1090 MHz and uses a 56-bit or 112-bit PPM data transmission. Each Mode-S interrogation will have a 24-bit address unique to the aircraft, together with a 24-bit parity check to validate the data.

Mode-S in its basic surveillance mode provides limited download formatted information known as DF0 for altitude-reporting aircraft identification known as the DF4 message and basic airframe information known as the DF11 message. Interrogation is repeated at a pulse-repetition frequency (PRF) of 50 Hz. Transponders will reply at the same rate. Once the interrogator has received the reply it will decode the Mode A, C or S from the demodulated information within each reception. Their are 25 different download-message formats, each one having a particular purpose. The major formats in use are shown below in Table 3. It be should be noted that DF0, DF4, DF5, DF11, DF16, DF20, DF21 and DF24 are at the present time used in civil aviation, with DF0 providing information for ACAS, whilst the DF17 format is used for the modern ADS-B system.

Downlink Format	Message Format	Content
DF0	Fig 14	Short Air to Air ACAS
DF4	Fig 14	Surveillance (roll call) Altitude
DF5	Fig 14	Surveillance (roll call) IDENT Reply
DF11	Fig 14	Mode S Only All-Call Reply (Acq Squitter if II=0)
DF16	Fig 15	Long Air to Air ACAS
DF17	Fig 15	1090 Extended Squitter
DF19	Fig 15	Military Extended Squitter
DF20	Fig 15	
DF21	Fig 15	Comm. B Altitude, IDENT Reply
DF22	Fig 15	Military use only
DF24	Fig 16	Comm. D Extended Length Message (ELM)

Table 3 - Mode-S Download Message Formats

Figure 13 show the typical modulation pattern of a reply that is comprised of two distinct parts: a preamble that is then followed by a block of data. The preamble is used to "wake up" the interrogator's receiver and decoder and consists of a pattern of four pulses each with a length of 0.5μs and spaced a T_0, T_0+1μs, $T_0+3.5$μs and $T_0+4.5$μs. The data block consists of either 56 or 112 bits with a length of either 56 or 112μs.

Figure 14 shows the content of a typical short message reply. Again there is an 8μs preamble but the 56 bit data block is divided into three parts. The first 5 bits comprise a format identifier, the next 27 bits a surveillance and control word, whilst the remaining 24 bits provide a parity check to ensure the quality of the data. The parity bits are calculated using a cyclic-redundancy code based on a modified G(x) polynomial.

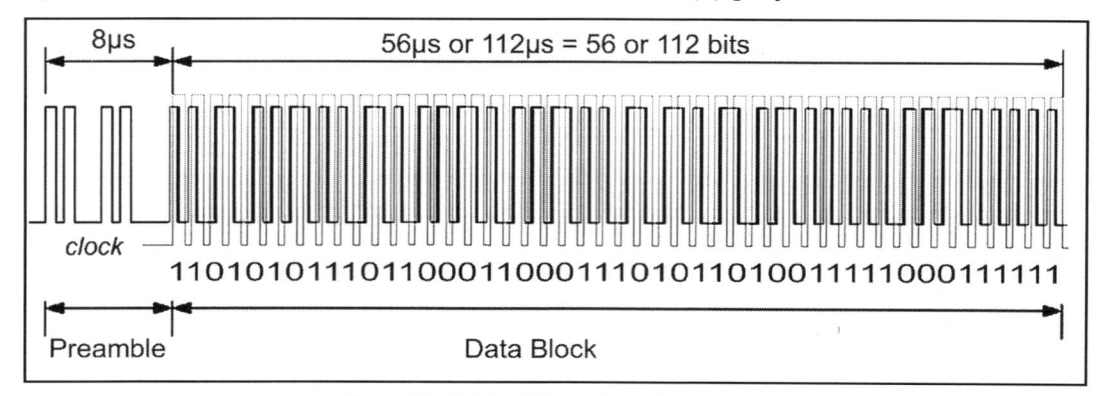

Figure 13 - Mode-S Typical Reply Encoding

Figure 14 - Mode S Short Message Reply

Figure 15 - Mode-S Long Message Reply

Figure 15 below shows the content of a typical long message reply. As in the short message reply there is a 8µs preamble but the data block now has 112 bits and is divided into four parts. Again the first 5 bits comprise a format identifier, the next 27 bits a surveillance and control word, but there is then a 56-bit message field and again a 24-bit parity check. The DF16 long message is used to communicate between aircraft using ACAS or TCAS units.

Figure 16 - Mode-S Exteneded Length Message Reply

Finally Figure 16 below shows the content of a typical of a reply with an extended-length message data block as its communication reply. Again, as in the short message reply there is a 8µs preamble followed by a 2-bit format identifier, a 6-bit communications control code and then an 80-bit message field and finally again a 24-bit parity check.

Squitters

A squitter is a reply-format transmission from an aircraft without the aircraft being interrogated. These unsolicited replies, or squitters, are used to provide the discrete address of a squittering aircraft fitted with TCAS 2-type systems, thus enabling ACAS and TCAS systems to detect and track an aircraft by using Mode S formats DF0 and DF16. The term "squitter" has its origins in the old Distance Measuring Equipment

(DME) systems where a DME ground station would broadcast unsolicited replies or squitters. When the airborne DME interrogator was in range, the squitter would be seen and the DME interrogator would then transmit a range interrogation and receive range replies from the DME ground station. This served to limit unnecessary transmissions over the air and optimised DME ground station handling capability. TCAS 2 systems use Mode S squitters in a similar fashion, but in these cases the TCAS only listens for the DF11 squitters that contain the transmitting aircraft's discrete address. This then reduces the need to interrogate the aircraft over the air. Once the discrete address has obtained it is placed on the TCAS 2 processors list of addresses for further ongoing tracking. Mode S technology has two types of squitters, the short 56-bit DF11 acquisition squitter and the extended 112-bit DF17 squitter.

ADS-B

The concept of DF17 extended squitter is similar to elementary and enhanced surveillance with one exception in that DF17 is a squitter and therefore does not need an interrogation. Therefore a DF17 message will report an aircraft's information whether or not a ground station or other aircraft is asking for it. A DF17 extended squitter is an important part of an **A**utomatic **D**ependent **S**urveillance-**B**roadcast (ADS-B) as a

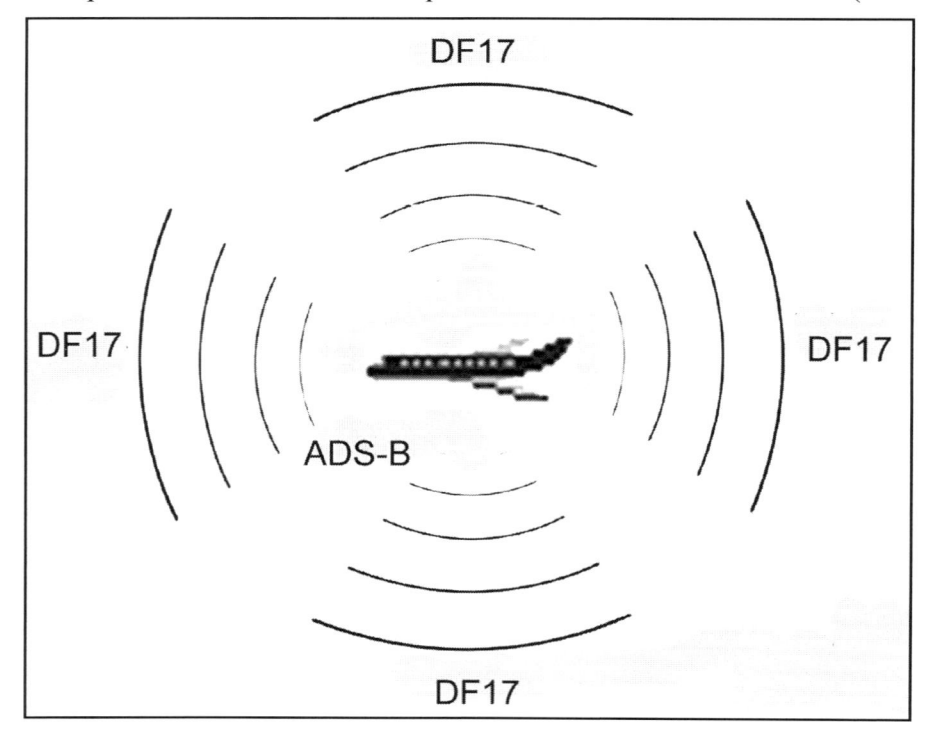

Figure 17 - The ADS-B DF17 Broadcast

6.67μsecs

DF17 extended squitter provides airborne position, surface position, extended squitter status, identity and category, as well as the aircraft's speed. The aircraft's airborne position includes the longitude and latitude of the aircraft, the barometric altitude, the GPS derived height and surveillance status. Its surface position is similar to its airborne position for the longitude and latitude of the aircraft but with the movement and heading of the aircraft on the round.

DF17 is the integral and working portion of ADS-B. The meaning of the acronym comes from the terms:

Automatic - There is no interrogation needed to start the data or squitter coming from the transponder.

Dependent - It relies on onboard navigation and broadcast equipment to provide information to other ADS-B users.

Surveillance - It is a means of automatic surveillance and traffic coordination.

Broadcast - The message is transmitted freely to the world.

Having now covered some of the basic technologies of detecting aircraft and tracking their positions, we can now move on to the PlanePlotter program and see how it uses these technologies to the benefit of the aviation enthusiasts.

Figure 18 - Air Traffic Control Radar Tower

Chapter 3
Radios and Virtual Radars

PlanePlotter can receive raw and/or decoded data from a number of different sources.

VHF

For receiving ACARS transmission, almost any VHF receiver or scanner tuned to an appropriate ACARS channel in AM mode and for aircraft within VHF range are suitable. Recommended products include the Signal R532/R53, the Unidens USC230, UBC-30XLT, USC230, UBC3500XLT & Bearcat UBC180XLT, the Yaesus FT-817 or the Yupiteru MVT range. Those receivers can also be used to receive local Air Traffic Control (ATC) transmissions both to and from aircraft.

HF

For both HFDL plotting and HF SelCal decoding a HF receiver with USB mode and fine-tuning steps.

ADS-B & Mode-S

To process and display ADS-B and Mode-S information, there are a number of virtual radar boxes readily available in the commercial market place.

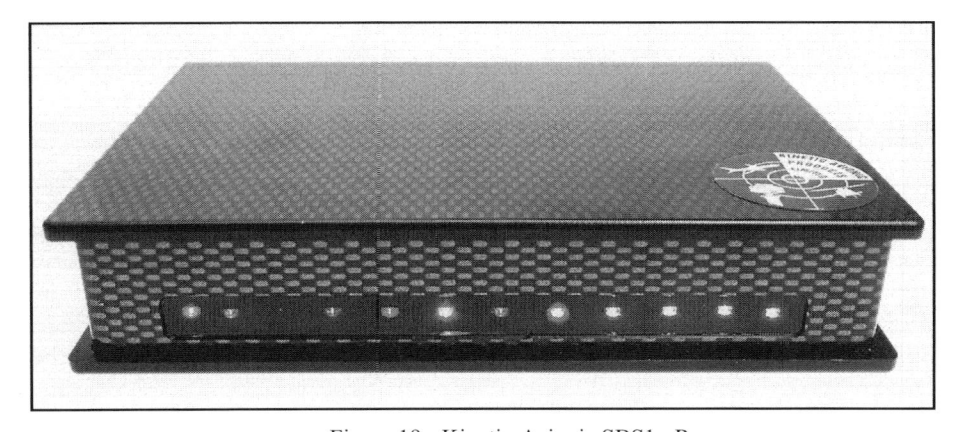

Figure 19 - Kinetic-Avionic SBS1-eR

Kinetic-Avionic SBS1-eR

The Kinetic-Avionic SBS1-eR is a Mode-S receiver that has its own BaseStation software for processing and displaying received and decoded software. The SBS1-eR also has a built-in VHF receiver that can be used to receive, but not decode, ACARS messages. The SBS1-eR is a popular virtual radar box and well supported by PlanePlotter. More details can be found on the manufacturer's website at *www.kinetic-avionics.co.uk*.

PlaneGadget Radar

A relatively new product is the PlaneGadget Radar (PGR) and is a quality virtual-radar receiver at an affordable price than other receivers on the market, but with all the features required to interface with PlanePlotter.

Figure 20 - PlaneGadget Radar

The PlaneGadget Radar has been designed to interface with PlanePlotter and it is PlanePlotter that does all the processing and decoding of received data. It comes complete with its own magnetic base antenna already fitted with 2 meters coax and a USB lead to connect it to the PlanePlotter PC. The PGR also supports PlanePlotters Multilateration.

The PlaneGadget Radar is CE Marked and has been tested under the Electromagnetic Compatibility Directive 2004/108/EC to standards, EN55022:2006 +A1:2007 (Emissions) and EN55024:1998 + A1:2001 + D2:2003 (Immunity). On the emissions side, the product passes both the industrial and the more stringent consumer-emissions levels. More details can be found on the manufacturer's website at *www.radargadgets.com*.

AirNav Systems Radar Box

The AirNav Systems virtual radar is quite an expensive product and has been around for a number of years and can be found in use around the world. Further details can be found on AirNav web site at *www.airnavsystems.com/radarbox*.

Figure 21 - AirNav Systems RadarBox

Aurora Mode-S Receiver

This is a medium-priced product that should be available in the virtual radar marketplace early in 2011. Early information from the manufacturer suggests that it will interface to the PlanePlotter software although it has its own software.

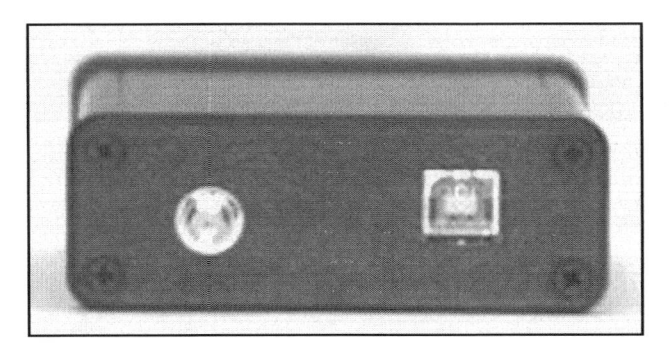

Figure 22 - Aurora Mode-S Receiver

Home Build

It is possible for users to build their own ADS-B receivers and decoders. A kit is available for RxControl Mode-S receiver system from *rxcontrol.free.fr/PicADSB/index.html*.

Figure 23 - Qatar Airways Airbus A340-642X
Photo Courtesy and Copyright of Ian Kirby

Chapter 4
Software Installation

PlanePlotter has been designed to work on the Microsoft Windows platform family of operating systems including Windows XP™, Windows Vista™ and Windows 7™ and should run without difficulty on any modern PC or Laptop with a minimum of a Pentium-type processor and at least 4Mb RAM is recommended to display large charts.

One or more of the radios or virtual radar boxes described previously in Chapter 3 can be installed before or after installing the PlanePlotter software program. If a radio or virtual radar is already installed then the software installation may be a little quicker.

Preliminary Steps

It is highly recommended that a few preliminary steps be taken before downloading and installing the PlanePlotter software. First create a new folder on the PC called C:\PlanePlotter. Then download the PlanePlotter Support Files from www.planeplotters.com/support.zip. This compressed file contains five folders that should be unzipped into the C:\PlanePlotter folder providing the result shown in Figure 24 below. The Chart Files folder contains a number of prepared and calibrated charts to get the user started and the Log Files folder contains two empty databases for later use, the Photo Files folder allows the user to save their own aircraft pictures. The remaining two folders are for some add-on programs that will be discussed later in the book.

For those readers who already have PlanePlotter installed prior to reading this book, they should consider adopting this common file folder directory structure as it will make maintenance of the PlanePlotter software much easier.

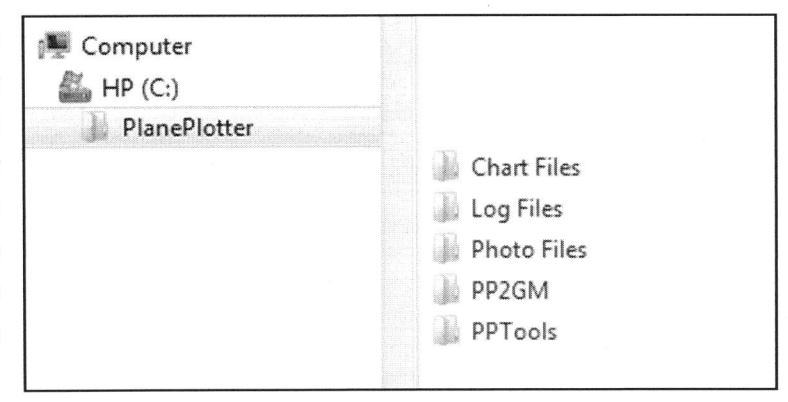

Figure 24 - PlanePlotter Program and Installation Folders

Download and Install the PlanePlotter Software

The latest version of the PlanePlotter software can be downloaded from the COAA PlanePlotter website at *http://www.planeplotter.com*, note that both English and French language versions of the program are available.

The software should be downloaded to the PCs Downloads Folder if using Windows Vista™ or Windows 7™ operating systems. If using Windows XP™, create a download folder called C:\Zip if one does not already exist.

Once the file has been safely downloaded close the Internet connection and then run the program by opening the folder where the downloaded file has been stored by double clicking the filename which will be of a similar format to *planeplotter5_5_2_9.exe*.

Figure 25 - PlanePlotter Program Setup Wizard

After accepting the license agreement, Figure 26, select the folder where the PlanePlotter program and its supporting files will be installed. **Do not accept** the default folder but ensure that *C:\PlanePlotter* is specified as the target destination folder as shown in Figure 27 on the next page.

If the folder already exists then a dialog box may be displayed as shown in Figure 28. The user should just click the *Yes* button to continue.

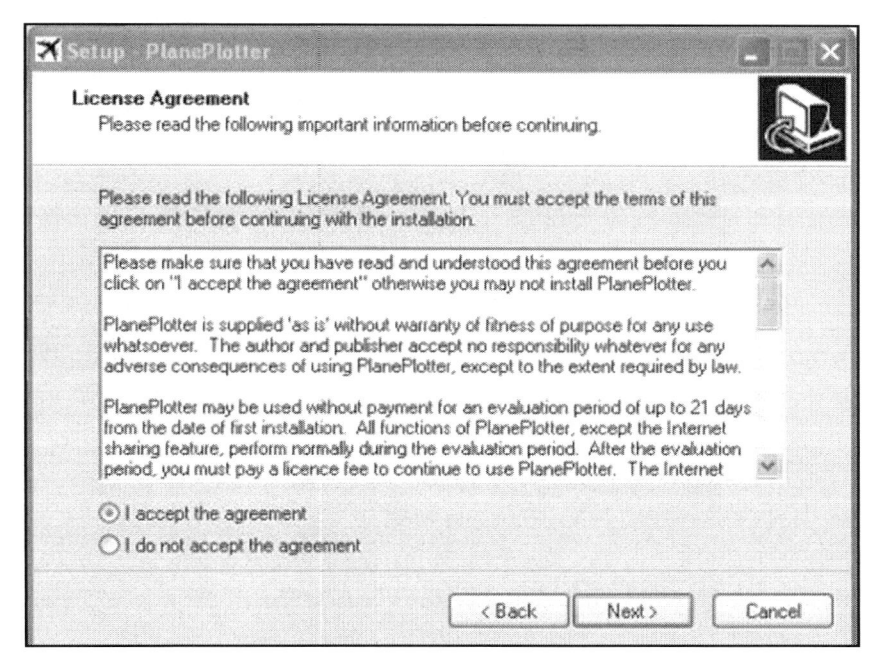

Figure 26 - PlanePlotter License Agreement

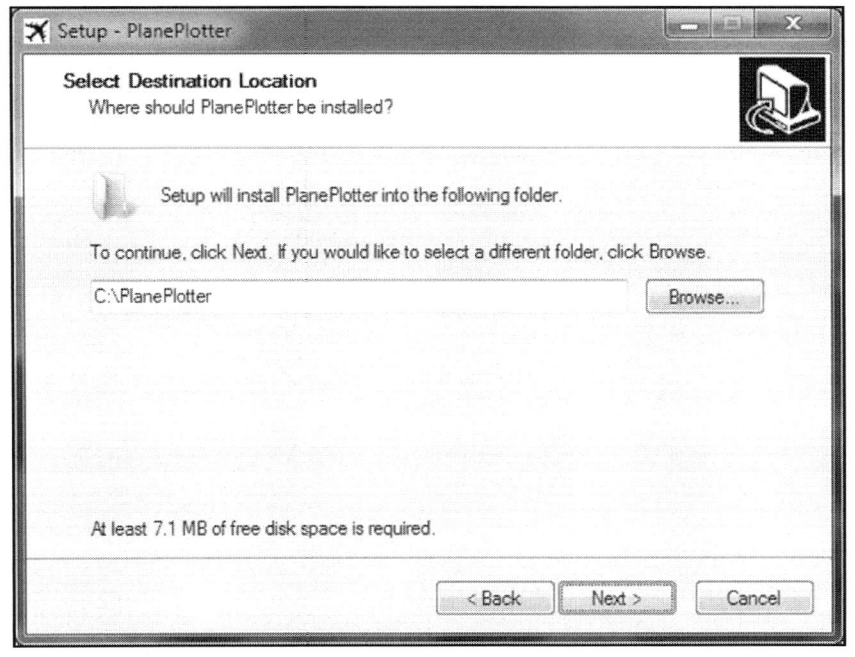

Figure 27 - PlanePlotter Program Destination Location

If the PlanePlotter folder already exists the dialog shown in Figure 27 will be displayed in which case the *Yes* button should be pressed. Otherwise a new folder will be created.

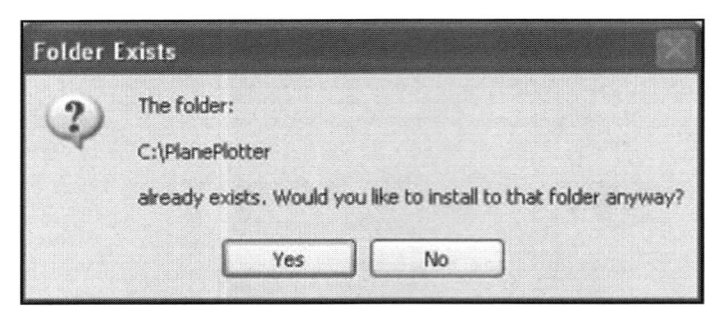

Figure 28 - Folder Exists Dialog

The installation will then create the three main sub folders for Chart Files, Log Files and Image Files under the main Planeplotter Folder as shown below.

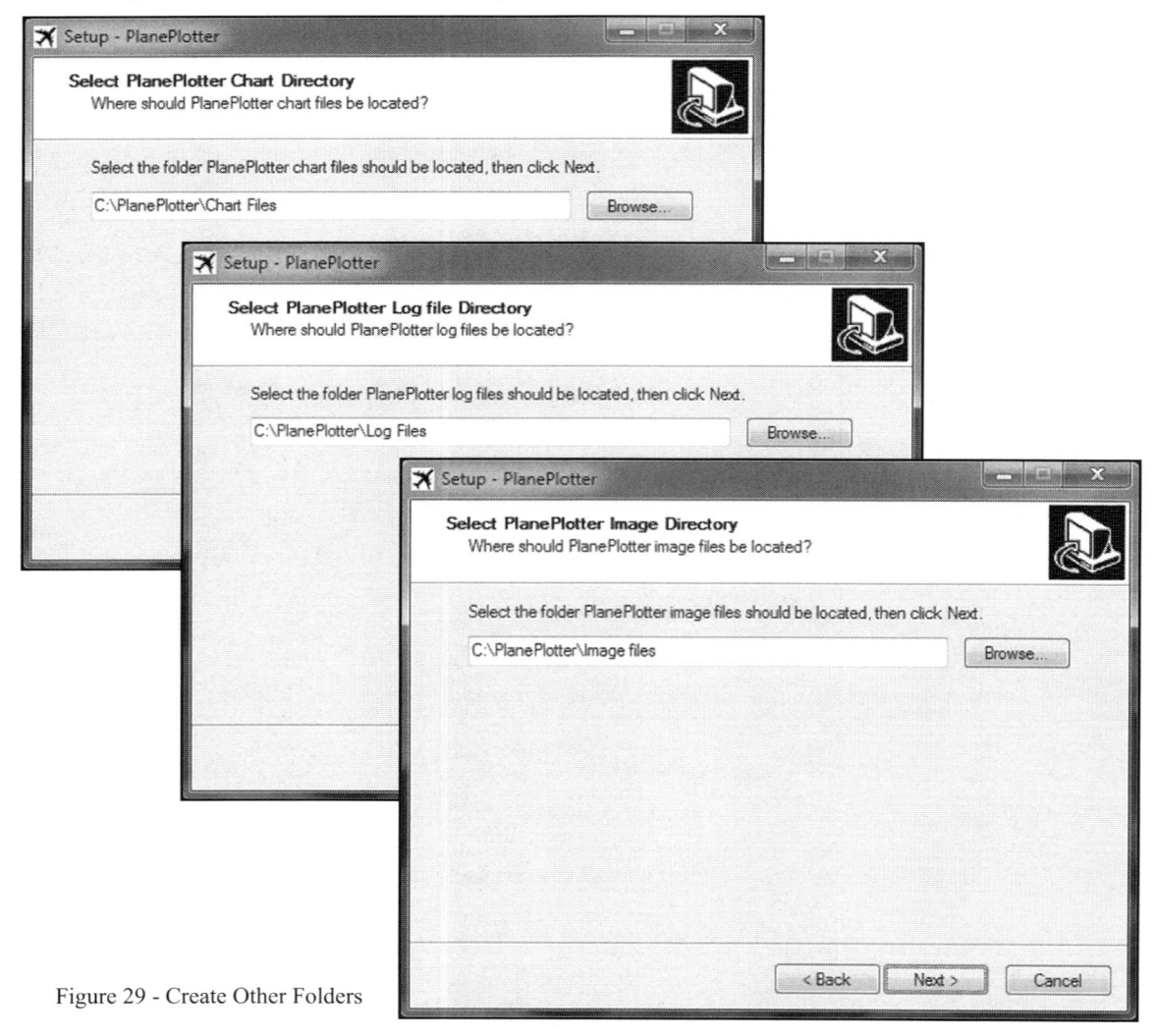

Figure 29 - Create Other Folders

Software Installation

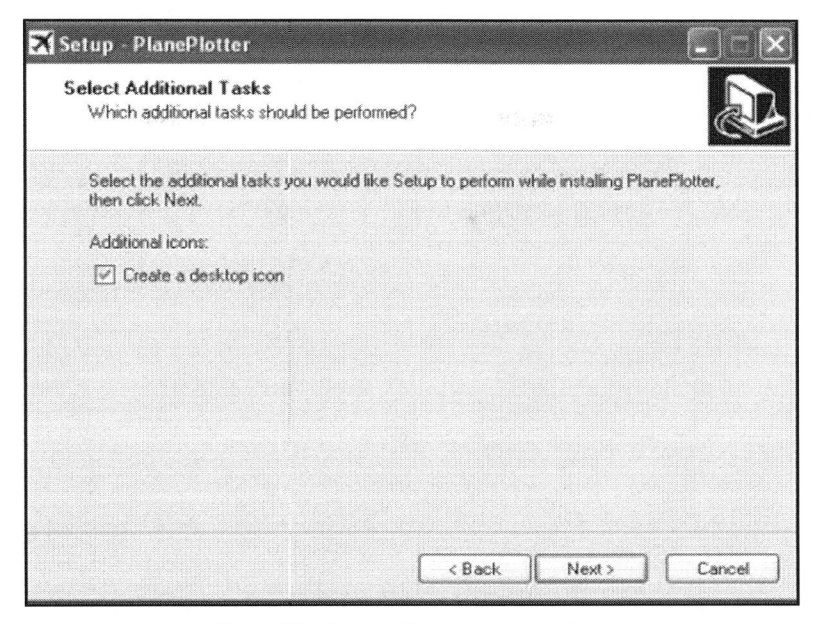

Figure 30 - Check 'Create a desktop icon'

The installation will continue step-by-step until completed and then the PlanePlotter program will start and a small dialog will be displayed stating that the software as yet is still unlicensed and asking whether the user wishes to register the installation immediately or continue to use it for a trial period of 21 days as shown in Figure 33.

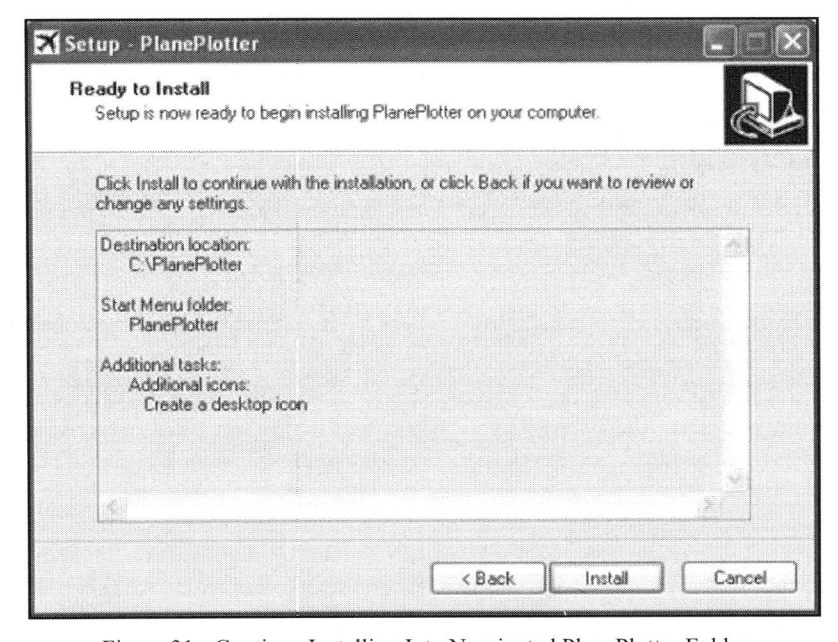

Figure 31 - Continue Installing Into Nominated PlanePlotter Folder

As this is the first time that the PlanePlotter program has been executed, it is possible that the PCs Firewall and/or protection devices may display a warning similar to that shown in Figure 32. The user should take the appropriate action to allow to the PlanePlotter program to continue.

Figure 32 - Firewall Check

Registration

To register the PlanePlotter software the user should click on the *www.coaa.uk/ COAAreg.htm* button in the dialog box as shown overpage in Figure 34.

Figure 33 - Trial Period or Registration

Software Installation

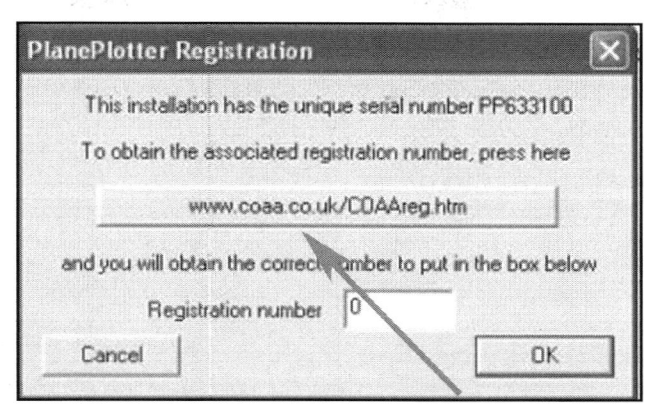

Figure 34 - Registration Dialog

After clicking on that button the users Internet Browser will open the COAA Internet registration page which is easily completed. After the registration form is complete a second screen will be displayed to collect the registration fee of 25€ normally using the Paypal method.

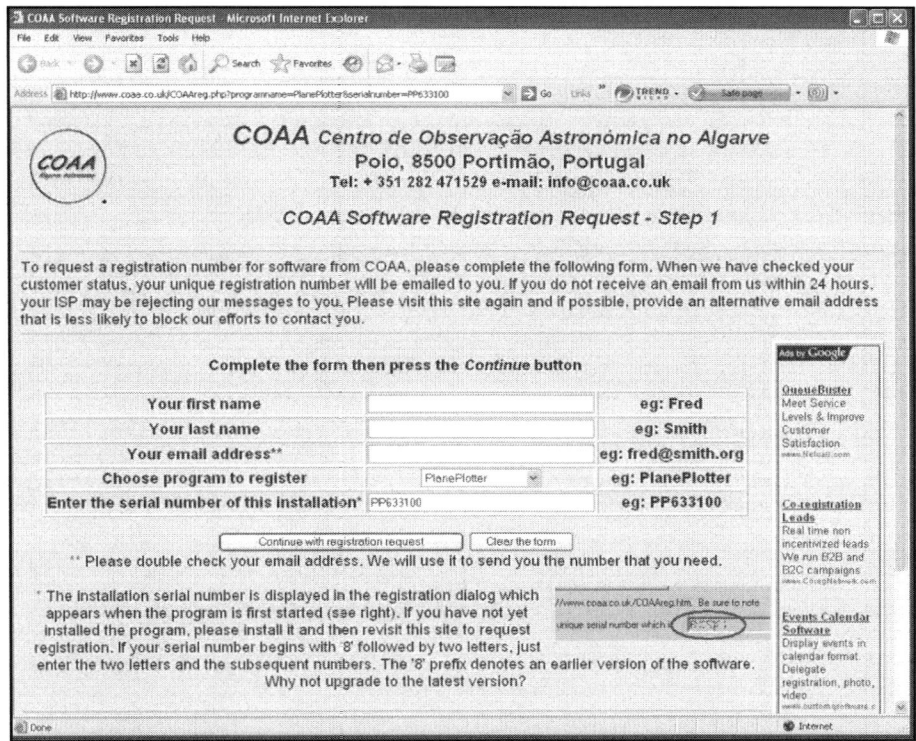

Figure 35 - PlanePlotter Software Registration Screen.

Once this stage has been completed it is recommend that the user should close and then restart the PlanePlotter before following the *First Time Use* instructions in Chapter 5.

Figure 36 - F17C
Photo Courtesy and Copyright of Kevin Daws

Chapter 5
First Time Use & Setup Wizard

It was recommended in Chapter 4 that the folder structure for PlanePlotter should be as shown in Figure 37 and no apologies are made for emphasising this point once again. By adopting this structure it will make life far easier in the future when dealing with all aspects of using the PlanePlotter program. Version 5.5.2.9 of PlanePlotters allows the user

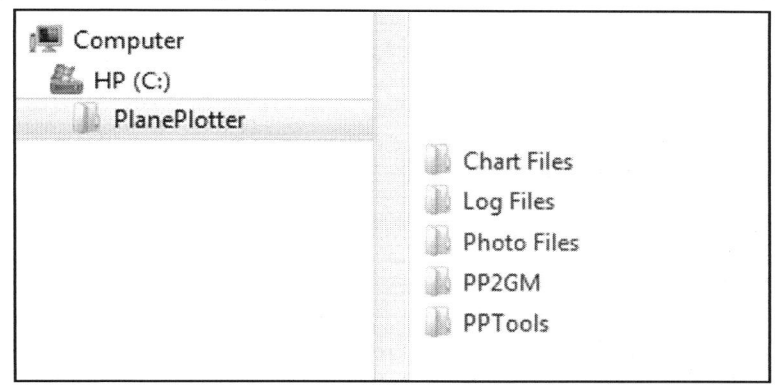

Figure 37 - Recommended PlanePlotter Folder/Directory Structure

to specify where the main Planeplotter program is installed to and also the folder for the Chart, Log & Image files. For earlier versions folders where created in the *%APPUSER%* area of the computers hard disk normally seen as *C:\Documents and Settings\Username\ AppData\COAA\PlanePlotter*, where *Username* may be *Administrator, Fred, Joe* or whatever is used on the user's PC. This can be a problem for many users especially later on when using many of the PlanePlotter program features, so it is really important that the following folders and their content should all be moved from their existing locations:

C:\Documents and Settings\Username\AppData\COAA\PlanePlotter\chart files
C:\Documents and Settings\Username\AppData\COAA\PlanePlotter\log files
C:\Documents and Settings\Username\AppData\COAA\PlanePlotter\photo files
to
C:\ PlanePlotter\chart files
C:\PlanePlotter\log files
C:\ PlanePlotter\photo files

Then check that the *C:\ PlanePlotter\chart files* folder contains these files:

C:\ PlanePlotter\chart files\General.out
C:\ PlanePlotter\chart files\globe.clb
C:\ PlanePlotter\chart files\globe.jpg

Once this has been done the user should start the PlanePlotter program and then select *Options, Directories* from the Main Menu Bar. The select and set each of the following in turn:

PP Log files" to point to "C:\PlanePlotter\log files
PP Chart files" to point to "C:\PlanePlotter\chart files
PP Photo files" to point to "C:\PlanePlotter\photo files

The other four directories can be left until later.

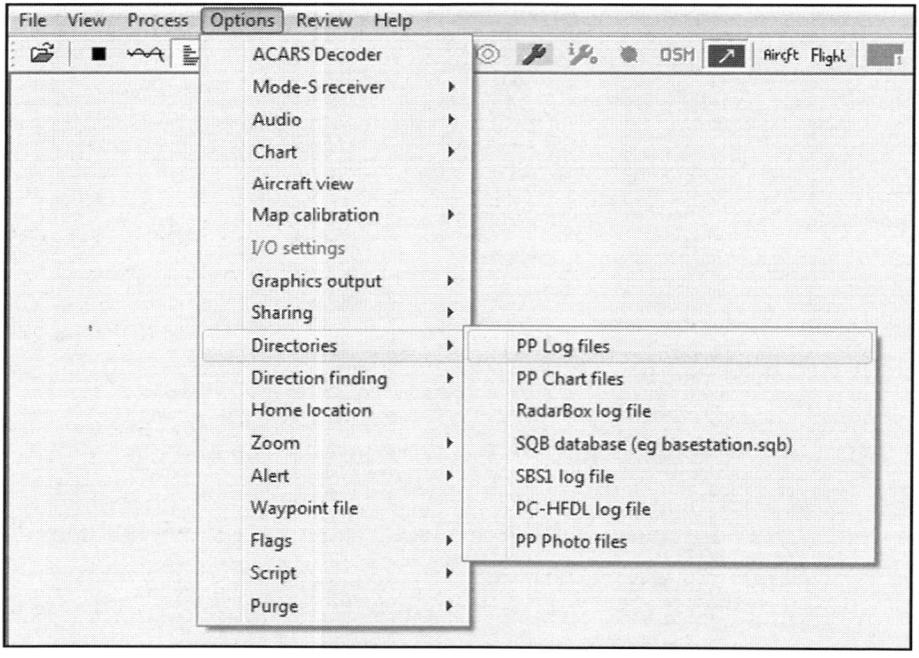

Figure 38 - Defining the PlanePlotter Folder locations

Once this has been done and completed then the settings should be saved. This is done by selecting from the main menu Bar *File, Save Restore point* and then clicking on the letter A.

Note: As will be discussed in more detail later there are four Restore Points: A, B, C, D each of which can save and store a complete set of settings including the folder locations, current chart and all the defined options. Each saved Restore Point can be restored at any time.

The Setup Wizard

The Setup Wizard will be launched automatically as part of the installation program. It can however be started again at any time by selecting *Help, Setup Wizard* from the Main Menu bar shown below in Figure 39.

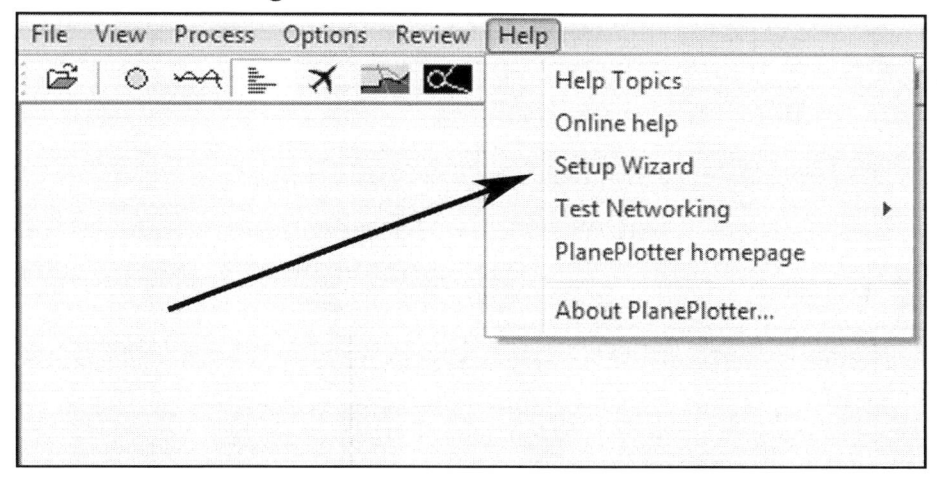

Figure 39 - Defining the PlanePlotter Folder locations

The Wizard provides an easy method of setting up all the initial options especially those associated with the type of Receiver to be used with PlanePlotter.

Rather than describe each step of the Wizard a flow chart showing the steps has been created and is shown on the following pages.

It is recommended that the reader has these pages near at hand as they run the Setup Wizard.

Once all the necessary steps of the Wizard have been completed a dialog box as shown below in Figure 40 will be displayed and the PlanePlotter program will then be ready to start processing aircraft data.

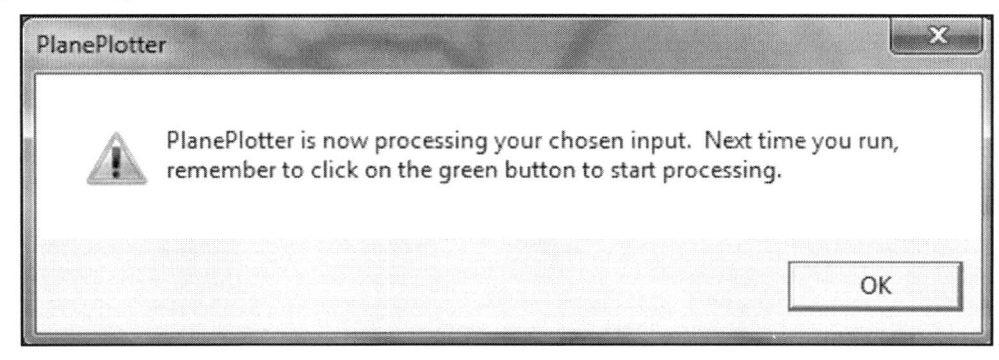

Figure 40 - Setup Wizard Completion Dialog

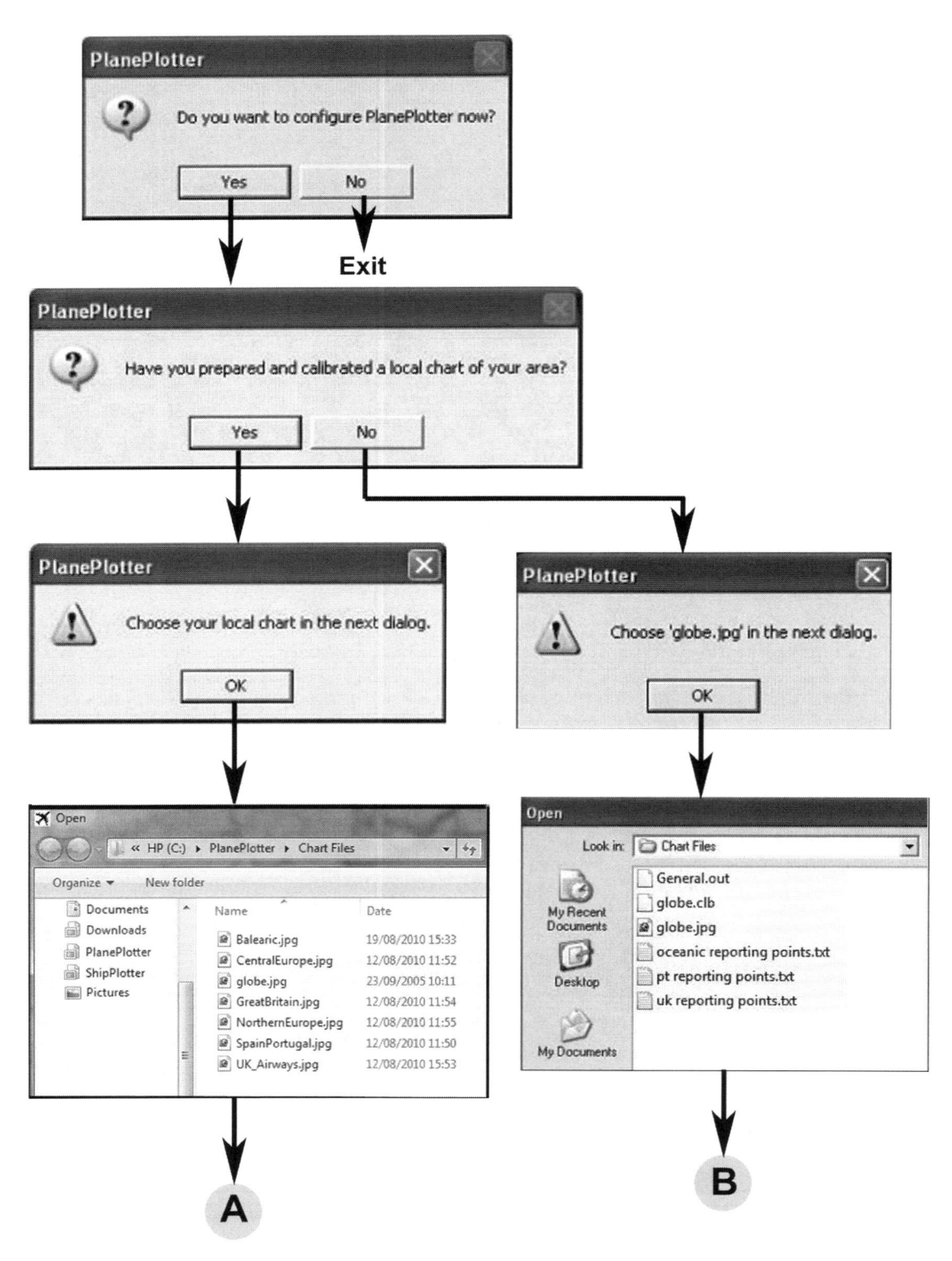

PlanePlotter User Guide

First Time Use & Setup Wizrd

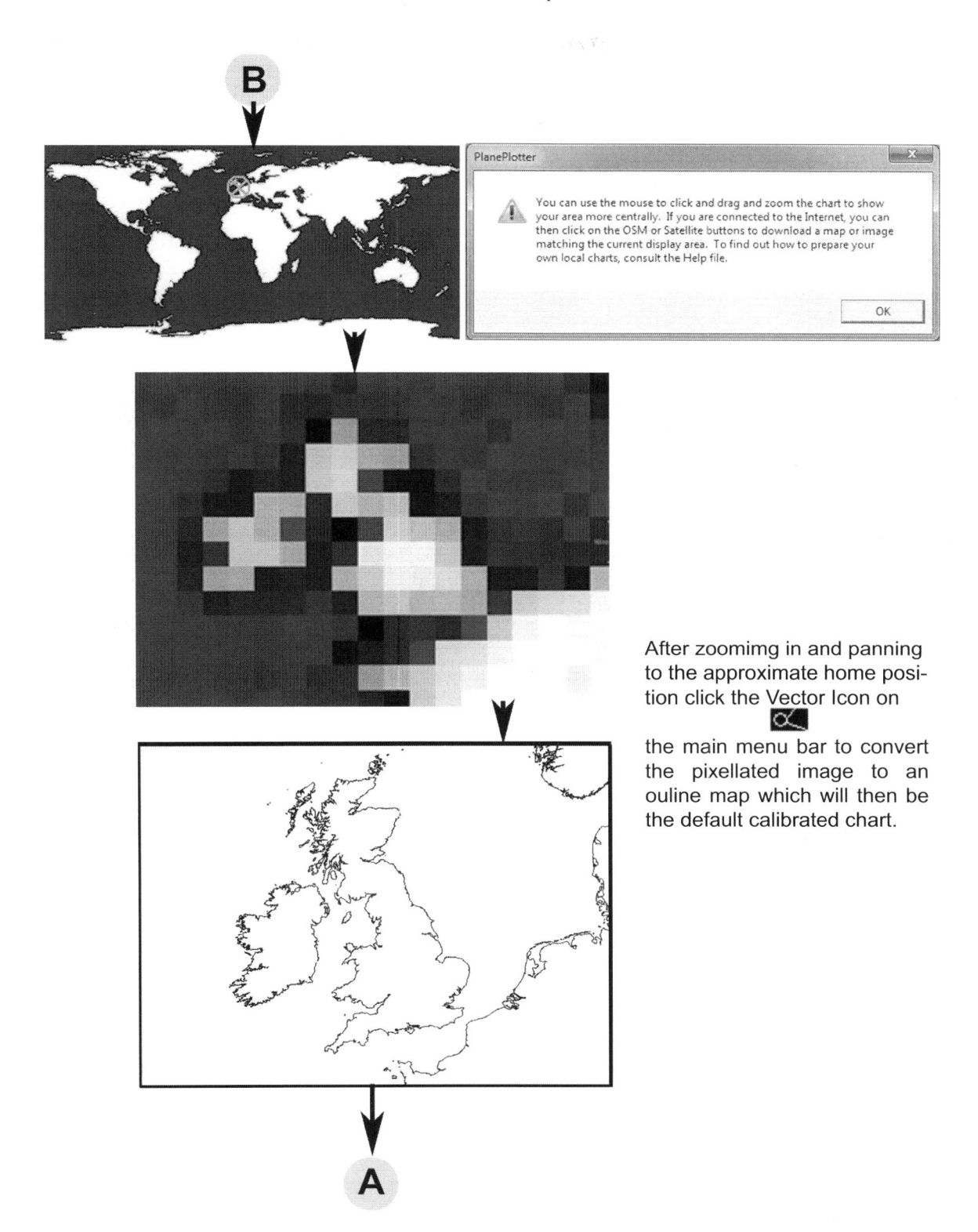

After zoomimg in and panning to the approximate home position click the Vector Icon on

the main menu bar to convert the pixellated image to an ouline map which will then be the default calibrated chart.

35

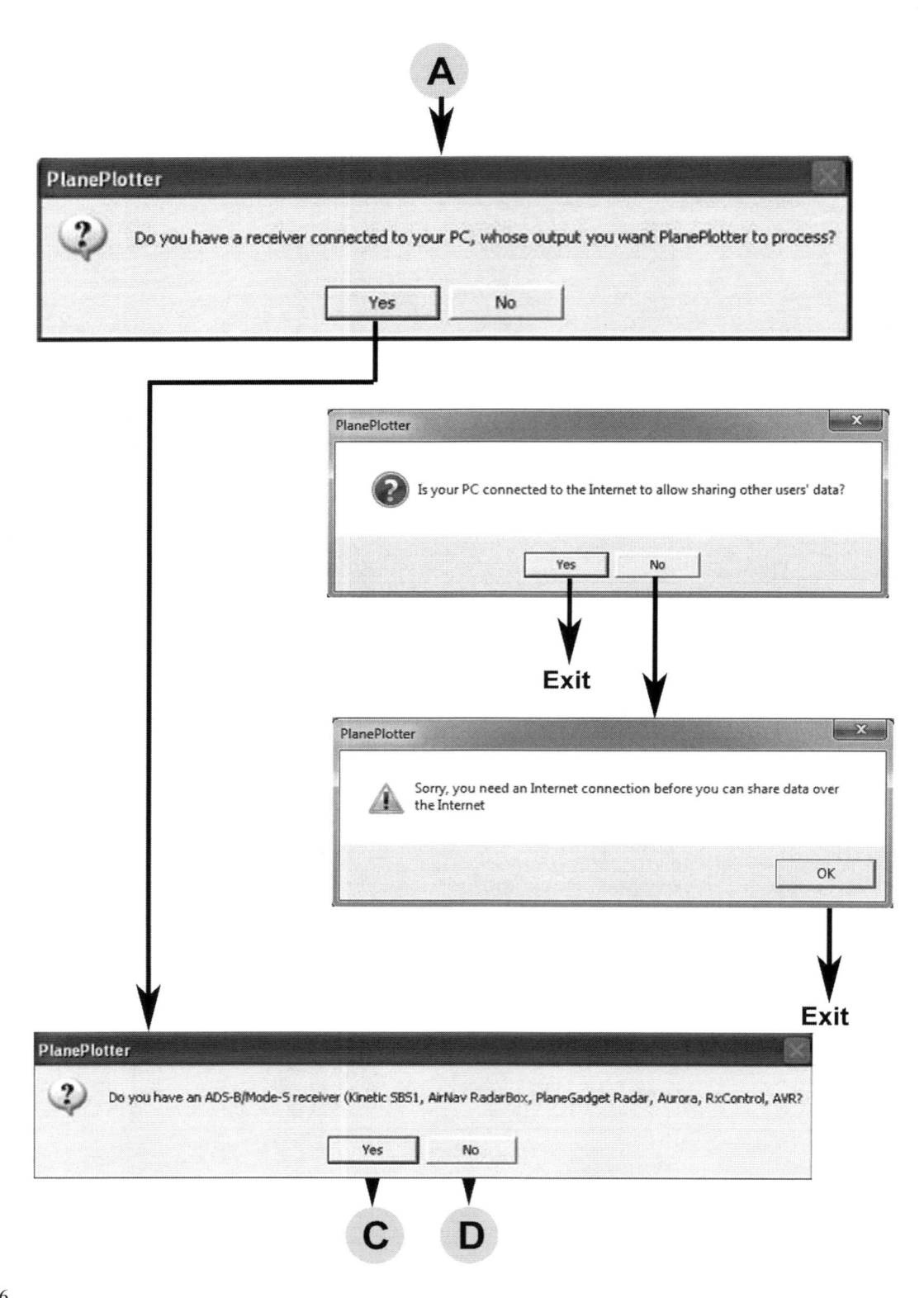

PlanePlotter User Guide

A

PlanePlotter
Do you have a receiver connected to your PC, whose output you want PlanePlotter to process?
[Yes] [No]

PlanePlotter
Is your PC connected to the Internet to allow sharing other users' data?
[Yes] [No]

Exit

PlanePlotter
Sorry, you need an Internet connection before you can share data over the Internet
[OK]

Exit

PlanePlotter
Do you have an ADS-B/Mode-S receiver (Kinetic SBS1, AirNav RadarBox, PlaneGadget Radar, Aurora, RxControl, AVR?
[Yes] [No]

C   **D**

First Time Use & Setup Wizrd

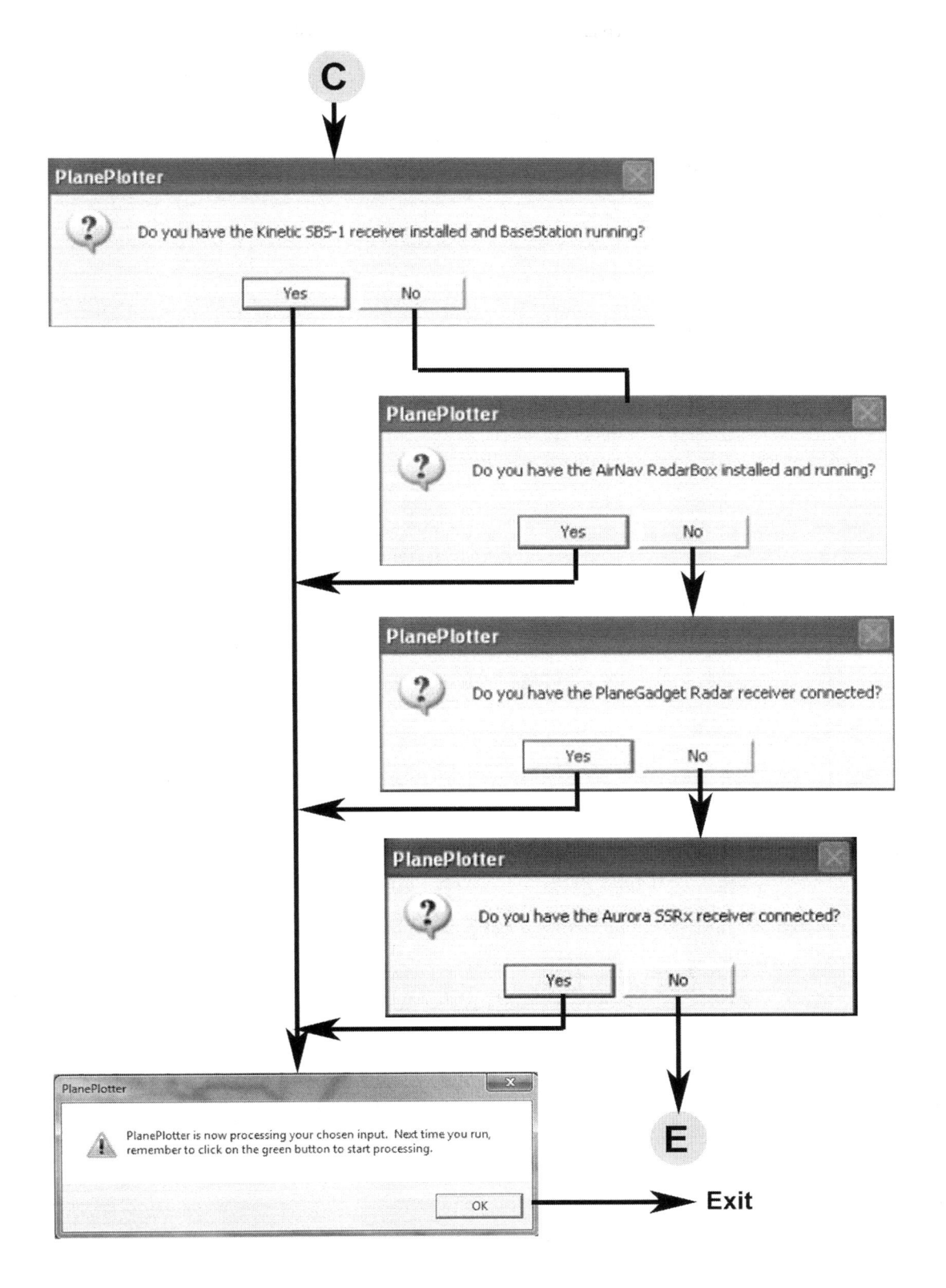

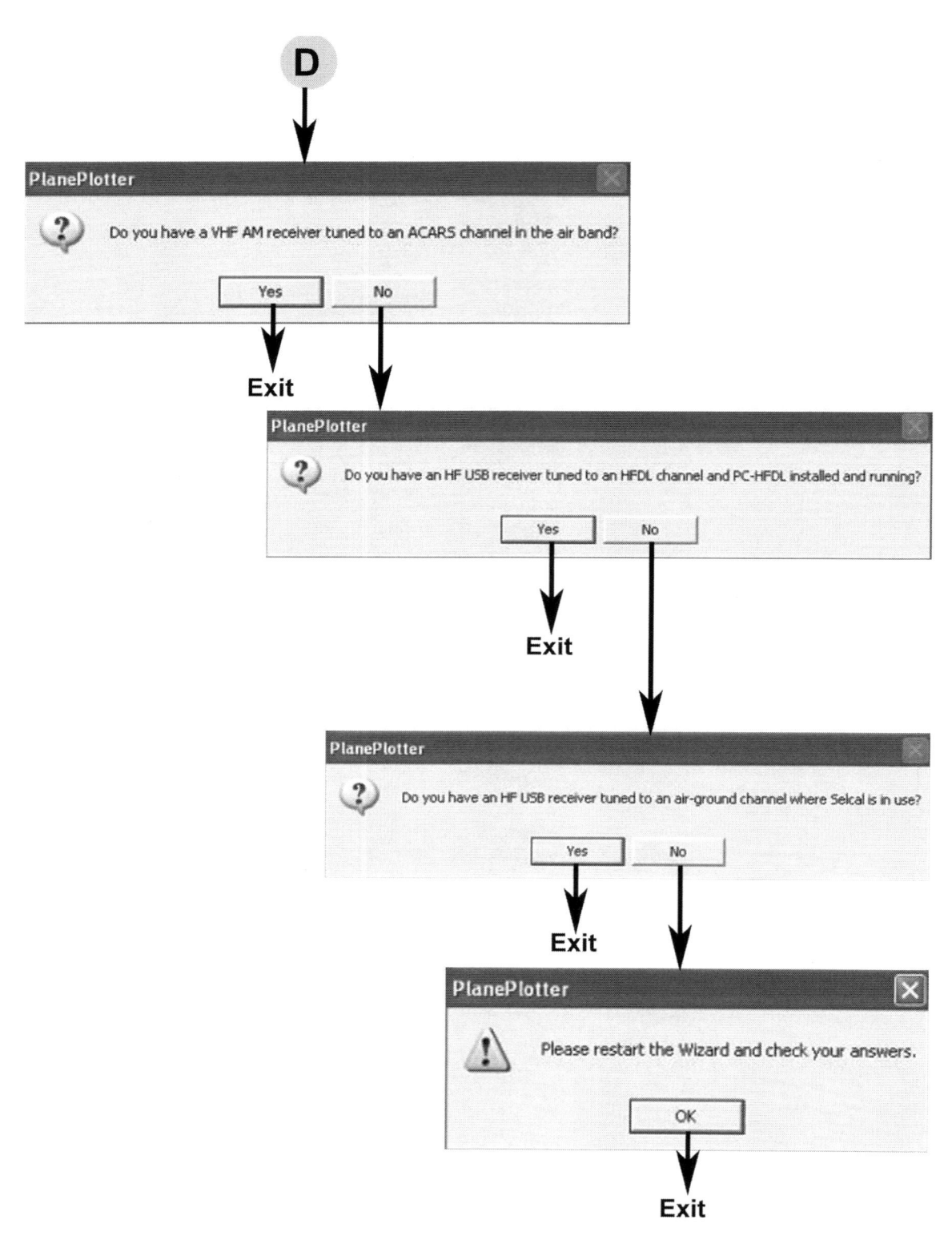

PlanePlotter User Guide

D

PlanePlotter

? Do you have a VHF AM receiver tuned to an ACARS channel in the air band?

[Yes] [No]

Yes → Exit

PlanePlotter

? Do you have an HF USB receiver tuned to an HFDL channel and PC-HFDL installed and running?

[Yes] [No]

Yes → Exit

PlanePlotter

? Do you have an HF USB receiver tuned to an air-ground channel where Selcal is in use?

[Yes] [No]

Yes → Exit

PlanePlotter

⚠ Please restart the Wizard and check your answers.

[OK]

→ Exit

First Time Use & Setup Wizrd

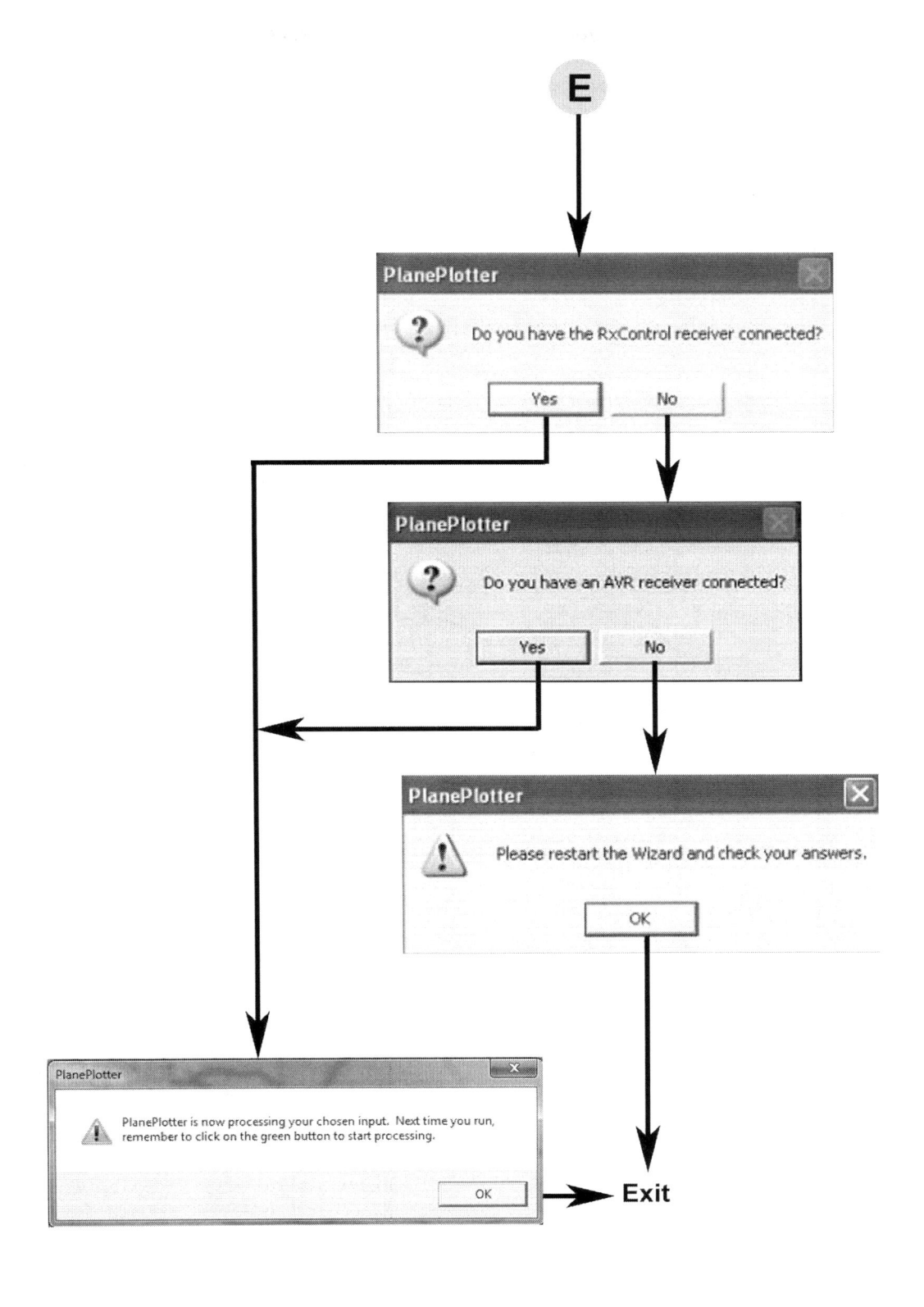

E

PlanePlotter
Do you have the RxControl receiver connected?
[Yes] [No]

PlanePlotter
Do you have an AVR receiver connected?
[Yes] [No]

PlanePlotter
Please restart the Wizard and check your answers.
[OK]

PlanePlotter
PlanePlotter is now processing your chosen input. Next time you run, remember to click on the green button to start processing.
[OK] → **Exit**

Figure 41 - EasyJet
Photo Courtesy and Copyright of Trevor Marshall

Chapter 6
PlanePlotter Menus

As can be seen in Figure 43 over page the PlanePlotter program has a comprehensive Menu structure to enable the user to easily deal with all the features and facilities available in the program. This chapter describes in detail each of the Menu commands.

File Menu Commands
The File menu offers the following commands:

Open Chart
This command is used to load a new PlanePlotter chart background. Charts or maps are graphic BMP or JPG files and must be accompanied by a CLB calibration file.

Pos rep (Ctrl-P)
Enthusiasts monitoring air traffic control frequencies are able to hear aircraft position reports especially for oceanic traffic on HF receivers. The typical format for such reports is of the form *Identity, position/waypoint, time at report point, altitude* quite often followed by *next position/waypoint and estimated time for that point*. The *Pos rep* menu option opens a small entry dialog box to enable the user to enter a manual position report. The format of such an entry is *id pos time alt pos time* where
   id is the aircrafts callsign
   pos is the aircrafts position or waypoint in the form 40N01E or 42N15W or a waypoint name.
   time is the 24 hour clock time as a four digit number e.g. 2359

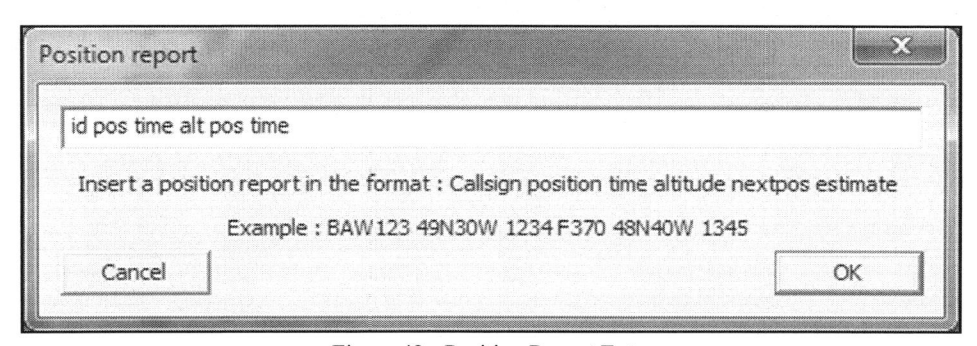

Figure 42 - Position Report Entry

PlanePlotter User Guide

File

- Open chart
- Pos rep
- Report
  - Save report
  - Setup report
- Save Restore point
- Load Restore pont
- Factory default settings
  - A
  - B
  - C
  - D
- Exit

- Toolbar
- Expand toolbar
- Status Bar
- Signal
- Messages
- Aircraft
- Aircraft list display options
- Chart
- Chart display options
- Always on top
- My sky
- Copy view position Ctrl-C
- Paste view position Ctrl-C
- Best chart (F7)
- Minimise Mlat window

- All aircraft
- Chart Mlat only (F10)
- Chart Desig only (Shift-F10)
- Chart 'Interested'
- Chart 'Not Interested'

- All aircraft
- All aircraft with flags
- Aircraft with positions
- Aircraft without positions
- Aircraft with Mlat possible
- Aircraft 'Interested'
- Aircraft 'Not Interested'

View

- ACARS Decoder
- Mode-S Receiver
- Audio
- Chart
- Aircraft view
- Map calibration
  - Add calibration point
  - Save calibration data
  - Clear calibration
  - Display calibration points
  - Display geo grid
- I/O settings
- Graphic output
- Sharing
  - Setup
  - Enable
- Directories
  - PP Log files
  - PP Chart files
  - RadarBox log file
  - SQB database
  - SBS1 log file
  - PC-HFDL log file
  - PP Photo files
- Direction Finding
  - DF setup
  - DF test
- Home location
- Zoom
  - Zoom in
  - Zoom out
- Alert
- Waypoint file
  - Alert Zone
  - Alert Shell
    - Clear
    - Add
    - Save
    - Load
- Flags
- Scripts
  - Enable
  - Define
    - Alert Shell Setup
    - Alert Shell Enable
- Purge
  - Define
  - Run
  - Purge flights
  - Purge routes

Process

- Start
- Stop

Options

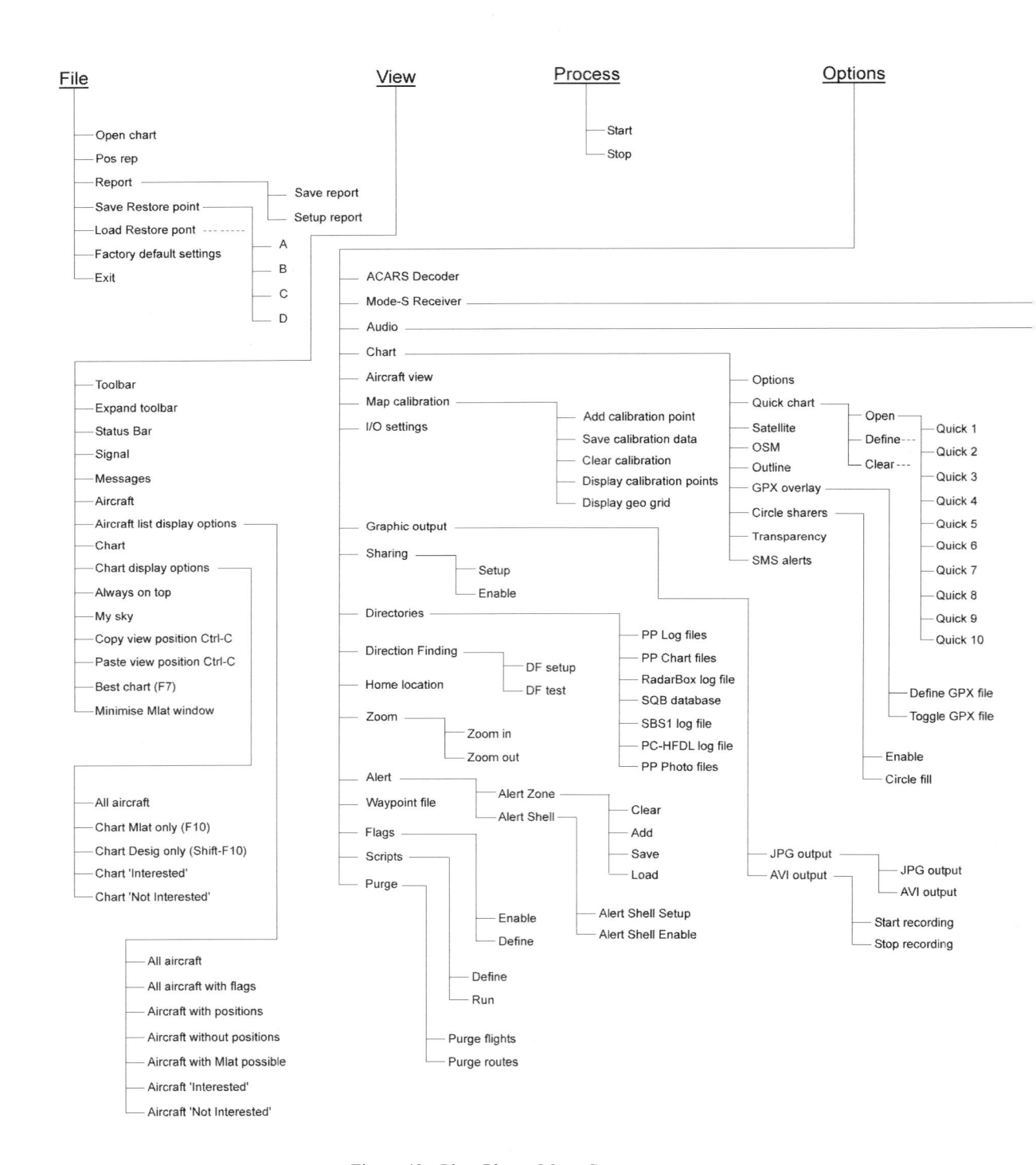

- Options
- Quick chart
  - Open
    - Quick 1
  - Define
    - Quick 2
  - Clear
    - Quick 3
    - Quick 4
    - Quick 5
    - Quick 6
    - Quick 7
    - Quick 8
    - Quick 9
    - Quick 10
- Satellite
- OSM
- Outline
- GPX overlay
  - Define GPX file
  - Toggle GPX file
- Circle sharers
  - Enable
  - Circle fill
- Transparency
- SMS alerts
- JPG output
- AVI output
  - JPG output
  - AVI output
  - Start recording
  - Stop recording

Figure 42 - PlanePlotter Menu Structure

PlanePlotter Menus

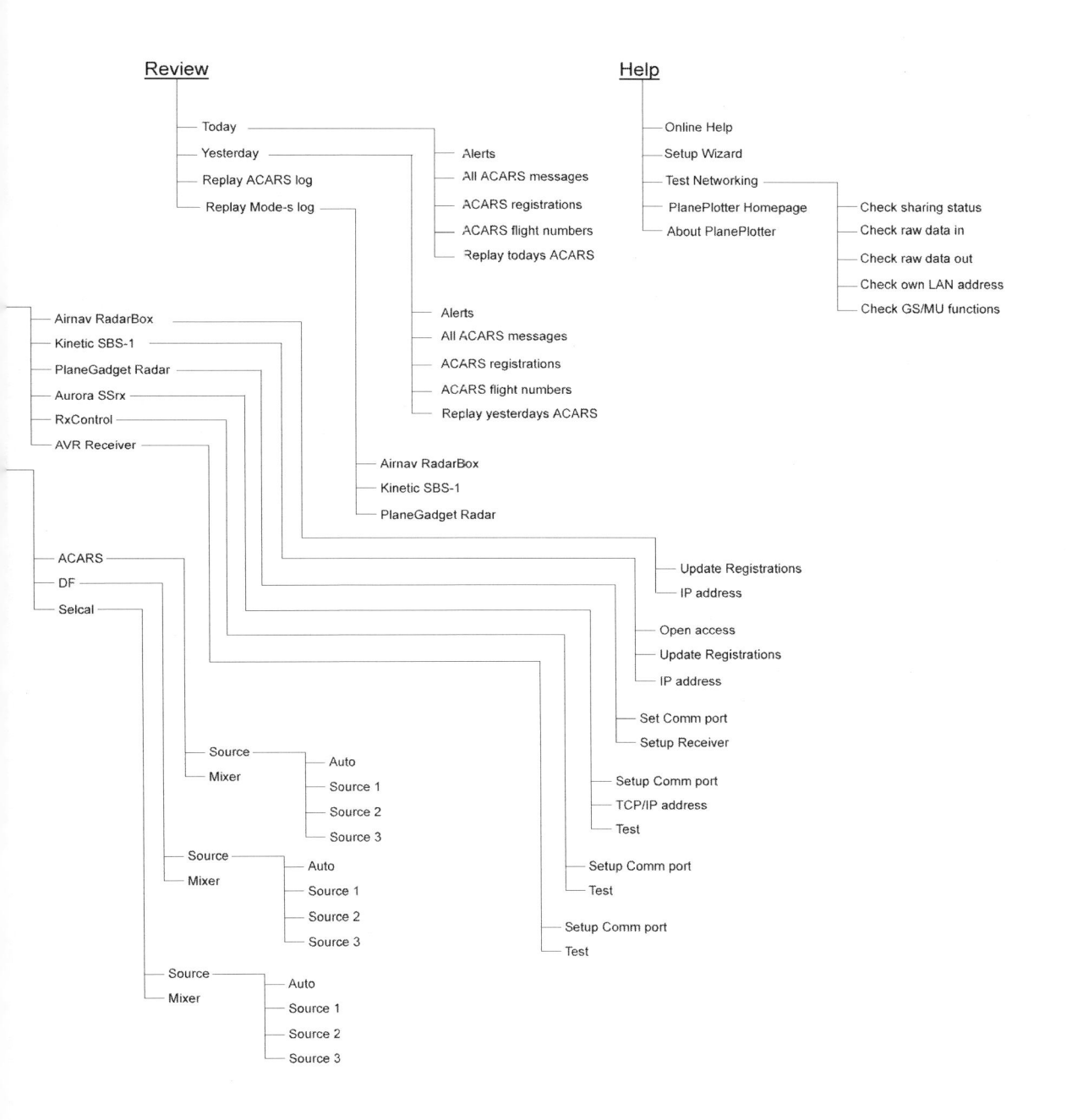

Review
- Today
- Yesterday
- Replay ACARS log
  - Alerts
  - All ACARS messages
  - ACARS registrations
  - ACARS flight numbers
  - Replay todays ACARS
- Replay Mode-s log
  - Alerts
  - All ACARS messages
  - ACARS registrations
  - ACARS flight numbers
  - Replay yesterdays ACARS
  - Airnav RadarBox
  - Kinetic SBS-1
  - PlaneGadget Radar

Help
- Online Help
- Setup Wizard
- Test Networking
  - Check sharing status
  - Check raw data in
  - Check raw data out
  - Check own LAN address
  - Check GS/MU functions
- PlanePlotter Homepage
- About PlanePlotter

- Airnav RadarBox
  - Update Registrations
  - IP address
- Kinetic SBS-1
- PlaneGadget Radar
- Aurora SSrx
  - Open access
  - Update Registrations
  - IP address
- RxControl
  - Set Comm port
  - Setup Receiver
- AVR Receiver
  - Setup Comm port
  - TCP/IP address
  - Test

- ACARS
  - Source
    - Setup Comm port
    - Test
  - Mixer
    - Auto
    - Source 1
    - Source 2
    - Source 3
- DF
  - Source
    - Setup Comm port
    - Test
  - Mixer
    - Auto
    - Source 1
    - Source 2
    - Source 3
- Selcal
  - Source
  - Mixer
    - Auto
    - Source 1
    - Source 2
    - Source 3

43

alt is the aircrafts altitude either a flight level e.g. F250 or height in feet e.g. 250000

pos (2nd instance) is the aircrafts next position or waypoint

time (2nd instance) is the estimated 24 hour clock time at the next position

for example

ken120 40n01e 0935 f250 41n02w 1020

or

ken120 alpha 0935 f250 sigma 1020

Once the entry has been made the details can be seen in the Aircraft View display. Double clicking on the entry will switch the display to a Chart View where an icon will be seen for the entry. When the next point and time have been included in the entry PlanePlotter will cleverly use the two times to determine the speed and course of the aircraft.

If a waypoint name is used then PlanePlotter will search the waypoints.txt file for the waypoint coordinates. Note that this file is stored in the main program application folder directory and can be edited by the user to add waypoints in their own geographical area.

Report

Save Report

Creates and saves the report with the parameters setup. The report will be saved in the Log Files folder with a name like *pp_report_12345.txt*.

Setup Report

The Report option enables the user to specify the data fields to be included in a report file. To include a data field the corresponding box should checked. The bottom check box allows the user to generate a report only for those aircraft which have been defined by a flag. See *Menu Options, Flag, Define* a feature used to restrict reports to aircraft or fleets of particular interest.

Save Restore point

This command allows the user to save a current configuration as one of four Restore Points A, B, C or D.

Each saved restore point contains all the current settings within the program including the chart, the i/o settings and all the defined options including alerts, waypoints, flags, directory

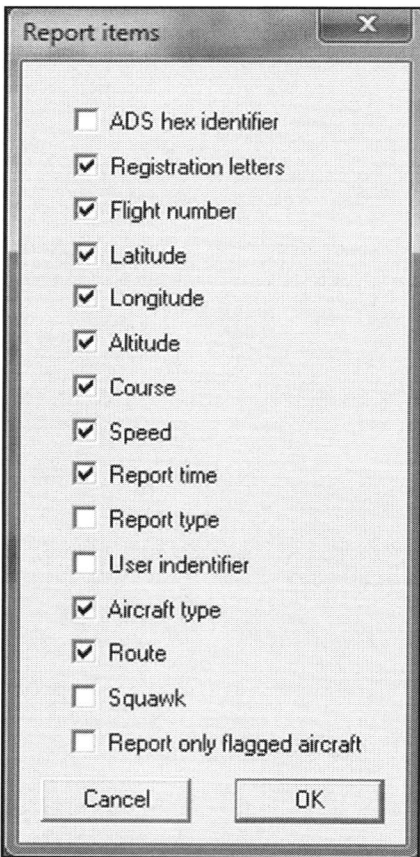

Figure 43 - Report Setup

locations, sharing setup. If any Quick Charts have been defined then these are also saved.

Load Restore point

This command is used to reset the current configurations back to one of the four saved Restore Points. Note that no warning is given that the current configuration all existing settings and data will be overridden! For this reason it is recommended that before loading a Restore Point that the current configuration is saved as Restore Point D and that Restore Point D be used as a safety net.

Factory default settings

This command restores all the settings to the default settings provided with a new install of the PlanePlotter program. It is recommended that the Setup Wizard is run immediately after invoking this command as it is only intended as a last resort to obscure problems that may arise by some options being incorrectly set by the user option.

Exit

This command will end the current PlanePlotter session.

View Menu Commands

The View menu offers the following commands:

Toolbar

This command is used to show or hide the toolbar at the top of the PlanePlotter screen.

Expand Toolbar

Each of the ABCD buttons can be hidden by a *Shift-Left Click*. The Expand Toolbar command is used to reveal any hidden buttons.

Status Bar

This command is used to display and hide the Status Bar at the foot of the PlanePlotter screen. The Status Bar has five areas of interest. The left area of the status bar describes actions of menu items as the arrow keys are used to navigate through the menus or to describe the actions of toolbar buttons as they are pointed to. The second area shows the Geo Co-ords of the current cursor position. The next the number of messages received then the name of vessel bring tracked with its speed, course and time of last message and finally the current UTC time.

Signal

This command is used to display a graph of the raw audio signal that PlanePlotter is processing when receiving ACARS messages. This view is used to ensure that audio signal from the ACARS receiver is correctly connected and that the volume level is correctly set. Note that processing must have been started for anything to be shown in this view.

Messages

This command is used to display any recently received and decoded ACARS or HF SelCal messages. Again note that processing must have been started for anything to be shown in this view.

Aircraft

This command is used to display a table of the aircraft that have currently been processed by PlanePlotter. The table columns list the ICAO hex code, aircraft registration, flight number, latitude, longitude, altitude, course, speed, aircraft type, route, squawk code, time of the latest position report, report type, the uploading sharer who last uploaded the data, interested tag, user Tag, and a string of the last 5 users who have uploaded the data.

The information provided by the data displayed in the table is enhanced by the use of mouse key clicks and a keyboard shortcut and these are discussed below.

The entries in the table are initially sorted in the order that they are received in the current PlanePlotter session. However the table can be sorted by any of the column headings, in ascending or descending order, by using a mouse *left-click* on

ICAO	Reg.	Flight	Lat.	Long.	Alt.	Course	Speed
740728	JY-AYH	RJA110	41.04230	3.65450	34000'	103.1°	497.0kts
342488	EC-JYC	TLY401T	0.00000	0.00000	0'	0.0°	0.0kts
3414C9	EC-HQJ	VLG3211	41.05450	1.74470	8075'	2.4°	260.2kts
4B161E	HB-IJU	EDW102	41.51420	3.50100	36000'	75.5°	520.5kts
020009	CN-RNB	RAM960	0.00000	0.00000	19550'	0.0°	0.0kts
4CA25A	EI-DHE	RYR8037	40.47400	1.35513	16875'	26.1°	405.2kts
4A08E8	YR-BGH	ROT413	39.55740	-0.64429	36950'	299.7°	421.5kts
44A847	OO-JBG	JAF283	38.79350	-0.31687	37000'	176.5°	405.8kts
3425D8	EC-KEC	JKK5207	40.53900	0.96393	19325'	52.0°	450.3kts
3946E4	F-GRXE	AFR454H	40.76800	3.15333	37000'	182.1°	386.3kts
342590	EC-KFI	VLG2112	40.40934	0.65437	30800'	42.6°	529.4kts
342181	EC-JCG	ANE8907	0.00000	0.00000	11650'	0.0°	0.0kts
3424D5	EC-KDG	VLG2012	39.72200	-0.05278	37050'	235.6°	379.2kts
341141	EC-ISY	PVG7846	41.30580	2.10130	38975'	256.3°	388.1kts
4404E1	OE-LEF	BER64Z	39.41540	0.79137	22200'	92.2°	467.3kts
3C4881	D-ABDA	BER19Z	40.74590	2.71260	29200'	181.8°	420.2kts
451E8A	LZ-WZB	WZZ633L	39.58320	0.69586	19350'	275.6°	386.9kts
400AF7	G-EZAM	EZY2TB	39.96300	0.19070	33350'	36.8°	514.4kts
4CA736	EI-EBM	RYR5374	41.50560	3.25253	38000'	78.6°	507.9kts
4CA27C	EI-DHO	RYR9PY	41.06330	3.73660	37000'	189.6°	390.4kts
4CA56E	EI-DWI	RYR8517	41.56820	2.74658	37000'	181.5°	383.1kts

Aircraft View Table

the column heading label. Any aircraft data that is received after the sorting process will be added to the end of the list and not in the sorted order.

A *right click* on one of the aircraft will initiate a lookup of the aircraft registration letters through an external database.

With a *Ctrl-right click* on one of the aircraft PlanePlotter will open up the aircraft information dialog box where the user can edit the properties of the aircraft.

For mice with a mouse wheel button then a *middle click* on an aircraft will designate the aircraft. Unlike the *left click* option described below this does not change to view to Chart View.

A *left click* on one of the aircraft will open the current chart or outline and centre it on the chosen aircraft. That aircraft will then become the designated aircraft and dependant on setting in "Options, Chart, Options" the chart will continue to centre itself on the designated aircraft.

A *Ctrl-left click* is available only to *Master Users* and is used to initiate the Multilateration process for position-less aircraft. Green text denotes that three or more raw data Ground Stations received the aircraft at the last refresh cycle and that a position fix is probable whilst Orange text denotes that two raw data Ground Stations received the aircraft at the last refresh cycle but that a position fix is not likely although but a position curve is possible. (Multilateration is discussed in further detail in Chapter 11).

Type	Route	Squawk	Time	Rep	Sh.	Int	U-tag	Sharers
A321	LEMD-OJAI	5503	16:05:31UTC	6	e3			e3cIaTe3cI
SW4	*-*	0000	15:58:45UTC	6	pA			pAiZpAiZ
A320	GCXO-LEBL	0000	15:57:59UTC	6	pA			pApA
A320	*-*	5376	16:05:35UTC	6	cI			cIaTcIaTcI
B735	GMMN-LEBL	6474	15:59:18UTC	6	*1			
B738	LEIB-LEBL	0000	16:02:50UTC	6	*1			pA pA
B737	*-*	6771	16:01:27UTC	6	pA			pApApApApA
B738	*-*	7162	15:58:03UTC	6	pA			pApApApApA
A320	GCLP-LEBL	6354	16:06:29UTC	6	*1			pA pA
A319	*-*	7644	16:01:28UTC	6	aT			aTcIaTcIaT
A320	LEMG-LEBL	3740	16:06:24UTC	6	*1			pA pA
CRJ2	LEIB-LEAL	1240	16:00:07UTC	6	pA			pApApA
A320	LEBL-LEGR	0000	16:04:52UTC	6	pA			pApApApApA
B752	*-*	4007	15:59:08UTC	6	pA			pAcIcIcIaT
A320	*-*	0000	16:05:03UTC	6	pA			pApApApApA
A320	*-*	2575	16:04:40UTC	6	cI			cIcIcI
A320	*-*	0000	16:05:03UTC	6	pA			pApApApApA
A319	*-*	5544	16:06:28UTC	6	*1			pApApApA
B738	LEMD-LIPE	5512	16:05:26UTC	6	cI			cIaTcIaTcI
B738	*-*	1103	16:04:38UTC	6	cI			cIaTcIaTcI
B738	ENRY-LEPA	0735	16:00:13UTC	6	cI			cIaTcIaTaT

A keyboard shortcut *Ctrl-F* open a dialog box to enable the table to be searched for a particular aircraft ICAO hex code or the registration letters, the flight number or a sharer code.

Aircraft list display options
The list of aircraft displayed can be filtered in several ways by choosing one of the following commands.

All aircraft
This command displays a table of all aircraft that are currently known to PlanePlotter

All aircraft with flags
This command displays a table of flagged aircraft and is only available when the user has defined and enabled flags. See *Menu Options, Flags, Enable*.

Aircraft with positions
This command displays a table of the aircraft that are currently known to PlanePlotter and have known positions.

Aircraft without positions
This command displays a table of the aircraft that are currently known to PlanePlotter and do not have known positions.

Aircraft with Mlat possible
This command displays a table of the aircraft that are currently known to PlanePlotter and have the possibility of having their positions determined by Multilateration. The filtering is based on the presence of Ground Stations in the sharer history string and is only a guide to the possibility of successful Multilateration.

Aircraft 'Interested'
This command displays a table of aircraft that the user has tagged with the "Interested" tag in the SQB database. The tag can be set/reset in the aircraft information dialog.

Aircraft 'Not Interested'
This command displays a table of aircraft that the user has not tagged with the "Interested" tag in the SQB database.

Chart
This command is used to display the positions of aircraft on the currently selected chart which by the way should have an associated calibration file.

To assist in identifying position reports from ACARS messages or manually entered position reports aircraft are plotted with a *red symbol* whilst position reports from real time Mode-S/ADS-B messages are plotted with a *yellow symbol*. If the report is from a delayed or expiring Mode-S/ADS-B messages then an *orange symbol* is used. Position reports from in HFDL messages are plotted with a *cyan symbol* and *blue symbols* are used to plot the next intended waypoint with *green symbols* being used to plot previous positions included in an ACARS AMDAR weather report.

Magenta symbols are used to plot waypoints that are contained within ACARS flight plan messages.

Master Users may also see *white symbols* which show *position-less* aircraft from Mode-S reports without an ADS-B message and whose position has been determined by the Multilateration process which is discussed in detail later in Chapter 11.

If an aircraft is known to be descending the nose of the plotted symbol is *brown* whilst if it known to be climbing the nose of the plotted symbol is *pale blue*.

The home location is displayed by an 'X' in a circle using the colour defined for DF vectors. Further information on customizing the home location symbol will be found later in *Options, Home location*.

When a Mode-S report includes speed and heading the aircraft symbol is a outline of an aircraft pointing in the angle of the heading. When no speed and direction are available then a simple triangular symbol is used.

When PlanePlotter receives a succession of Mode-S or HFDL messages with position reports the program will attempt to calculate the speed and heading from the vector between reports.

At the bottom right of the chart display are four small windows that give details of the designated aircraft, the coordinates of the cursor together with the distance and bearing from the either the home location or the designated aircraft to the cursor position The third window contains four numbers separated by slashes representing the number of aircraft received in this session, the number of messages received, the number of messages with positions and the number of raw data records in the buffer. The fourth window displays the PCs current time which should be synchronized to a standard source.

As with the Aircraft View data displayed on the chart is enhanced by the use of a number mouse key clicks and a keyboard shortcut.

Using a *left click and drag* the user can move the chart or outline to follow the cursor movement. Using a *Shift left click and drag* the user can move the label for the aircraft nearest to the mouse position will be moved accordingly and thin line will connect the displaced label to the aircraft symbol.

By using a *right click* on an aircraft an AirRep dialog box will be displayed to show the details for the chosen aircraft. In that dialog you can also enable *Flight deck view* for use with Google Earth and

Figure 44 - Chart AirRep

Contrail to enable the contrail feature for that aircraft regardless of the Contrail options specified in *Options, Chart, Options*.

A *double left click* on an aircraft symbol makes it the *designated aircraft* and places a ring around it on the chart. When the option is enabled in *Options, Chart, Options* the chart will automatically be centred on the designated aircraft. When an aircraft has been designated then the distance and bearing from the aircraft to the cursor is displayed in the status bar alongside the cursor coordinates. When no aircraft is designated, the distance and bearing from the home location to the cursor are displayed.

The user can search for a particular aircraft by pressing *Ctrl-F* to open a dialog box where the aircraft ICAO hex code, the registration letters or the flight number or just a part of the string can be entered. If the string is found then the first active aircraft in the internal data which matches the string will be centred on the chart and will become the designated aircraft and highlighted in red. Pressing *F3* will repeat the search starting at the currently designated aircraft and using the same search string.

By pressings *Ctrl-C* the position and scale of the chart view will be saved to the Clipboard in the form of a text string like this *PP view: 40.22307 1.234108 3.971* this string can be pasted back into a chart or outline view in PlanePlotter using *Ctrl-V* or *View, Paste view*. Doing so it will adjust the position and scale of the current view to match the one that was originally copied.

Further keyboard shortcuts include the use of the Function keys. *F7* searches the chart files directory for the largest scale map or chart that covers the current centre of the field. *F8* toggles the display of the GPX overlay when one has been defined. *F9* toggles the prediction option which is discussed later in *Options, Chart, Options*.

The *F10* function key toggles the option to display only Mlat aircraft on the chart, pressing the F10 key enables the option to be alternately turned on or off. *Shift-F10* toggles the option to display only the designated aircraft on the chart. Whilst *Ctrl-F10* toggles the option to display only flagged aircraft on the chart. And finally *F11* toggles the MySky display on or off.

Chart display options
All aircraft
The chart displays all aircraft currently being processed.
Chart Mlat only (F10)
Toggles the display of Mlat aircraft between showing all aircraft and showing only those aircraft that have had their positions determined by Multilateration.
Chart Desig only (Shift-F10)
Toggles the display of the designated aircraft to show only the aircraft designated or all aircraft or all aircraft
Chart 'Interested'
Toggles the display of aircraft to show only the aircraft tagged as "Interested" in SQB database or all aircraft.

Chart 'Not Interested'
Toggles the display of aircraft to show only the aircraft not tagged as "Interested" in SQB database or all aircraft.

Always on top
This command is used to display all the decoded AIS messages as they are received with he latest message always at the top of the list.

My sky
This command is used to provide a graphical interpretation of how the sky may be viewed within an arc of 90° and an elevation up to 60°.

Copy view position Ctrl-C
This command is used to save the position and scale of the chart view to the Clipboard in the form of a text string like this *PP view: 40.22307 1.234108 3.971.*

Paste View position Ctrl-V
This command will paste a text string of the form *PP view: 40.22307 1.234108 3.971* to adjust the position and scale of the current view to match the one that the text string was originally from.

Best Chart (F7)
This command will automatically load the most suitable chart from those charts saved in the Charts folder for the currently received data.

Minimise Mlat window
This command is used to minimise the progress window that is normally shown when Multilateration calculations are being performed.

Process Menu Commands
The Process menu offers the following commands:

Start
This command is used to start PlanePlotter processing any input signals. Depending on the options made in the I/O settings dialog the PlanePlotter program will process audio data from the receiver to decode ACARS messages or Mode-S data.

A PlanePlotter desktop program shortcut can initiate processing immediately by using command line with the argument */start* e.g. *C:\PlanePlotter\PlanePlotter.exe /start*.

Stop
This command stops any processing currently in progress.

Options Menu Commands

This menu provides a comprehensive set of options and is perhaps the most important menu for setting up how PlanePlotter processes, decodes and displays data. These options are reviewed in detail below but the reader is recommended to review the menu structure shown earlier in Figure 45.

ACARS Decoder

This option is used to set the ACARS decoder.

The *CRC correct* check box is used to ensure that the message is only processed if the Cyclic Redundancy Check (CRC) checksum is true. The CRC check is the only way of knowing that the message is correct although some users may wish to see corrupted messages.

A value can be entered in the *Bad parity count* to define the maximum extent to which a corrupted messages will be accepted.

Checking the *Overwrite reports* option lets the user choose whether to plot all received ACARS position reports or only the last position report decoded. The default is not to overwrite position reports.

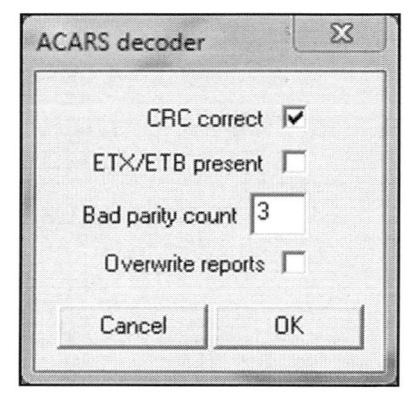

Figure 45 - ACARS Decoder

Mode-S receiver

PlanePlotter support six different families of Mode-S receiver.

RadarBox

There are two options for the AirNav RadarBox

Update registrations

This option is used to update the database with the registration letters of any aircraft received in an ACARS message or from an online look-up of the ICAO24 address in an ADS-B message.

IP address

This option is used only when the TCP input is selected and is used to add an IP address where a RadarBox is installed on another PC. When the RadarBox is on the same PC as PlanePlotter then the default IP address is *127.0.0.1*.

Kinetic SBS-1

There are three options for the Kinetic SBS-1.

Open access

This feature is no longer required as the recommended access method is via the Basestation TCP port.

Update registrations

This option is used to update the database with the registration letters of any aircraft received in an ACARS message or from an online look-up of the ICAO24 address in an ADS-B message.

As different aircraft sometimes fly the same flight number on different occasions then PlanePlotter only updates the database if the ADS-B message has been received from the aircraft within the last hour.

IP address

This option is used only when the TCP input is selected and is used to add an IP address where a Kinetic SBS-1 receiver is installed on another PC. If the BaseStation is running on the same PC then the default IP address of *127.0.0.1* is used.

When accessing a TCP server on another PlanePlotter installation then a non-standard IP port number can be used by placing a colon after the IP address followed by the port number as in *192.168.2.104:9774*.

PlaneGadget Radar

There are two options for the PlaneGadget Radar.

Setup Comm port

This option to define which Comm Port the PlaneGadget Radar is connected to.

If the Comm Port number is not known then it can be found using the *Windows Control Panel* in the *Device Manager* panel of the *Hardware & Sound* settings.

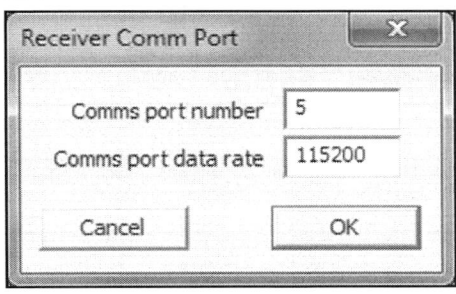

Figure 46 - PGR Setup Comm Port

Setup Receiver

The PlaneGadget Radar is provided with a Receiver Setup dialog box that enables the user to change the receiver gain setting and monitor the message error count. As the receiver is pretuned to the supplied antenna the user is advised not to alter these settings. However should they do so then an option is provided to restore the settings to a factory default.

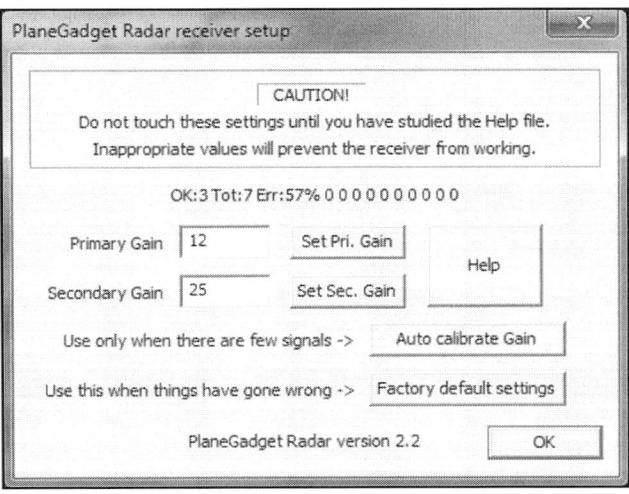

Figure 47 - PGR Setup Receiver

Aurora SSrx

There are three options for the Aurora SSrx.

Setup Comm port

This option is used to define which Comm Port the Aurora SSrx is connected to.

TCP/IP Address

This option is used only when the TCP input is selected and is used to add an IP address where an Aurora SSrx is installed on another PC.

Test

This can be used to test the link to the Aurora SSrx.

RxControl

The RxControl has two options.

Setup Comm port

This option is used to define which Comm Port the RxControl is using.

Test

This can be used to test the link to the RxControl receiver.

AVR Receiver

There are two options for an AVR.

Setup Comm port

This option is used to define which Comm Port the used.

Test

This can be used to test the link to the AVR receiver.

Audio

There are three possible audio message sources and each option has its own two similar settings.

ACARS

There are two options here.

Source

This option is used to define the audio source to be used to process the ACARS messages.

Mixer

This option to open the Direction Finding soundcard mixer dialog. It can be used to select the appropriate audio input and also to adjust the volume settings.

DF

Direction Finding (DF) also has two options.

Source
This option is used to define the audio source or soundcard to be used for Direction. When receiving and processing ACARS messages and also using the DF facility a different audio source must be used.

Mixer
This option to open the Direction Finding audio mixer dialog. It can be used to select the appropriate audio input and also to adjust the volume settings.

SelCal
SelCal has two similar options.

Source
This option is used to define the audio source to be used to process the ACARS messages.

Mixer
This option to open the SelCal audio mixer dialog. It can be used to select the appropriate audio input and also to adjust the volume settings.

Chart
There are nine groups to be set-up for Charts.

Options
Selecting Option will open the dialog box shown below. Each individual option will be described in the following pages.

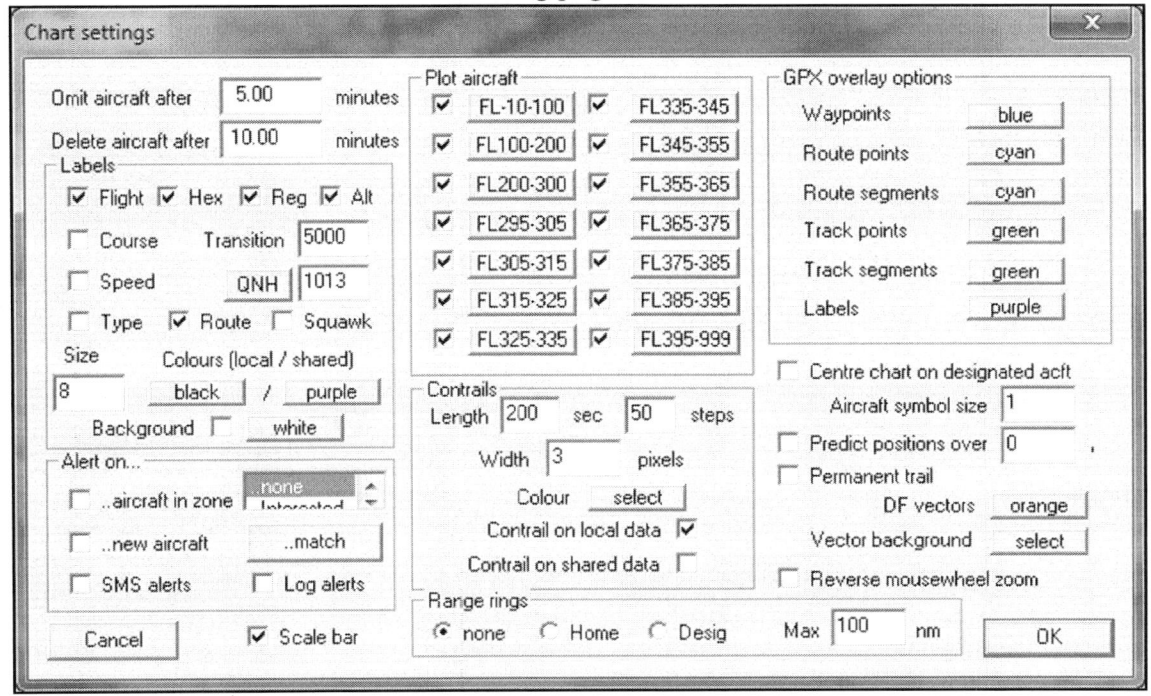

Figure 48 - Chart Settings

Omit after

The *Omit after* option determines the elapsed time from the message being received after which the aircraft will not be displayed.

Delete after

The *Delete after* option determines the elapsed time from the message being received after which the aircraft will be deleted. By entering an integer preceded by an upper case letter A the delete threshold will be reached at the given number of aircraft within PlanePlotter. e.g. *A150* will restrict the total number of aircraft to 150 by deleting the oldest aircraft in the list.

Labels

Aircraft symbols on a chart can have labels using a combination of identification, registration, flight number, altitude, course, speed, route, type and or squawk code.

The font size and colour of the labels can be set to different colours for aircraft received locally or for those received by sharing using the colour selector shown in Figure 49. The label may be transparent or have a coloured.

Alert on

Alerts can be switched on for four types of condition producing both a pop up warning and an audible alert. An aircraft raising an alert is shown by a circled on the chart.

Alert on any aircraft with an alert zone.

An alert is raised when an aircraft a defined zone.

Alert on any new aircraft.

An alert is raised when a previously unknown aircraft is detected.

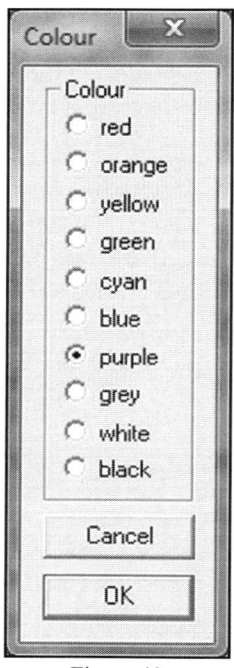

Figure 49
Colour Selector

Alert on an aircraft with the "Interested" tag set or unset.

An alert is raised when aircraft marked with one of three options is detected. The option are *Neither*, *Interested* or *Not interested*. The *Interested/Not interested* state of an aircraft can be changed using the Edit button in the AirRep dialog.

Alert on any aircraft matching a user-defined list.

A user defined list as shown in Figure 50 can be created to define up to 50 aircraft registrations, flight numbers, ICAO24 numbers, squawk codes, aircraft types or User Tags, it is also possible to specify a particular flight such as LEPA-LEMD, using the dialog that appears when this option is selected. An alert will be raised any time that a message is received from an aircraft in the list. Wild cards can be used in the list. e.g. For example, [flt] EZY will alert on any EasyJet flight with a prefix such as EZY282 or EZY6204 and [reg]

G- alert on any aircraft registration with that prefix such as G-EUYE or G-SZTP.

Log Alerts

Alerts can be recorded in a log file by enabling this option. The log file will be saved and automatically named in the form *pp_alert_yymmdd.log*.

SMS Alerts

For the aircraft enthusiast a useful option is to send alerts as text messages to a designated mobile phone. However to use this feature the user must first nominate a mobile phone number and purchase message credits from PlanePlotter using the web page at ***http://www.coaa.co.uk/smsalert.php***. When a new alert is generated on the PlanePlotter screen then a text message is sent to the designated mobile phone. Each message sent decrements the credit of text messages so it wise not to configure PlanePlotter to send a large number of text messages. It is also advisable to turn off the *SMS alerts* option by unchecking the box when the feature is not needed.

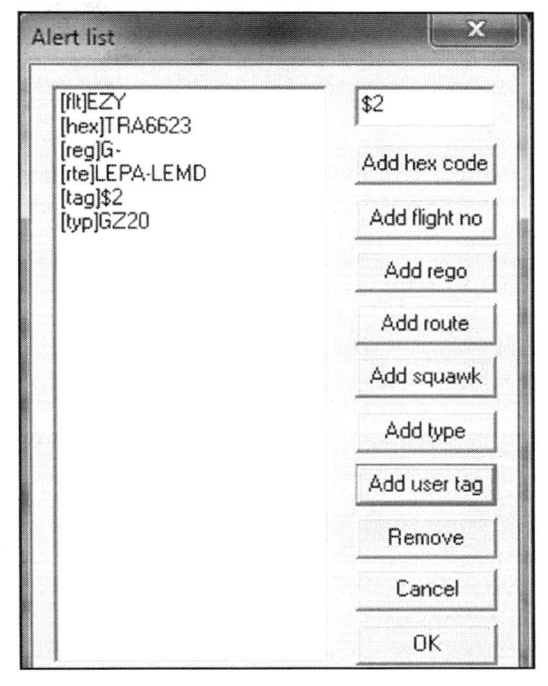

Figure 50 - Alert List

Plot aircraft

A group of 14 buttons and check boxes enables the user to display aircraft in particular height bands. This is useful for say separating say airfield traffic from over flying aircraft.

Note that ACARS and HFDL reports do not contain altitude but are given by PlanePlotter a dummy altitude of 33333. Thus a range of FL332-FL334, which are not normal cruising levels, could be defined and used to allow the user to turn on and off the Mode-S & ADS-B reports or ACARS & HFDL reports independently.

Contrails

There are a group of Contrail setting that are used to enable a short track history or contrail to be displayed for each aircraft plotted. A contrail is made up of a number of aircraft location points that have been recorded at uniform intervals over a specified duration. The width of the contrail in pixels and its colour be specified as can the number of steps and the time interval that the steps cover. Contrails can be enabled for local and/or shared traffic by checking the appropriate boxes.

GPX overlay options

Each overlay can use a different colour for better recognition on a chart. The colour selector is enabled by clicking on each button in turn to change the allocated colour.

Centre chart on designated aircraft

When this option is checked the displayed chart will be centred on the designated aircraft.

Aircraft symbol size

The size of symbols displayed on the chart can be made larger or smaller by selecting a 0.5 and 5.0.

Predict positions over ... feet

PlanePlotter can predict the future position of an aircraft based on the reported course and speed information from a Mode-S report. With prediction enabled a lower altitude limit can be entered below which prediction will not be implemented to avoid aircraft that are descending to land appearing to over fly the airport when prediction is applied after contact is lost.

Permanent trails

Check this option to leave a permanent trail of each flight track on the chart. You can always erase the trails by reloading the chart. The button alongside the option chooses the colour to be used for the permanent trail and for the contrail if any.

DF vector

When using PlanePlotters Direction Finding feature the colour of the vectors can be determined using a colour selector to display the DF measurements. The same colour will also be used for the range rings and scale bar when these have been selected.

Vector background

The background colour of a vector chart can be determined with a colour selector with this option.

Reverse mouse wheel

The way the mouse wheel zooms in and out of a chart can be changed with this option.

Scale bar

A scale bar can be plotted on the right side of the chart or outline display if this option is checked.

Range rings
Range rings centred on home location or centred on a designated aircraft can be displayed using these options The maximum range of the rings can be specified and the spacing of the rings depend on the scale of the chart.

Quick chart
PlanePlotter Quick Charts is a feature that allows the user to have readily available up to ten different charts. A currently displayed chart can rapidly replaced by another chart using just a single button press. The PlanePlotter toolbar has ten Quick Chart buttons and a different chart can be allocated to any of those buttons.

Figure 51 - Toolbar Quick Chart Buttons

Open
Selecting the Open option loads the nominated Quick Chart. But pressing a Quick Chart toolbar button is easier.

Define
This option is used to define the chart to which a button relates. A Quick chart can also be defined by holding down the control key whilst clicking a toolbar button.

Clear
This option is used to remove a Quick Link to a chart.

Satellite
This option is used to download from the Internet a satellite chart of the area covered by the current display. The downloaded chart is automatically saved in the charts folder and can used at any time..

OSM
This option will download an OpenStreetMap™ covering the current displayed areas. By holding down the Control key while pressing the OSM button an OSM map larger than the current screen area will be downloaded. This enables the user to drag the new map around without losing map detail.

Outline
This option draws an outline map covering the area covered by the current display.

GPX overlay
This option is used to first open a file dialog to define a GPX file that can be switched on or off as an overlay on the current chart.

Circle sharers
Enable
This option draws a circle around the location of each sharer that has reported a designated aircraft in the last minute or two when the sharers' locations is known.

Fill
When this option is selected the area contained within all the circles around the locations of each sharer that has reported the designated aircraft in the last minute or two will be shaded

Transparency
Quite often users may wish to place more importance on the display of aircraft symbols rather than on the detail of a chart. This option enables the user to fade a chart into the background by varying degrees. A factor of 0% denotes no change whilst factors between 1% and 99% will progressively lighten a chart when it is loaded. A factor of -100% denotes that the chart is to be faded to black. Note the transparency setting will not affect the currently loaded chart but only charts loaded after the setting is made.

SMS Alerts
This option allows the user to complete a web form to apply to use COAA's SMS alerts service, this service is only available to registered PlanePlotter users.

Aircraft view options
This option enables the user to select colours for the *View, Aircraft* screen. The background colour and the text colour for normal reports, designated aircraft and the prospects for Multilateration can all be set.

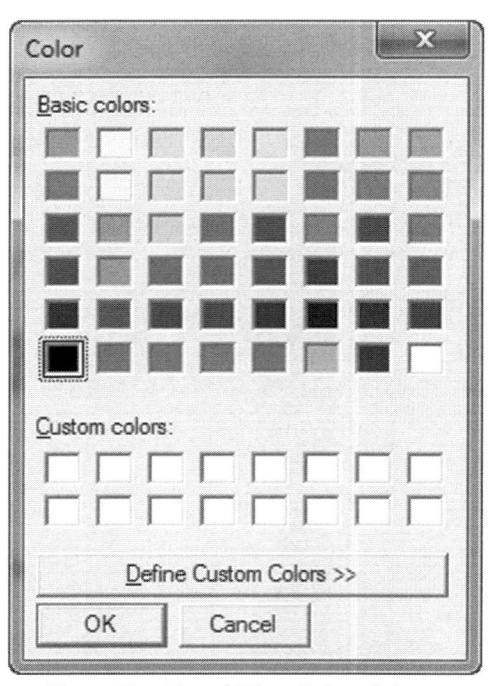

Figure 52 - Aircraft View Colour Selector

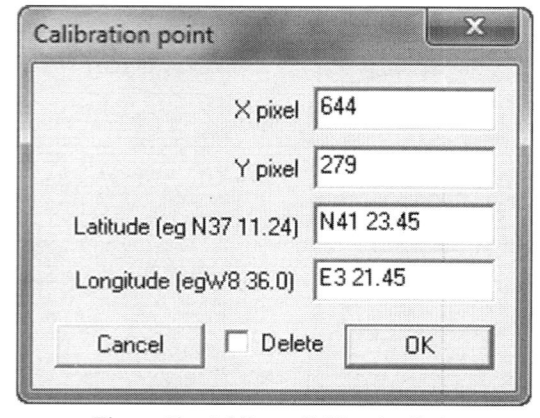

Figure 53 - Adding a Calibration Point

PlanePlotter Menus

Calibration

Add
This option to is used add a new calibration point to the current chart by clicking on a point with known coordinates and then entering the latitude and longitude of the point in the correct format. The correct format is quadrant letter, degrees, space, minutes and decimals of a minute. e.g. N41 35.21 or E14 17.52.

Save
Once the calibration points have been added his option is used to save the calibration file with the same name as JPG chart file but with the file extension CLB. e.g. thames.jpg and thames.clb.

Clear calibration points
This option is used to clear all the calibration points for the current chart.

Display calibration points
This option will display all the calibration points on the current chart. The user can right click on a calibration point to edit it or delete it.

Display geographical grid
This option will toggle the display of a geographical grid over the current chart and can be used to confirm that the calibration is correct.

I/O settings

This is perhaps the most important of all the option as it used to change the input and output parameters for PlanePlotter. When the option is elected a dialog box as shown in Figure 54 over page is displayed. Each of the various parameters is discussed below.

Input data
Checking the Mode-S reception will enable processing of messages from the receiver selected in the drop down list. PlanePlotter Master Users (MU) can make Multilateration (Mlat) requests by checking both *UDP/IP data from net* and the *Raw data for Mlats* boxes.

ACARS processing of messages received from a VHF receiver or scanner and connected to the PCs audio input is enabled by checking the *ACARS reception from audio input* box.

Direction finding using a suitable VHF receiver and antenna switch is enabled by checking the *DF from audio input* box and checking *HFDL with PC-HFDL* box enables HF ACARS. HF SelCal messages can be processed by checking the *HF Selcal* box.

Be aware however that those latter options use audio input and processing their inputs are mutually exclusive so selecting one option will disable others that cannot coexist with that option.

Output data

There are six parameters that the user may setup or change. The first four enable a number of different log files to be created.

Log Mode-S

This check box is used to enable the creation of Mode-S raw data messages. This log file is only created to enable the user to replay a sequence of received messages. It is not a readable text type log.

Log desig acft

This check box is used to enable the creation of Mode-S raw data messages only for a designated aircraft. Again the log file is only created to enable the user to replay a sequence of received messages.

Airmaster log format

This check box is used to enable an ACARS log file to be saved in Airmaster format.

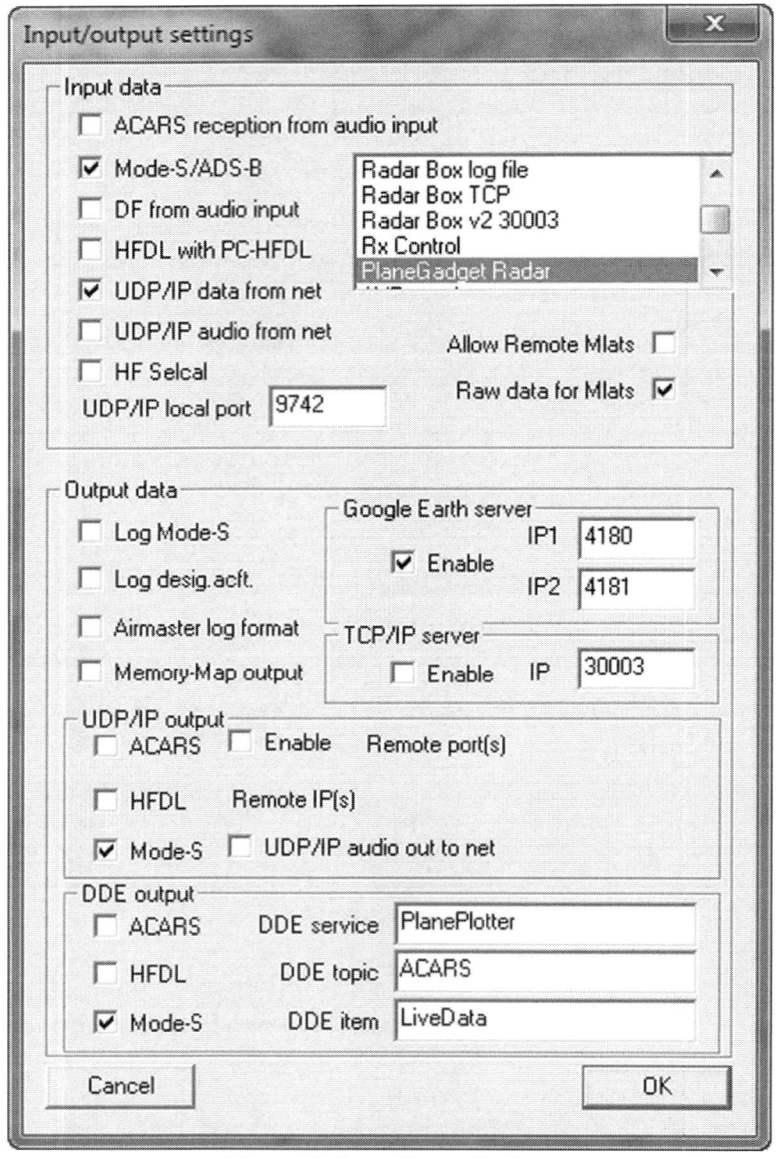

Figure 54 - Input/Output Settings

Memory map output

This check box enables support for the Memory-Map Navigator, a proprietary plotting program, map plotting program.

The next two parameters enable data to be exported and made available to other applications.

PlanePlotter Menus

Google Earth server

This box should be checked to enable PlanePlotter data to be exported to a Google Earth KML file. The default IP addresses should only be changed if there is a conflict with other servers installed on the same PC. Not that the two KML files created in the installation of the PlanePlotter and placed in the main program application folder should not be removed.

TCP/IP server

This box should be checked to enable a TCP/IP server that will output data in the SBS1 30003 format.

UDP/IP output

This section enables peer-to-peer output to other PlanePlotter installations through internetworking using UDP/IP datagrams.

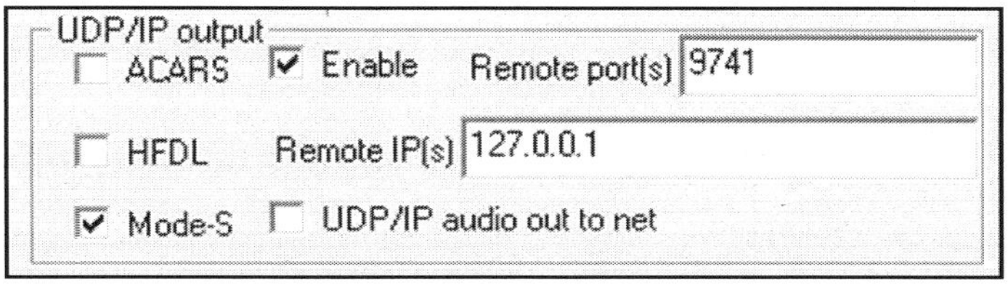

When this option is enabled the destination IP address and port number boxes will appear in the dialog box as shown in Figure 55. When the data is to be sent to more than one other destination multiple IP addresses and port numbers may be entered but the number of fields for IP address and port number must be the same. For example to send the data to three other applications on a local machine, the IP address field would be:

Figure 55 - UDP/IP output

127.0.0.1,127.0.0.1,127.0.0.1
and the port number field might be :
9741,9742,9743

Using the three check boxes the user can select Mode-S, ACARS or HFDL messages for forwarding to a remote machine. Additionally any locally received ATC audio may be forwarded to a remote PC, however one cannot send audio via UDP/IP if it is to be processed locally.

DDE server

PlanePlotter can act as a DDE server for other ACARS software. If the DDE client requires a service name, topic name or item name that are different from the defaults these can be changed in this dialog. If changes are made to the service name then PlanePlotter mist be restarted for the service to be registered with the new name. If the topic or item name is changed whilst a DDE connection is active the DDE connection must be closed and then reconnected with the new names.

The default values are

Service name : PlanePlotter
Topic name : ACARS
Item name : LiveData

The DDE outputs from ACARS, HFDL or Mode-S processing can be enabled or disabled independently of each other.

Graphical output

JPG output

Setup

This option is used to configure the automatic saving of JPG image files of the current PlanePlotter display. A file name and folder location should be entered in the dialog box. New images can automatically be created and saved at regular intervals based on the number of seconds specified for that. When the *Auto increment file name* box is checked a four character number will be appended to the file name as in *planeplotterimage0000.jpg, planeplotterimage0001.jpg* etc. Each image has date/time stamp caption inserted at the lower right corner of each image.

Enable

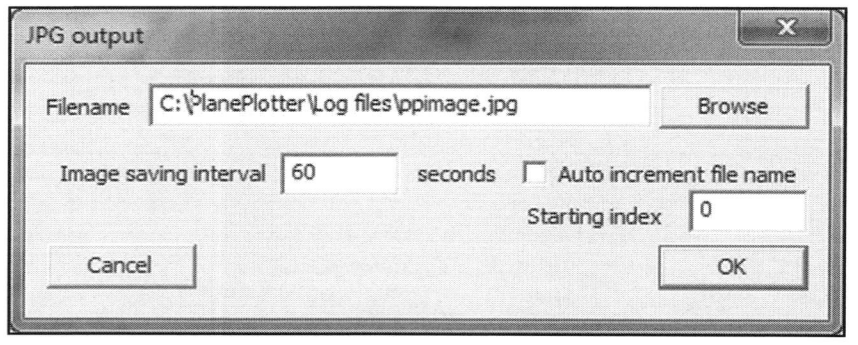

Figure 56 - JPG Output

Use this option to toggle the saving of JPG images of the PP window at regular intervals.

AVI output

Start recording

This option is used to start recording an AVI file of the current window that is saved in the log files folder. The current window is saved to the AVI file at 30 second intervals. The AVI file header defines a replay rate of 2 frames per second giving a time-lapse speed increase of 60 times. A UTC date/time stamp is overlaid in the lower right corner of each frame.

Stop recording

This option will stop recording an AVI file of the current window. AVI files can be replayed by double clicking the file name found in the log files folder.

Sharing
Setup
This option enables the user to define a sharing identifier and determine the types of message to be uploaded and/or downloaded.

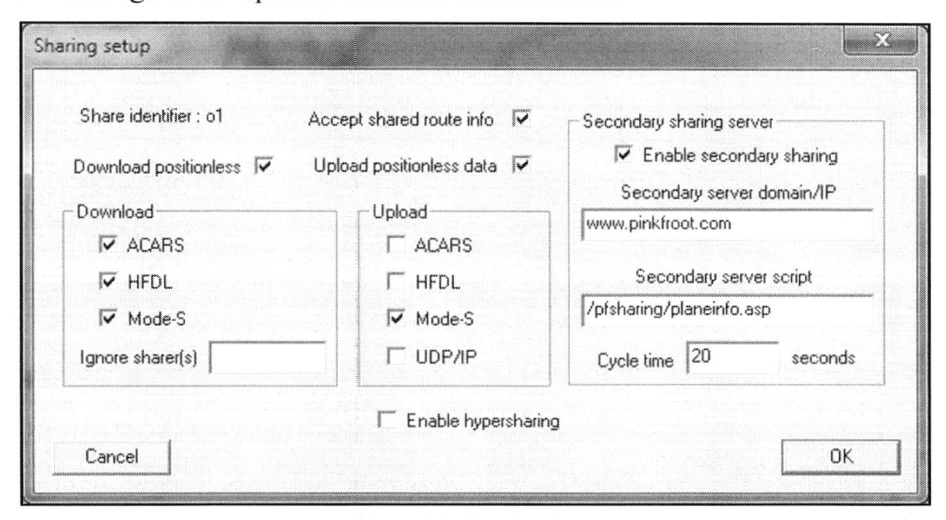

Figure 57 - Sharing Setup

Share identifier

The Share identifier is fixed and allocated by COAA. to registered users the first time they invoke sharing.

Download positionless information

Checking this box will enable receipt from the sharing server of aircraft reports that lack positions.

Download

Checking the appropriates boxes will permit the downloading of ACARS, HFDL and Mode-S traffic messages. Uncheck all three options will allow local received aircraft reports to be shared with other users but their data will not be downloaded. Data from particular sharers can be ignored by entering one or more share identifiers in the *Ignore share* dialog.

Accept shared route info

With option checked route information received via the sharing system will be used whether or not it is correct.

Upload positionless information

Checking this option data which may include altitude but not latitude and longitude will be uploaded to the sharing server along with the other messages.

Upload
Any combination of ACARS, HFDL and Mode-S traffic messages may be uploaded. Data will only be uploaded if both the UDP option and the specific message type options are enabled.

Secondary sharing server
Completing the entries in this section enables current data to be uploaded to sharing servers other than the COAA server.

Enable hypersharing
Hypersharing feature enables shared data to updated directly from suitable Ground Stations at a faster rate than normal typically a cycle of one per second and only operates for a designated aircraft.

Enable
This option is used to enable or disable sharing of position message data with other users. A toolbar button for Sharing has three states: *No sharing, Upload & Download* and *Upload only*. No sharing is denoted by the sharing button with a red line across it, Bilateral sharing is denoted by the button depressed showing two diagonal arrows whilst Upload only is denoted by the button with a single diagonal arrow.

Directories

PP Log files
This option to is used to define the folder where PlanePlotter saves the various log files created. Typically *C:/Planeplotter/Log Files*

PP Chart files
This option to is used to define the folder where PlanePlotter saves the various chart and calibration files created. Typically *C:/Planeplotter/Chart Files*

RadarBox log file
This option to define the folder where the software that controls the RadarBox receiver stores its log file. PlanePlotter uses this information to access the Mode-S data that the RadarBox is receiving. Note that RadarBox must be recording its log file in order for PlanePlotter to access the data. Typically *C:\Program files\AirNav\RadarBox\Recorder*

SQB database
This option is to define the folder and filename where the database file is saved. Typically *C:\Planeplotter\log Files\basestation.sqb or for those with a* SBS1 receiver *C:\Program files\Kinetic\BaseStation\basetstaion.sqb*

SBS1 log files
This option is used to define the directory where the BaseStation software that controls the SBS1 receiver saves its log files. Typically *C:\Program files\Kinetic\BaseStation*

HFDL files
This option is used to define the folder where PC-HFDL saves its log files.

Photo files
This option is used to define the folder where PlanePlotter looks for JPG photo files.

Direction finding
This option is used to set up the options for Direction finding with PlanePlotter. A descriptive tutorial on how PlanePlotter uses a unique passive Doppler direction finding method with the PC own audio system to generate and process the signals required for determining the direction of any incoming aircraft transmission can be found in PlanePlotters online help.

Home Latitude & Longitude
The latitude and longitude of the DF antenna where the home symbol is plotted on the chart is also the starting point of the direction finding vector display.

Home Altitude
The altitude of the DF antenna must be entered in metres.

Antenna switching frequency
This must be an integer number of cycles over 0.25 seconds but a frequency is 3000Hz is recommended.

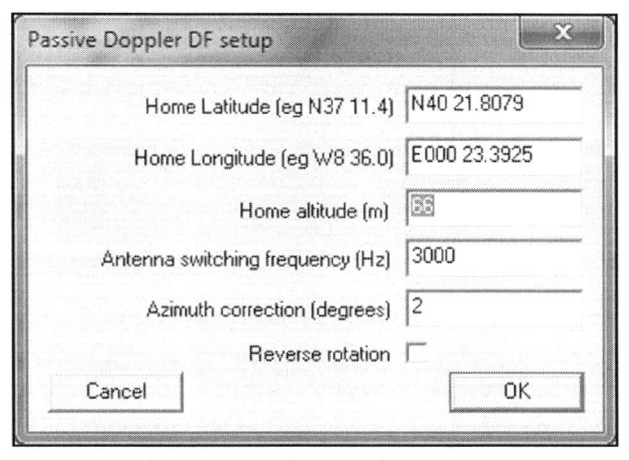

Figure 58 - Passive Doppler DF Setup

Azimuth offset
An azimuth offset may be entered to adjust the calibration of the DF antenna.

Reverse rotation
Checking this box will reverse the order of rotation of antenna elements

Test
This option is used to test the audio system for suitability with the direction finding feature.

Home location
This option is used to specify users home location. The values are identical to the home location Direction finding Passive Doppler DF setup shown in Figure 58. Data may be entered as decimal degrees e.g. *N40.363465* or degrees & decimal minutes e.g. *N40 21.8079*. A latitude entry must begin with a letter *N* or *S* and a longitude entry with a letter *E* or *W*.

A *Test* button is provided that opens a small window showing an OpenStreetMap centred on the coordinates entered. If the position is slightly wrong small adjustments to adjust the position can be made by a *Control-Left* click on the correct location. The cross representing the home location will then be moved to the new position and the coordinates ill be adjusted appropriately.

Zoom
In
This option is used to zoom in to the current chart.
Out
This option is used to zoom out of the current chart.

Alert Zone
The Alert zone is a user defined polygon outlined in red on the current chart. PlanePlotter can be set to check every aircraft regularly and to display and alert pop-up warning if any aircraft is detected in the Alert zone. It also generates an audible warning on such an event. Any aircraft causing an alert is circled on the chart for a few seconds after the alert sounds.

Clear
To clear an Alert zone
Add
This option is used to define a polygon on the chart. The zone is drawn on the chart by clicking points on the chart to create a series of vertices to construct polygon.
Save
This option is used to save the created zone polygon to a file. This option can be used several times to save various different alert zones although only one can be active at a time.
Load
This option is used to load an already defined and saved Alert zone.

Alert Shell
Alert Shell Setup
This option enables a shell command to be specified that will be executed when an alert is raised.
Alert Shell Enable
This option enables an alert shell command when an aircraft enters the current Alert zone.

Waypoint file
This option to is used specify then name and location of a text file containing a list of waypoint names and their associated coordinates. PlanePlotter will scan the file whenever it encounters a 5-letter waypoint name in a received message

and will convert the name into a coordinate to be plotted on the chart. The typical format for each waypoint entry is along the lines of the following example :
ABALO 32.33101 -18.13035
ABETO 40.42981 -8.05638
ABRAT 39.82168 -7.65425

Flags
Enable
This option is used to enable the display of flags in the Aircraft view.

Define
This option is used to specify the file defining the flags. Two example files are installed in the PlanePlotter application folder: rbflags.txt is typical of the flag images found in the AirNav RadarBox installation and *bsflags.txt* typical for Kinetics BaseStation.

Users can create their own flag files and by use specific hex codes for different aircraft. A flag file uses a fixed with each line containing the lowest hex code in the range, the highest hex code in the range, the complete file name and path for each flag applicable to those codes.
340000 37FFFF C:\Planeplotter\Flags\ES.bmp
3C0000 3FFFFF C:\Planeplotter\Flags\DE.bmp
400000 43FFFF C:\Planeplotter\Flags\GB.bmp

Script
Define
This option is used to define a script or batch file that will be run when the *Script* button on the toolbar is pressed.

Run
This option will run the script or batch file previously defined.

Review Menu Commands
The Review menu offers the following commands:

Today
The following command are used to review the current days activities.

Alerts
This command is used to review any Alerts detected and recorded in the current day.

All ACARS messages
This command is used to review all the ACARS messages received in the current day.

ACARS registrations
This command is used just to review any aircraft registrations received in today's ACARS messages.

ACARS flight numbers
This command is used just to review any aircraft flight numbers received in today's ACARS messages.

Replay today's ACARS
This command is used to replay all today's ACARS messages.

Yesterday
The following commands are used to review yesterdays activities.

Alerts
This command is used to review any Alerts detected and recorded yesterday.

All ACARS messages
This command is used to review all the ACARS messages received yesterday.

ACARS registrations
This command is used just to review any aircraft registrations received in yesterdays ACARS messages.

ACARS flight numbers
This command is used just to review any aircraft flight numbers received in yesterdays ACARS messages.

Replay yesterdays ACARS log
This command is used to replay all today's ACARS messages.

Replay ACARS log
This command is used to replay the full ACARS log.

Replay Mode-S log
These commands are used to replay the various receiver Modes-S logs.

AirNav RadarBox
This command is used to replay the Mode-S log of messages received by an AirNav RadarBox.

Kinetic SBS-1
This command is used to replay the Mode-S log of messages received by a Kinetics SBS-1 receiver.

PlaneGadget Radar
This command is used to replay the Mode-S log of messages received by a PlaneGadget Radar receiver.

Help Menu Commands

The Help menu offers the following commands:

Help Topics
This command is used to launch PlanePlotters online help and tutorial system.

Registering PlanePlotter
This command is used to register the current installed copy of PlanePlotter.

Check for new version
This command is used to check for a newer version of the software.

About PlanePlotter
This command is used to display the copyright notice and version number of the installed copy of PlanePlotter.

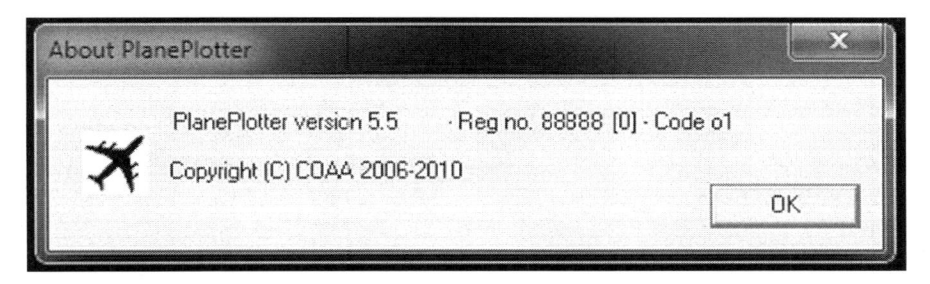

Figure 59 - Spitfires NG-D & PZ-I at Eslöv Airport
Photo Courtesy and Copyright of Lars Magnusson

Chapter 7
The PlanePlotter Toolbar

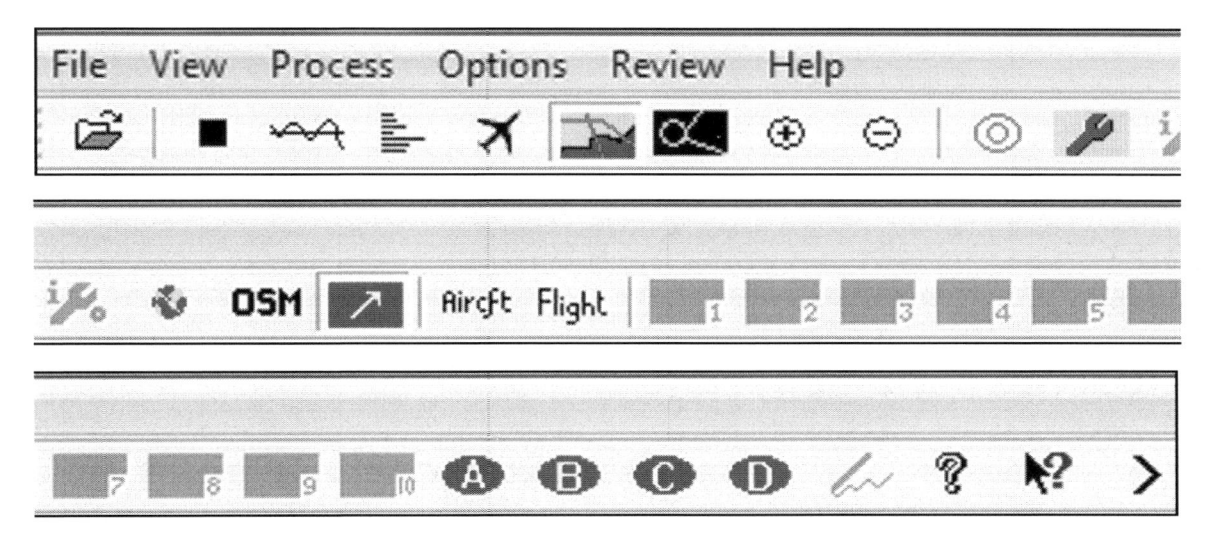

Figure 60 - PlanePlotter Toolbar (shown in 3 parts to fit on page)

The Toolbar is displayed across the top of the PlanePlotter application window and below the menu bar. The toolbar contains 36 buttons providing quick mouse access to many of the tools used in PlanePlotter and complimenting the PlanePlotter menus. The Toolbar can be hidden or revealed using the View menu keystrokes *ALT, V, T*. Some, but not all, individual Toolbar buttons can be hidden by using a *Shift-Click*. Buttons that have been hidden can be restored to the toolbar by clicking on the > button symbol at the right of the toolbar.

File
Opens a new chart file

Start
This button is normally a green coloured circle. By clicking the button processing of messages will commence and the button image will change to a black square.

Stop
Click this button to suspend processing. The image will revert to a green circle.

Signal
Clicking this button will change the screen display to a graph of the raw audio signal that PlanePlotter is processing when receiving ACARS messages. This view is used to ensure that audio signal from the ACARS receiver is correctly connected and that the volume level is correctly set. Note that processing must have been started for anything to be shown in this view.

Messages
This command is used to display any recently received and decoded ACARS or HF SelCal messages. Again note that processing must have been started for anything to be shown in this view.

Aircraft
This command is used to display a table of the aircraft that have currently been processed by PlanePlotter as described on Page 46 in the previous chapter.

Chart
Clicking this button will change the screen display to a chart.

Outline
Clicking this button will change the screen display to a vector outline chart.

Zoom In
Use this button to zoom in towards the centre of a chart.

Zoom Out
Use this button to zoom outwards from the centre of a chart.

Calibrate
This button is used to add a new calibration point to a chart. The user should first click the button and then move the cursor to a location where the geographical coordinates are known. Then click on that point and enter the coordinates.

Chart Options
This button is a blue spanner on a green background. When the button is pressed a Chart Settings dialog box will open as shown in Figure 61.

Omit after
> The *Omit after* option determines the elapsed time from the message being received after which the aircraft will not be displayed.

Delete after
> The *Delete after* option determines the elapsed time from the message being received after which the aircraft will be deleted. By entering an integer preceded by an upper case letter A the delete threshold will be reached at the given number of aircraft within PlanePlotter. e.g. *A150* will restrict the total number of aircraft to 150 by deleting the oldest aircraft in the list.

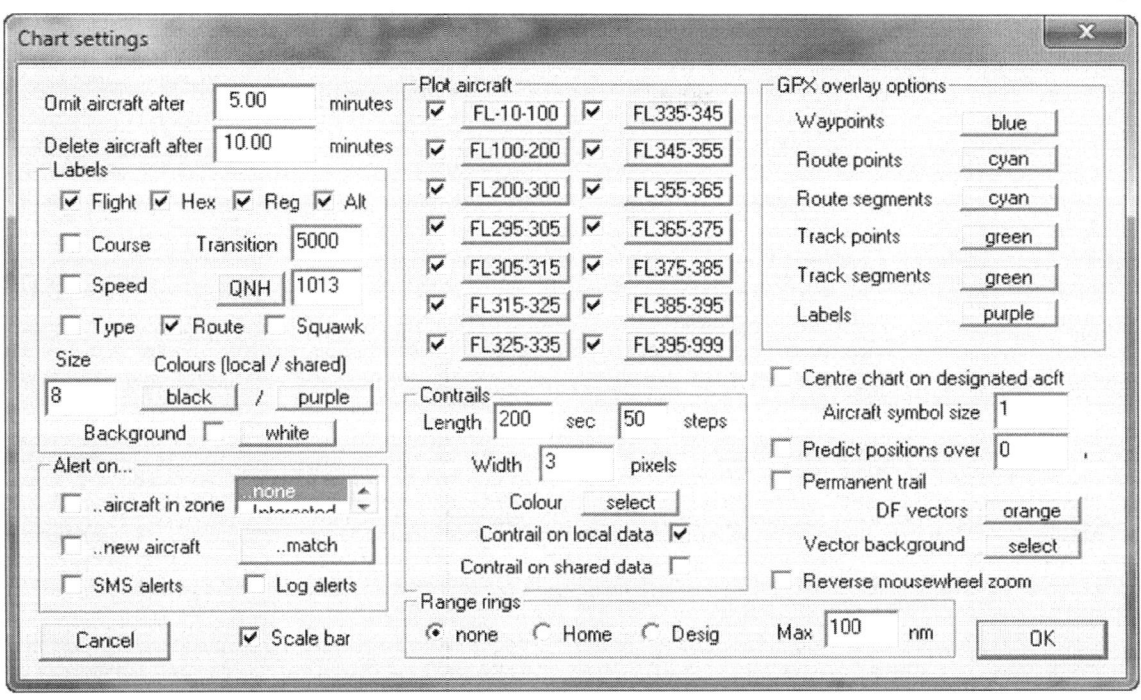

Figure 61 - Chart Settings

Labels

Aircraft symbols on a chart can have labels using a combination of identification, registration, flight number, altitude, course, speed, route, type and or squawk code. The font size and colour of the labels can be set to different colours for aircraft received locally or for those received by sharing using the colour selector shown in Figure 62. The label may be transparent or have a coloured.

Alert on

Alerts can be switched on for four types of condition producing both a pop up warning and an audible alert. An aircraft raising an alert is shown by a circled on the chart.

Alert on any aircraft with an alert zone.

An alert is raised when an aircraft a defined zone.

Alert on any new aircraft.

An alert is raised when a previously unknown aircraft is detected.

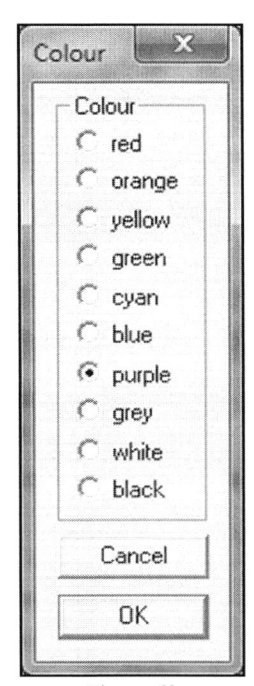

Figure 62
Colour Selector

Alert on an aircraft with the "Interested" tag set or unset.

An alert is raised when aircraft marked with one of three options is detected. The option are *Neither*, *Interested* or *Not interested*. The *Interested/Not interested* state of an aircraft can be changed using the Edit button in the AirRep dialog.

Alert on any aircraft matching a user-defined list.

A user defined list as shown in Figure 63 can be created to define up to 50 aircraft registrations, flight numbers, ICAO24 numbers, squawk codes, aircraft types or User Tags, it is also possible to specify a particular flight such as LEPA-LEMD, using the dialog that appears when this option is

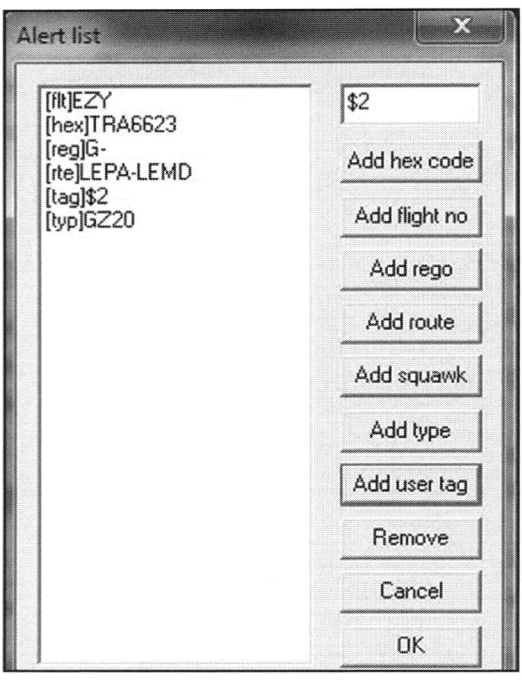

Figure 63 - Alert List

selected. Wild cards can be used in the list. e.g. For example, [flt] EZY will alert on any EasyJet flight with a prefix such as EZY282 or EZY6204. An alert will be raised any time that a message is received from an aircraft in the list.

Log Alerts

Alerts can be recorded in a log file by enabling this option. The log file will be saved and automatically named in the form *pp_alert_yymmdd.log*.

SMS Alerts

For the aircraft enthusiast a useful option is to send alerts as text messages to a designated mobile phone. However to use this feature the user must first nominate a mobile phone number and purchase message credits from PlanePlotter using the web page at ***http://www.coaa.co.uk/smsalert.php***. When a new alert is generated on the PlanePlotter screen then a text message is sent to the designated mobile phone. Each message sent decrements the credit of text messages so it wise not to configure PlanePlotter to send a large number of text messages. It is also advisable to turn off the ***SMS alerts*** option by unchecking the box when the feature is not needed.

Plot aircraft

A group of 14 buttons and check boxes enables the user to display aircraft in particular height bands. This is useful for say separating say airfield traffic from over flying aircraft.

Note that ACARS and HFDL reports do not contain altitude but are given by PlanePlotter a dummy altitude of 33333. Thus a range of FL332-FL334, which are not normal cruising levels, could be defined and used to allow the user to turn on and off the Mode-S & ADS-B reports or ACARS & HFDL reports independently.

Contrails

There are a group of Contrail setting that are used to enable a short track history or contrail to be displayed for each aircraft plotted. A contrail is made up of a number of aircraft location points that have been recorded at uniform intervals over a specified duration. The width of the contrail in pixels and its colour be specified as can the number of steps and the time interval that the steps cover. Contrails can be enabled for local and/or shared traffic by checking the appropriate boxes.

GPX overlay options

Each overlay can use a different colour for better recognition on a chart. The colour selector is enabled by clicking on each button in turn to change the allocated colour.

Centre chart on designated aircraft

When this option is checked the displayed chart will be centred on the designated aircraft.

Aircraft symbol size

The size of symbols displayed on the chart can be made larger or smaller by selecting a 0.5 and 5.0.

Predict positions over ... feet

PlanePlotter can predict the future position of an aircraft based on the reported course and speed information from a Mode-S report. With prediction enabled a lower altitude limit can be entered below which prediction will not be implemented to avoid aircraft that are descending to land appearing to over fly the airport when prediction is applied after contact is lost.

Permanent trails

Check this option to leave a permanent trail of each flight track on the chart. You can always erase the trails by reloading the chart. The button alongside the option chooses the colour to be used for the permanent trail and for the contrail if any.

DF vector

When using PlanePlotter's Direction Finding feature the colour of the vectors can be determined using a colour selector to display the DF measurements. The same colour will also be used for the range rings and scale bar when these have been selected.

Vector background

The background colour of a vector chart can be determined with a colour selector with this option.

Reverse mouse wheel

The way the mouse wheel zooms in and out of a chart can be changed with this option.

Scale bar

A scale bar can be plotted on the right side of the chart or outline display if this option is checked.

Range rings

Range rings centred on home location or centred on a designated aircraft can be displayed using these options The maximum range of the rings can be specified and the spacing of the rings depend on the scale of the chart

I/O Settings

This button will be greyed out if processing has been enabled. To make changes to the setting the *Stop* button must be pressed to stop any processing of messages.

Input data

Checking the Mode-S reception will enable processing of messages from the receiver selected in the drop down list. PlanePlotter Master Users (MU) can make Multilateration (Mlat) requests by checking both *UDP/IP data from net* and the *Raw data for Mlats* boxes.

ACARS processing of messages received from a VHF receiver or scanner and connected to the PCs audio input is enabled by checking the *ACARS reception from audio input* box.

Direction finding using a suitable VHF receiver and antenna switch is enabled by checking the *DF from audio input* box and checking *HFDL with PC-HFDL* box enables HF ACARS. HF SelCal messages can be processed by checking the *HF SelCal* box.

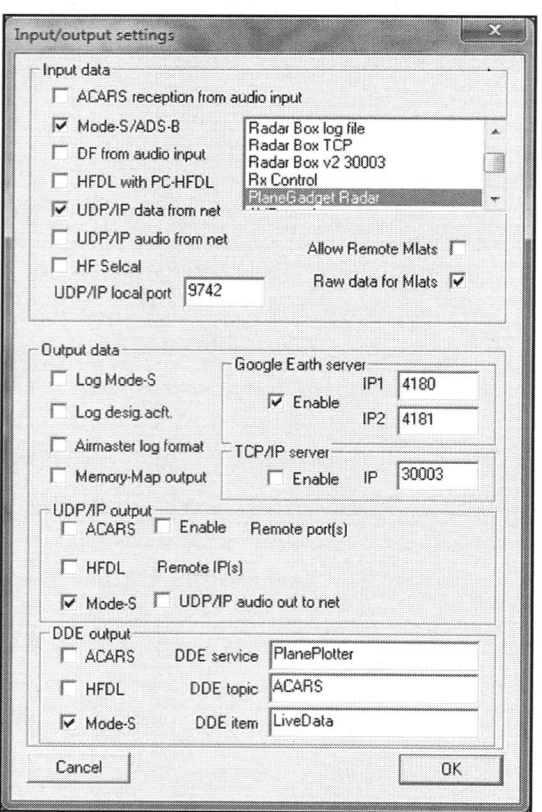

Figure 64 - Input/Output Settings

Be aware however that those latter options use audio input and processing their inputs are mutually exclusive so selecting one option will disable others that cannot coexist with that option.

Output data

There are six parameters that the user may set-up or change. The first four enable a number of different log files to be created.

Log Mode-S

This check box is used to enable the creation of Mode-S raw data messages. This log file is only created to enable the user to replay a sequence of received messages. It is not a readable text type log.

Log desig acft

This check box is used to enable the creation of Mode-S raw data messages only for a designated aircraft. Again the log file is only created to enable the user to replay a sequence of received messages.

Airmaster log format

This check box is used to enable an ACARS log file to be saved in Airmaster format.

Memory map output

This check box enables support for the Memory-Map Navigator, a proprietary plotting program, map plotting program.

The next two parameters enable data to be exported and made available to other applications.

Google Earth server

This box should be checked to enable PlanePlotter data to be exported to a Google Earth KML file. The default IP addresses should only be changed if there is a conflict with other servers installed on the same PC. Not that the two KML files created in the installation of the PlanePlotter and placed in the main program application folder should not be removed.

TCP/IP server

This box should be checked to enable a TCP/IP server that will output data in the SBS1 30003 format.

UDP/IP output

This section enables peer-to-peer output to other PlanePlotter installations through internetworking using UDP/IP datagrams.

When this option is enabled the destination IP address and port number boxes will appear in the dialog box as shown in Figure 64. When the data is to be sent to more than one other destination multiple IP addresses and port numbers may be entered but the number of fields for IP address and port number must be the same. For example to send the data to three other applications on a local machine, the IP address field would be:

127.0.0.1,127.0.0.1,127.0.0.1
and the port number field might be :
9741,9742,9743

Using the three check boxes the user can select Mode-S, ACARS or HFDL messages for forwarding to a remote machine. Additionally any locally received ATC audio may be forwarded to a remote PC, however one cannot send audio via UDP/IP if it is to be processed locally.

DDE server

PlanePlotter can act as a DDE server for other ACARS software. If the DDE client requires a service name, topic name or item name that are different from the defaults these can be changed in this dialog. If changes are made to the service name then PlanePlotter mist be restarted for the service to be registered with the new name. If the topic or item name is changed whilst a DDE connection is active the DDE connection must be closed and then reconnected with the new names.

The default values are
Service name : PlanePlotter
Topic name : ACARS
Item name : LiveData

The DDE outputs from ACARS, HFDL or Mode-S processing can be enabled or disabled independently of each other.

Satellite

On clicking this button a satellite image will be downloaded to replace the current chart and covering the same area.

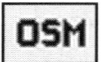

OSM

On clicking this button an OSM *OpenStreetMap*™ will be downloaded to replace the current chart and to cover the same area.

Share

This button will display one of three states. Each click of the button will change the state cycling through each one. The first state indicated by two arrows with a line through the button indicates that sharing is disabled. The second state is indicated by a single upwards facing arrow that indicates that the users local data will be uploaded to the sharing server. The third and final state is indicated by two arrows to show that the users local data will be uploaded to the sharing server and data from other users will be downloaded to the users computer.

PlanePlotter ToolBar

Aircraft
Clicking this button opens a list of the flight numbers in todays log file. A double click on a flight number will create a text document displaying all the ACARS messages from that flight number that appear in the log file for today.

Flight
Clicking this button opens a list of the aircraft registrations in todays log file. A double click on a flight number will create a text document displaying all the ACARS messages from that flight number that appear in the log file for today.

Quick Charts

There are a total of ten Quick Chart buttons. To start off with these buttons are normally greyed out. When a chart is allocated to one of the buttons the button is coloured. When the mouse pointer moves a defined Quick Chart button the name of the saved chart will be displayed. Quick Charts are saved with the current configuration but when a new configuration is opened any previously saved Quick Charts will not loaded. However when a Restore point is saved any defined Quick Charts will be saved at the same time.

Quick Charts are especially usefully in saving Outline charts that may cover the same geographical area but have different outlines, waypoints and TMAs.

A B C D

The ABCD buttons are used to quickly load a previously saved Restore point. Care should be taken when loading a Restore point as doing so will override all the current configuration. It is recommended that Restore point D be used as a safety net. Before loading a Restore point save the current configuration as Restore point D to ensure that one can always revert to a known configuration.

Script
Clicking this button will initiate a previously defined shell command. When no shell command has been defined the button will be greyed out.

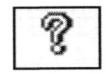

Help
Clicking this button will launch the Online Help and Tutorial screen.

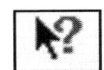

About
Clicking this button will display the *About PlanePlotter* dialog box which includes the software version and users *Share ID*.

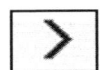

>
Clicking this button will restore any hidden toolbar buttons.

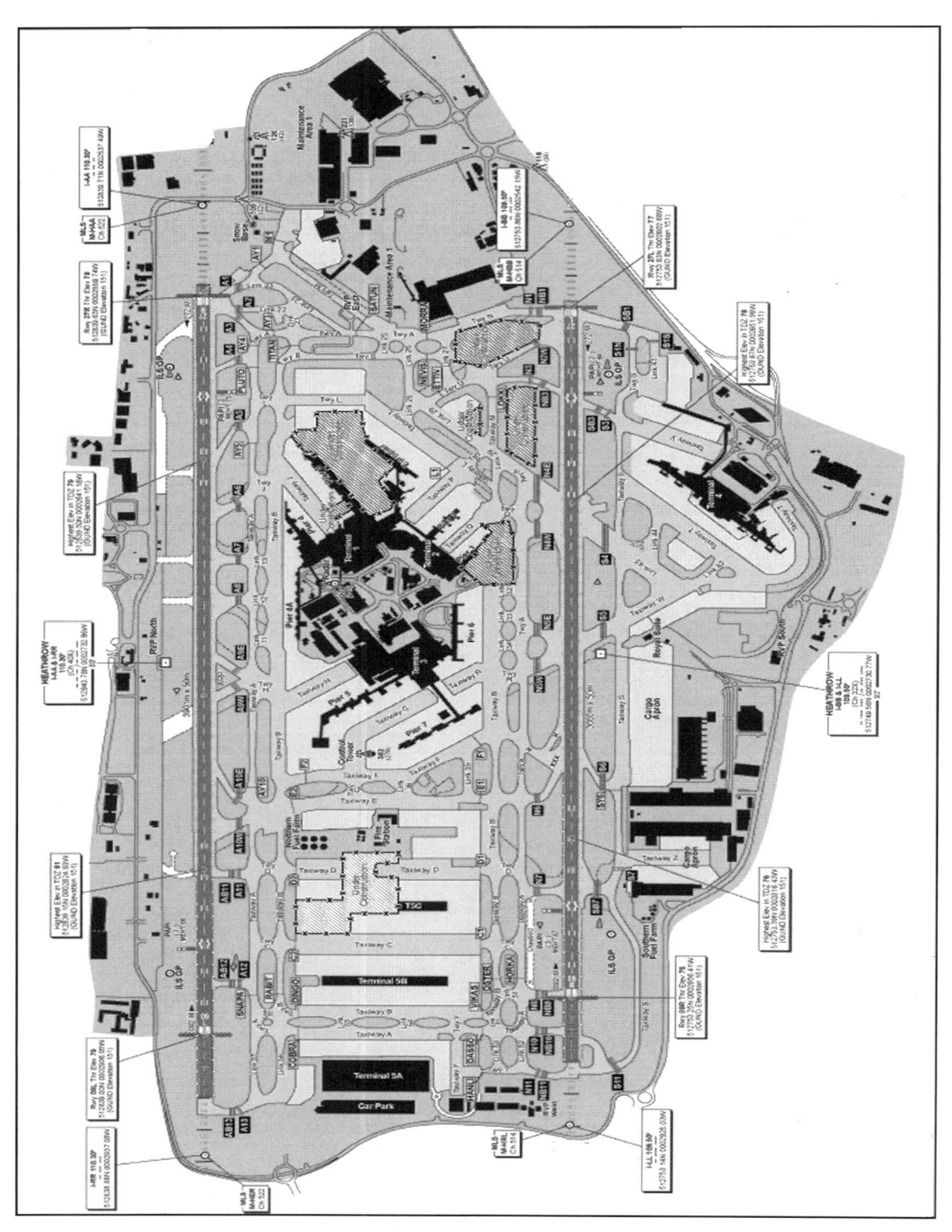

Figure 65 - LHR Airport

Chapter 8
Charts, Maps & Outlines

A chart may be defined as *a map for a very particular purpose such as shipping or aeroplanes/airplanes showing information useful for that purpose and ignoring most other information.* In topology as *a subspace of a manifold used as part of an atlas* or in general *a graphical presentation of something.*

In PlanePlotter a multitude of different charts may be used, the type of chart being very much a user choice. Some users prefer a simple geographical map type of chart showing roads, railways, rivers & coastlines, towns and airports etc. Others may prefer something resembling an air traffic controllers radar type display showing airways and TMA roses etc. All of these are possible with PlanePlotter

To use a chart or map as the base for the graphic display in PlanePlotter it must first be in the correct format and it must also be calibrated so that the software correctly associates chart pixels with geographical coordinates.

PlanePlotter uses charts in the familiar JPEG graphics file format. PlanePlotter can download OpenStreetMaps™ (OSM) and satellite images from the Internet. Both OSM charts and satellite images are self-calibrating but other charts must first be calibrated and this is discussed later.

Paper charts may be used as the source for a PlanePlotter chart but must first be scanned and then edited with a graphics editor program. Such a program should be used to size and orientate the chart to fit the users screen and of course be legible. Once this has been done the file should be saved in the *C:\PlanePlotter\Charts folder* in a JPG format.

An alternative to scanning a paper chart is to use a screen capture program to capture a chart image from a mapping program such as Google Maps, Multimap or better still Google Earth. A word of caution however – charts prepared from Google map and Multimap images may be difficult to calibrate, as they will have been stretched in one direction or the other to make them more suitable for road navigation and they ignore the curvature of the earth.

Creating your first chart

When PlanePlotter is used for the first time a user will have to create their own chart. Doing so is straightforward using the method described below.

First from the menu bar select *File, Open chart* or use the keyboard shortcut *Ctrl-O*. The select *globe.jpg*. The chart should be similar to that shown below in Figure 66. Note that crossed circle symbol indicates the home location. It is assumed that the user will have already set-up their home location as discussed earlier. For this example we will assume that the home location is in the town of Ryde on the Isle of Wight in Southern England.

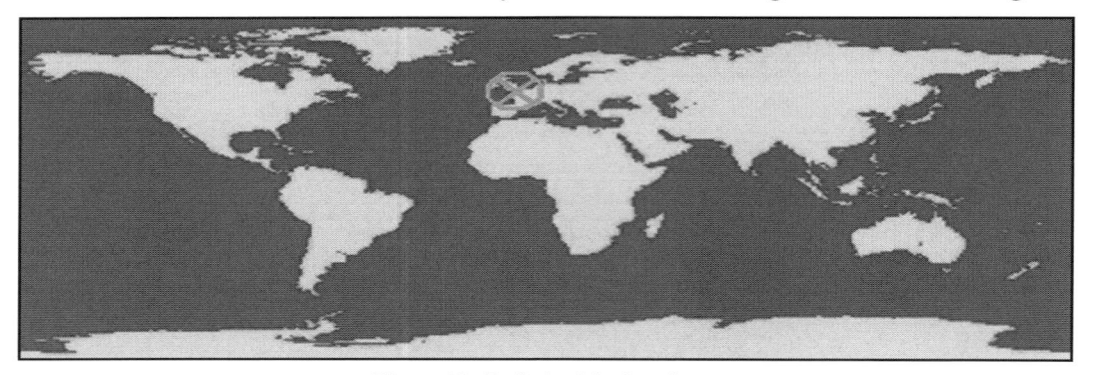

Figure 66 - Default globe.jpg chart

Then zoom into the home location using *Zoom in* button and the cursor to pan across the screen. The result will be a very pixellated screen but this should not cause any concern at this stage.

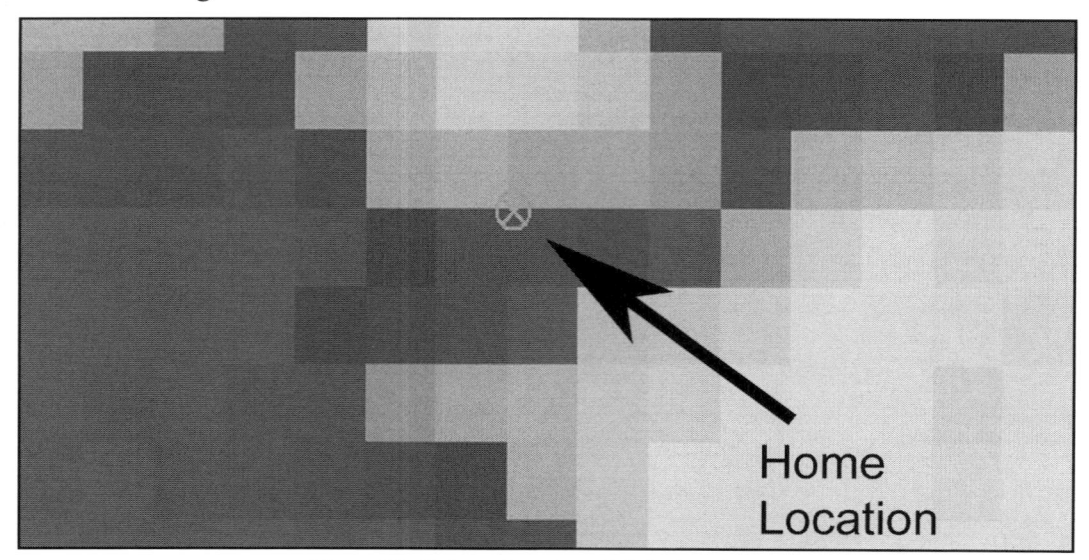

Figure 67 - Enlarged portion of globe.jpg chart

Click on the *OSM* button [OSM] to produce a chart similar to that shown in Figure 68, it may be necessary to zoom in a little further and pan until the desired area is achieved.

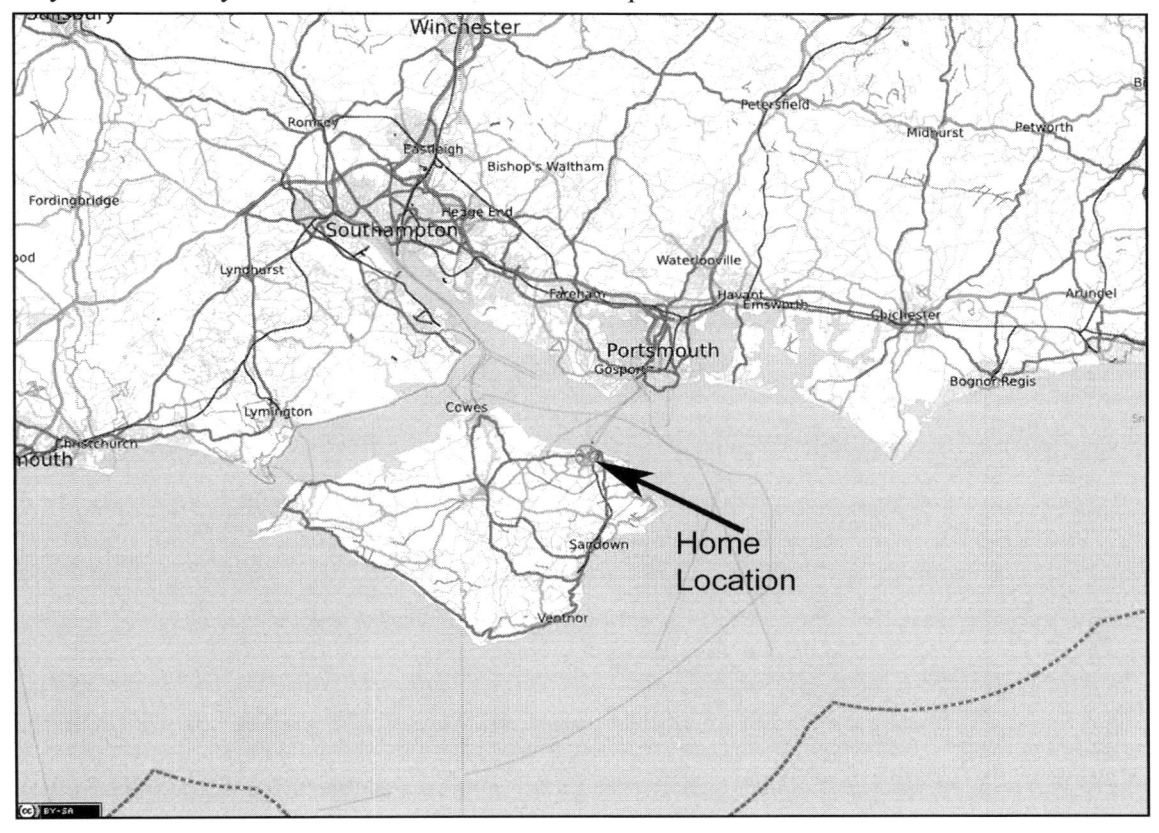

Figure 68 - OSM chart

PlanePlotter will automatically save the new chart that has just been created in the default Chart Files folder. The name of the chart can be seen in the top of the PlanePlotter screen display as shown in Figure 69. In this particular case it is 100928130409.jpg

Figure 69 - Chart name

Closing PlanePlotter at this stage will save the current configuration with that chart.

By zooming out one click and then selecting from the menu bar *Options, Map calibration, Display calibration points* the user will see that a calibration point has been added to each corner of the chart. At this stage the user may wish to use the Windows file manager to rename the file as *home.jpg*. An additional file will have been created and named *100928130409.clb*, this is the calibration file to accompany the chart and this also should be renamed, in this case as *home.clb*.

PlanePlotter should now be restarted. At this stage a dialog may be displayed informing the user that the program could not find the *100928130409* file. This is because the file was renamed as *Home*. The user should now use the menu command *File, Open chart* to open the chart *home.jpg*.

At this point the use is recommended to save the current configuration with the recently created chart by menu command *File, Save Restore point, A*.

Next click on the *Outline* button to change the OSM chart to an outline chart similar to that shown in Figure 70. Note that it will be displayed with a green outline on a black background.

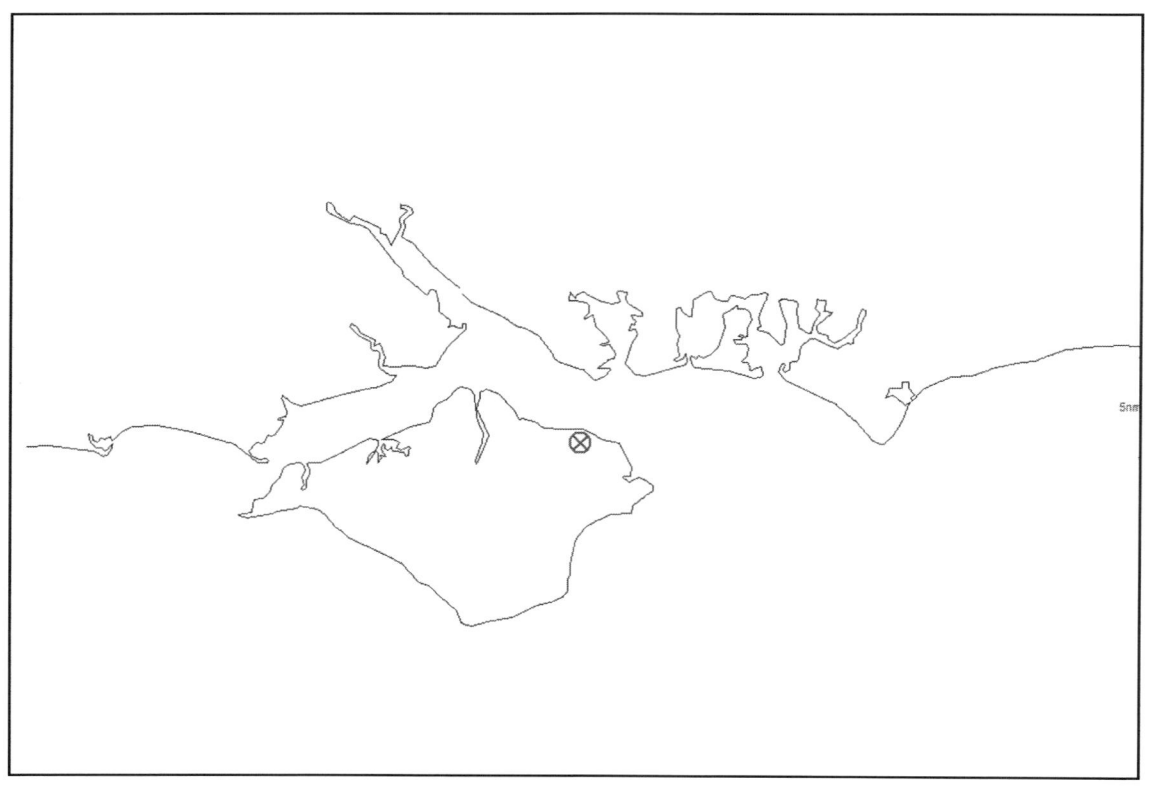

Figure 70 - Outline Chart

From the menu use the command *File, Chart, Quick charts, Define, Quick 2* to save the outline as a Quick Chart. Then revert back to the original OSM by clicking the *Chart* button and then use the command *File, Chart, Quick charts, Define, Quick 1* to save it. Now again use the menu command *File, Save Restore point, A*. Once this is done then user can start PlanePlotter with both an OSM and Outline of their home area readily available.

Charts, Maps & Outlines

Figure 71 - Satellite Chart

Next click on the *Satellite* button to change the chart to a satellite image similar to that shown in Figure 71. Then use the command *File, Chart, Quick charts, Define, Quick 3* to save the outline as a Quick Chart.

Once again use the menu command *File, Save Restore point, A* as this will provide the user with three variants of chart all covering the same geographical area centred on the home location.

Chart Generators

As mentioned earlier a user can create charts by scanning paper charts and then manipulating the images with a graphic editor. An alternative to this is to use one of the chart generators found on the Internet. Two examples of these are the *ShipPlotter Map Generator* found at *http://emit.demon.co.uk/map.php* and the *NoniMapView* found at *http://aeguerre.free.fr/Public/Windows/NoniMapView/EN/* . The first is a web application that runs in the user's web browser to produce both a JPG image that must be downloaded and also a table of calibration data which must be copied and pasted into a text file and saved as a CLB file. NoniMapView is a Java utility that must be installed onto the user's PC, this is done by double-clicking the file name *NoniMapView.exe.jnlp* in *http://aeguerre.free.fr/Public/Windows/NoniMapView/EN*. Note that Java must already be installed on the user's PC to execute the installer and then run the utility.

Figure 72 shows the ShipPlotter Map Generator screen with input boxes used to enter the size of chart required, the home coordinates and the radius of the map. Different styles of chart can be selected by changing the entry in the Visual box.

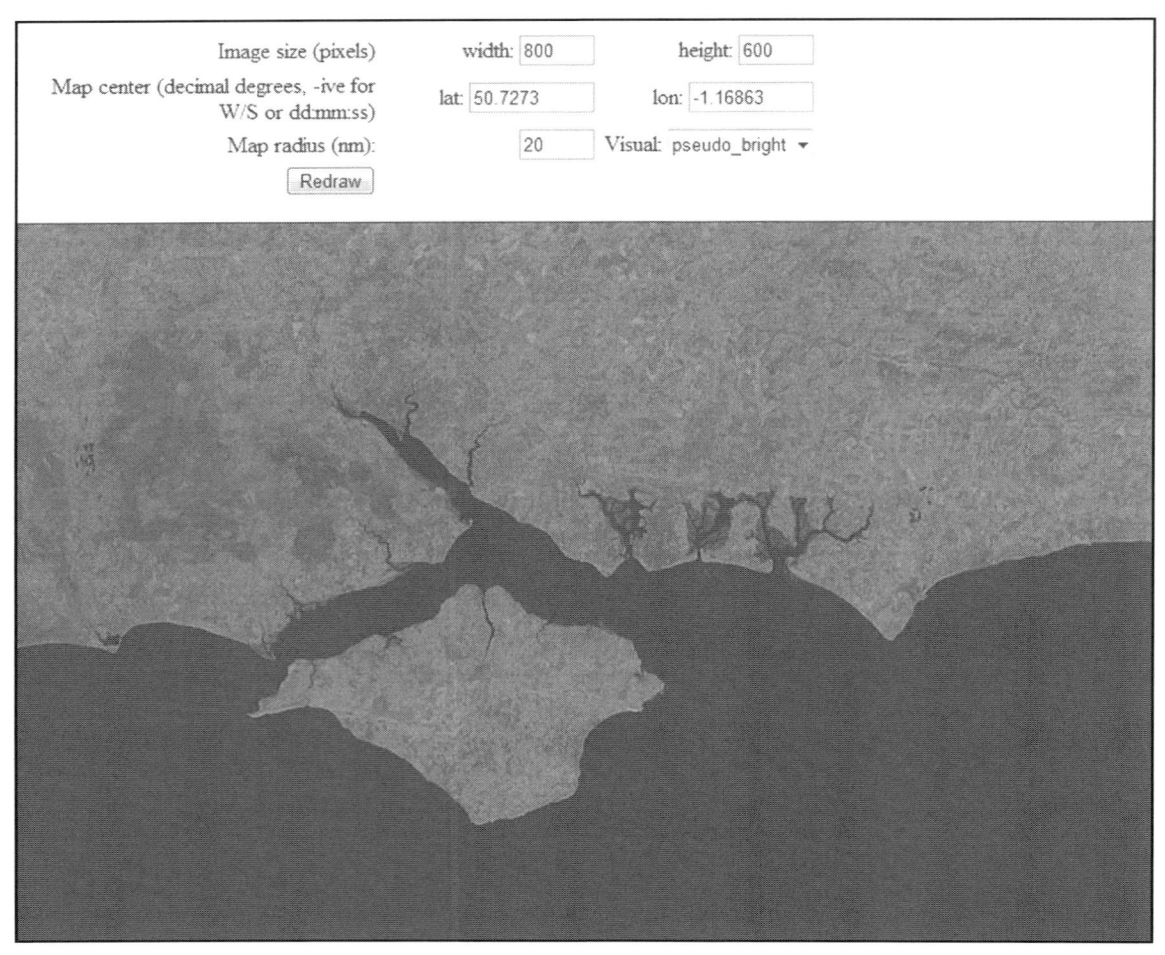

Figure 72 - ShipPlotter Map Chart Generator

Note that both the generators discussed here required the latitude and longitude to be entered in Decimal Degrees. A handy utility to convert between the various formats can be found at *http://www.directionsmag.com/site/latlong-converter*.

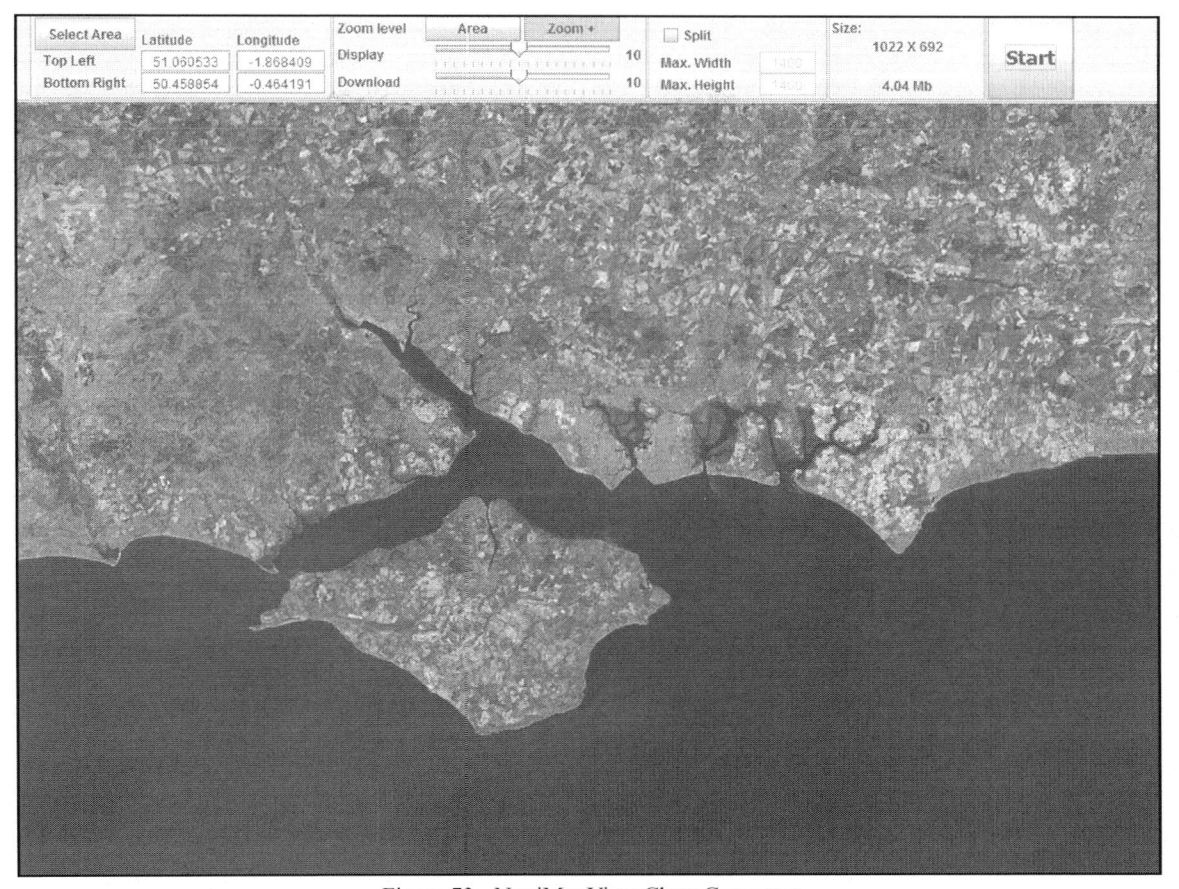

Figure 73 - NoniMapView Chart Generator

On the opening the NoniMapView the first task is to click on the Main Display dropdown and select the required source. Then drag the map about with the cursor and use the mouse wheel or Display slider to zoom in or out to define the area required. Alternatively the coordinates of the top left and bottom right corners of the chart can be entered. Then select the resolution with the Download slider which does not have to be the same as the Display slider but is good place to start. The Size window indicates the actual size of image that will be created, be careful as it is very easy to accidentally start a gigabyte download that may not load into PlanePlotter unless there is sufficient memory in the computer. Then Click Start to open a file dialog where a suitable name can be entered and then save it to the Chart Files folder. As well as saving a JPG image a calibration *.MAP* file will also be saved in the folder. The *.MAP* file is compatible with PlanePlotter *.CLB* files and PlanePlotter will open both JPG file and associated *.MAP* file when that chart is loaded. Note that as NoniMapView image files are high resolution files the user may like to reduce their pixel depth with a graphics image editor to create smaller files.

Calibration

Calibration is the process of relating the *x y* positions of pixels on a chart image to unique geographical coordinates. To ensure accuracy across a chart a number of reference points must be chosen and marked on the chart. Finally the calibration data that has been created must be saved in a calibration file with the same name prefix as the original chart.

From the Menu bar select *Options, Map Calibration, Add Calibration point* as this option to is used add a new calibration point to the current chart. Then by clicking on a point with known coordinates the latitude and longitude of the point can be entered as degrees decimal or degrees & minutes but a N, S, W or E must precede the angle to indicate the quadrant. This process should be repeated for several points, a minimum of at least four and around the chart with them well spaced around the chart.

Once sufficient calibration points have been added to the chart the user should select *Options, Map Calibration, Add Calibration point* to save the calibration file with same prefix name as the chart but with the file extension *.CLB*.

Should the user wish to create a new calibration file then selecting *Options, Map Calibration, Clear Calibration* will clear all the calibration points associated with the current chart.

The user can view individual calibration point data associated with a chart by using *Options, Map Calibration, Display Calibration points* when all the calibration points on the current chart will be displayed a red cross within a red circle. By right clicking on a displayed calibration point the user may edit the point or delete it.

A calibration file can be edited with a standard text editor such as Notepad.

```
Point00,xy,0,0,in,deg,51,3.64,N,1,52.24,W
Point01,xy,200,0,in,deg,51,3.64,N,1,31.18,W
Point02,xy,400,0,in,deg,51,3.64,N,1,10.12,W
Point03,xy,600,0,in,deg,51,3.64,N,0,49.05,W
Point04,xy,800,0,in,deg,51,3.64,N,0,27.99,W
Point05,xy,0,150,in,deg,50,53.64,N,1,52.24,W
Point06,xy,200,150,in,deg,50,53.64,N,1,31.18,W
Point07,xy,400,150,in,deg,50,53.64,N,1,10.12,W
Point08,xy,600,150,in,deg,50,53.64,N,0,49.05,W
Point09,xy,800,150,in,deg,50,53.64,N,0,27.99,W
Point10,xy,0,300,in,deg,50,43.64,N,1,52.24,W
Point11,xy,200,300,in,deg,50,43.64,N,1,31.18,W
Point12,xy,400,300,in,deg,50,43.64,N,1,10.12,W
Point13,xy,600,300,in,deg,50,43.64,N,0,49.05,W
Point14,xy,800,300,in,deg,50,43.64,N,0,27.99,W
Point15,xy,0,450,in,deg,50,33.64,N,1,52.24,W
Point16,xy,200,450,in,deg,50,33.64,N,1,31.18,W
Point17,xy,400,450,in,deg,50,33.64,N,1,10.12,W
Point18,xy,600,450,in,deg,50,33.64,N,0,49.05,W
Point19,xy,800,450,in,deg,50,33.64,N,0,27.99,W
Point20,xy,0,600,in,deg,50,23.64,N,1,52.24,W
Point21,xy,200,600,in,deg,50,23.64,N,1,31.18,W
Point22,xy,400,600,in,deg,50,23.64,N,1,10.12,W
Point23,xy,600,600,in,deg,50,23.64,N,0,49.05,W
Point24,xy,800,600,in,deg,50,23.64,N,0,27.99,W
```

Figure 74 - Calibration File for chart in Fig 72

Third Party Charts

PlanePlotter users are most fortunate in having access to a large range of charts created by air enthusiasts for use with PlanePlotter and some of the other virtual radar products on the market.

One source of charts is the Yahoo PlanePlotter user group on the Internet at http://groups.yahoo.com/group/planeplotter/. By searching through the Group's Files section a wide variety of prepared charts can be found.

A better source can be found at the Manchester TMA Overflights Group website who provide charts for the UK, Europe and the rest of the World. There website address is *http://www.mantma.co.uk/pp_maps_uk.html*.

An example of a chart available from ManTMA is the UK ATC Airspace chart shown in Figure 75. The chart image is shown inverted for printing has a black ground on screen.

Figure 75 - UK ATC Airspace Chart
Courtesy of John Locker & ManTMA

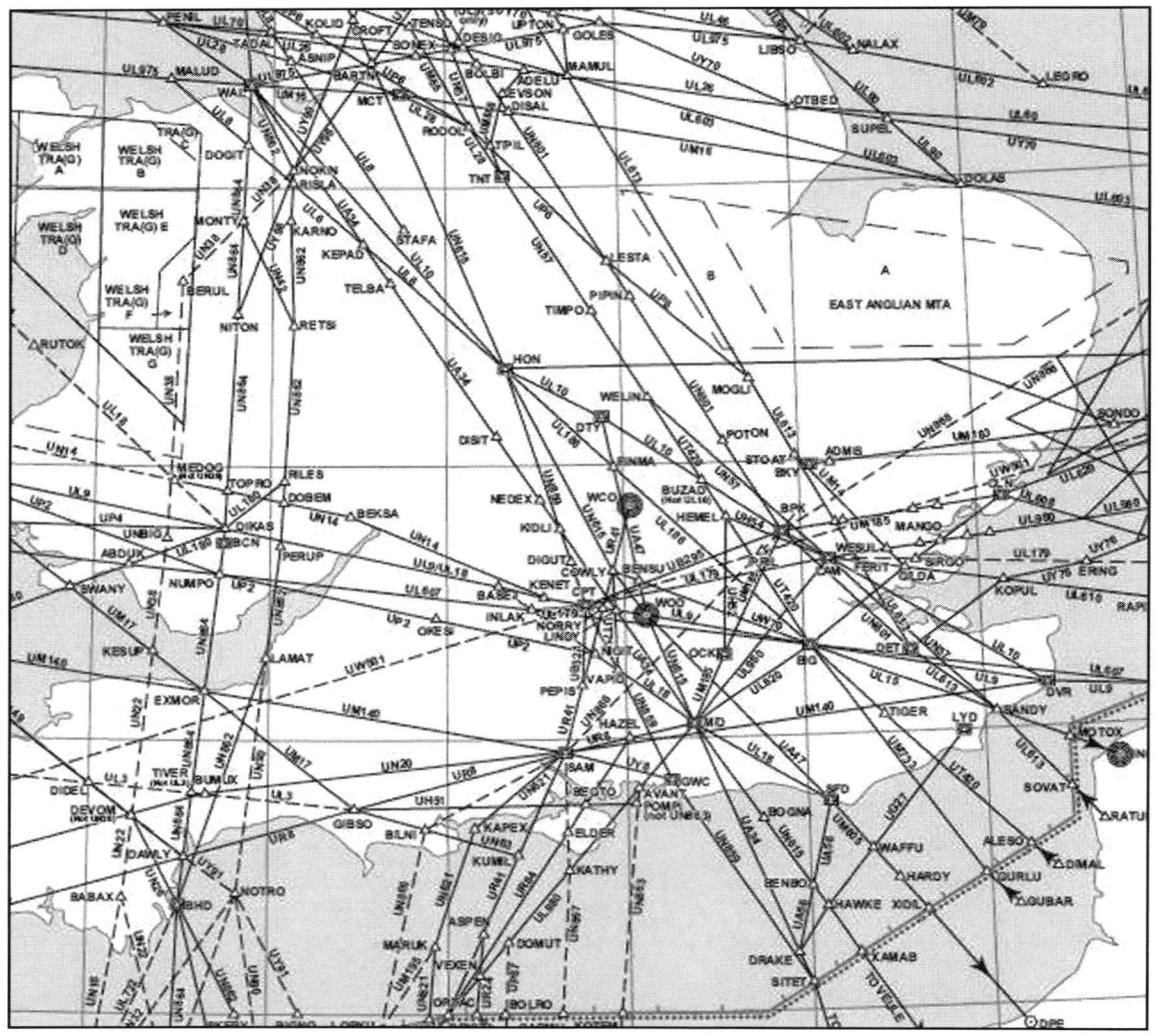

Figure 76 - UK Airway Map Style Chart
Courtesy of John Locker & ManTMA

Another example of a chart that can be downloaded is the map style chart of part of the UK shown in Figure 76 above. The original chart has been edited to provide additional information. The chart has a white mainland with blue sea and blue lines for the airways. Many PlanePlotter users try to emulate real radar displays by inverting the image to have a black background with different colours for the airways and reporting points. This is very easy to do using any of the popular graphic image editing programs such as Adobe's Photoshop. As most downloaded charts are provided with their own calibration files a little caution should be exercised when editing a chart as the calibration data points may no longer relate to the original coordinates.

Outlines

Outlines are simple vector drawings that are used to display coastlines, rivers, county borders, airways and of particular interest airports, their runways and other features. A typical example of an outline that can be used in PlanePlotter is shown in Figure 78 and can easily be compared with the detail shown in the OSM in Figure 77.

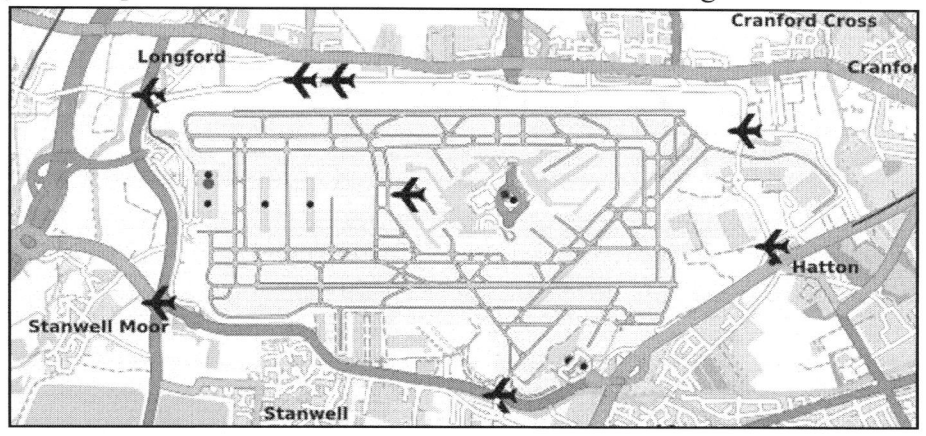

Figure 77 - Part of an OSM Chart showing London Heathrow Airport

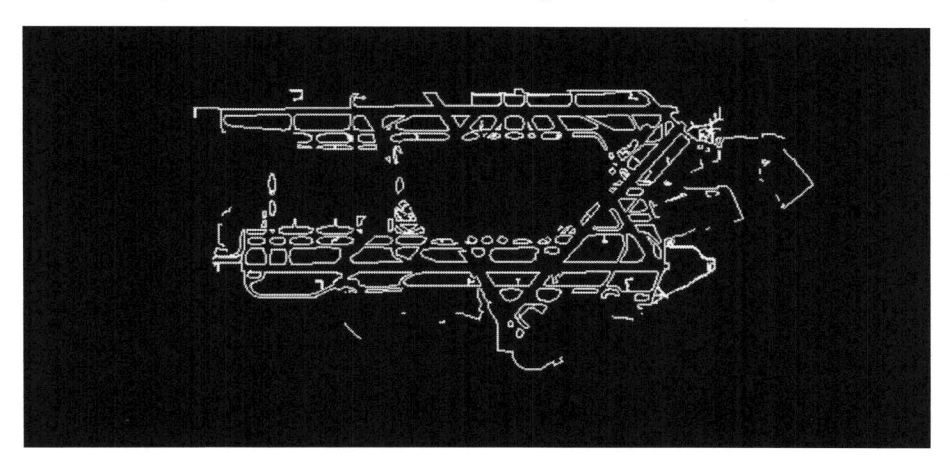

Figure 78 - An Outline of London Heathrow Airport

Fortunately for the PlanePlotter user a wide variety of outlines have already been created and are readily available on a number of Internet web sites. One of these if Yahoo PlanePlotter Group at *http://groups.yahoo.com/group/planeplotter/files/Map Files/* but by best the source is the SBSResources site at *http://www.sbs-resources.com/* which contains the some very popular Hi-Res outlines including ones for Airfields, Air Traffic Zones, Approach Paths, Control Areas, Control Zones, Danger Areas, Holds, Lower & Upper Airways, Military Zones, Terminal Manoeuvring Areas and VOR Roses.

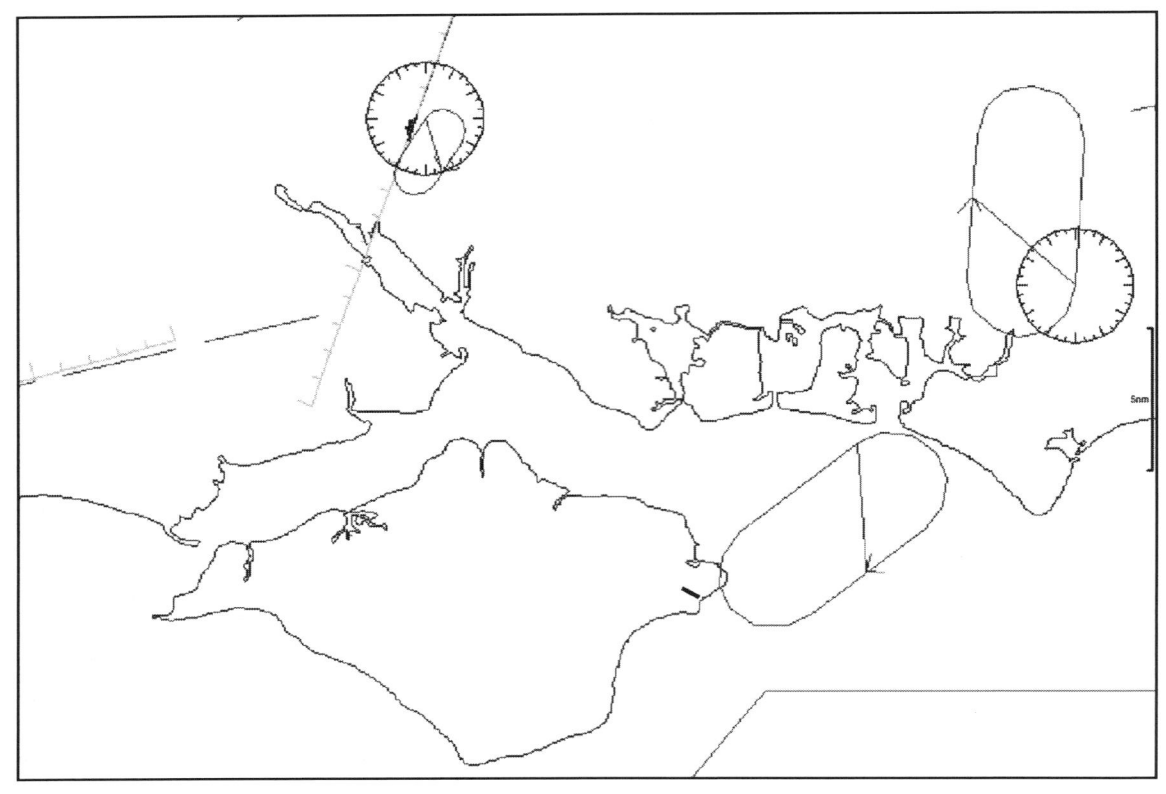

Figure 79 - Chart composed of several different outlines.

Figure 79 shows our basic coastal outline map with a number of different outlines added including airports, holds and VOR roses.

Although there is a wealth of prepared outlines readily available to a user occasionally one may wish to create their own individual outline. Jordan's Outline Maker found at http://www.planeplotters.com/outline_generator/ is a simple to use web based utility.

An example of using this is shown in Figure 81 where an outline of Bembridge Airport has been created. Once the outline has been completed the text below the image should be copied and then pasted into a new text file as shown in Figure 80. Two new lines of text should be inserted at the top of the file. First the name of the outline in squiggly brackets, secondly a Type number indicating the outline colour.

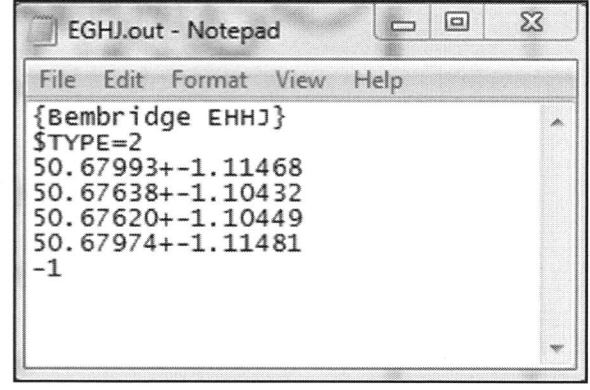

```
{Bembridge EHHJ}
$TYPE=2
50.67993+-1.11468
50.67638+-1.10432
50.67620+-1.10449
50.67974+-1.11481
-1
```

Figure 80 - EGH.out text file

Charts, Maps & Outlines

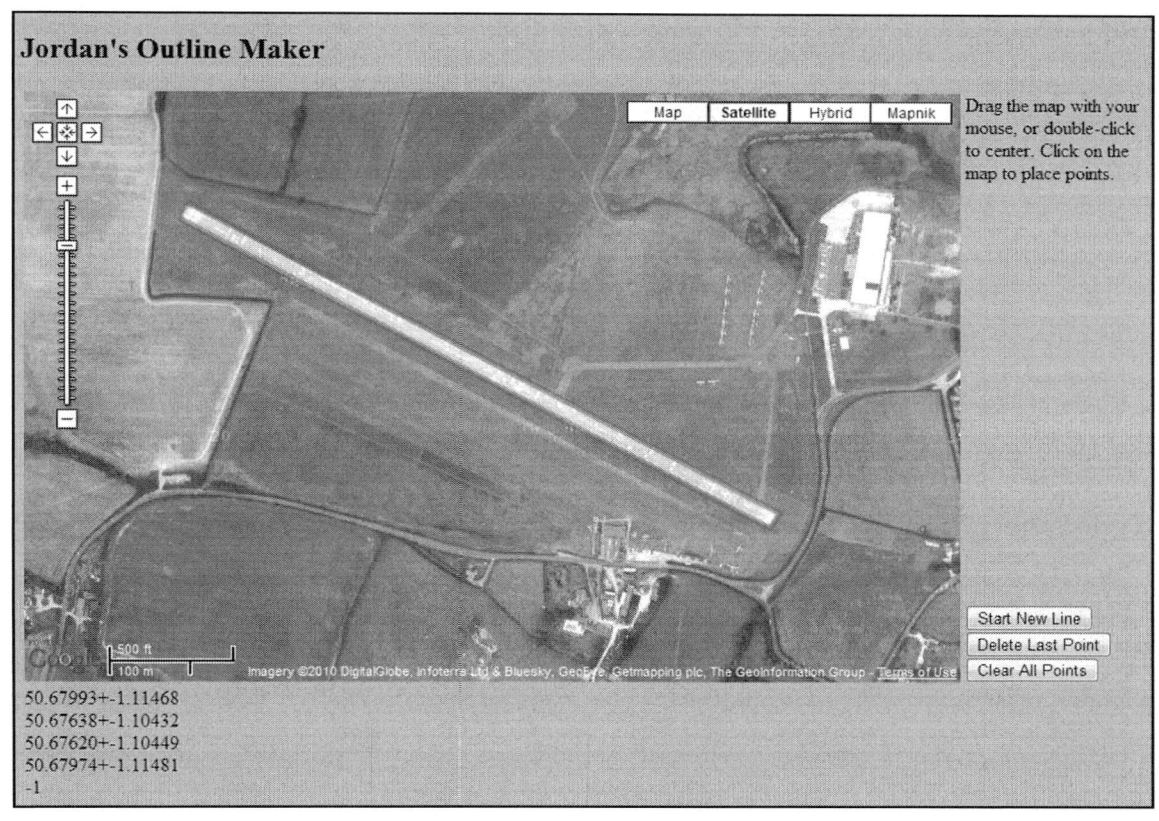

Figure 81 - Jordan's Outline Maker

Finally a word on the colour of an outline. Most outlines have default colours, for example coastlines are normally green, airfields yellow and airport approaches red. The PlanePlotter use may easily change the colour of any outline by simply placing the cursor on top of an outline and then making a Ctrl-click.

This will then open a colour selector as shown in Figure 82. Clicking on a colour in the palette will then change the colour of that particular outline.

If the user wishes to retain the changed colour then the current configuration should be saved with the File menu option *Save Restore point*.

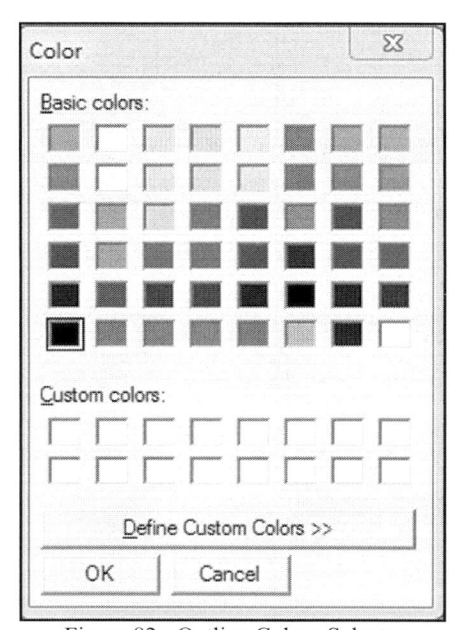

Figure 82 - Outline Colour Selector

95

GPX Overlays

PlanePlotter can display waypoints, tracks or routes over the current chart using data from a user-defined GPX format file.

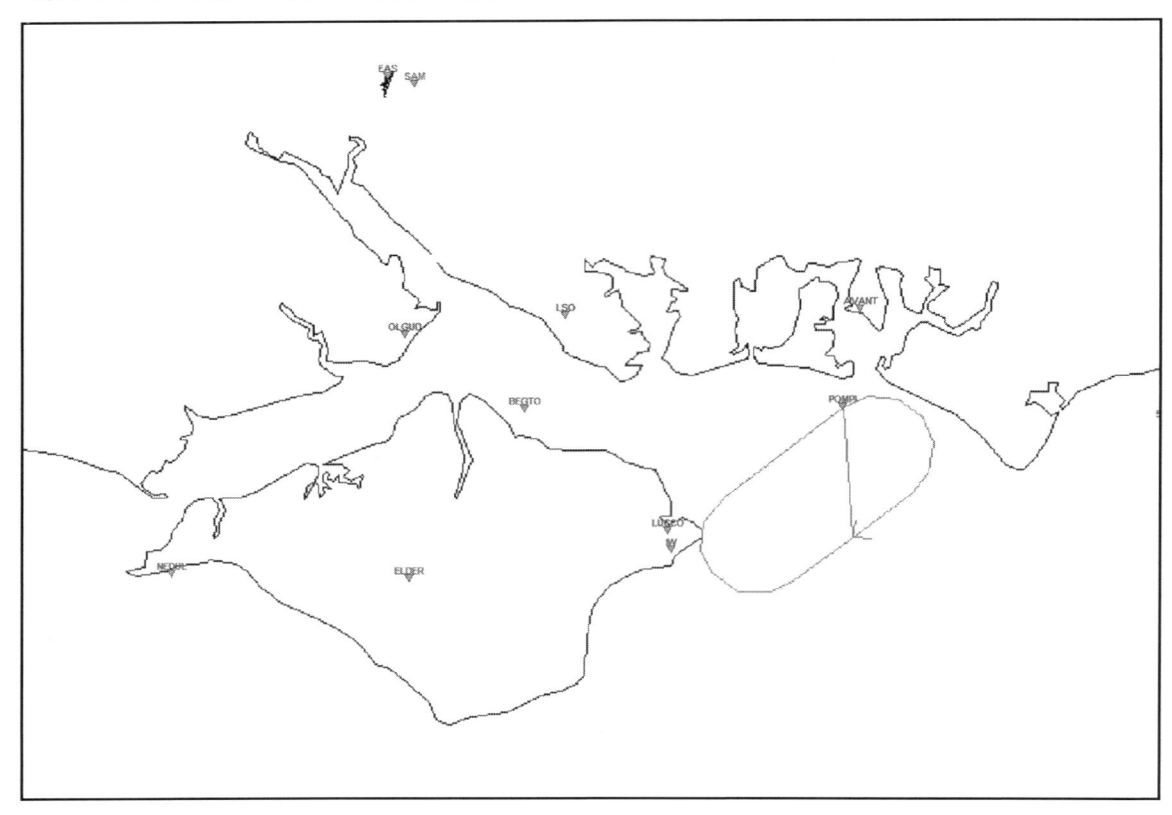

Figure 83 - Otline Chart with GPX Overlays

GPX overlays can be used with any of the different types of chart previously discussed in this chapter

Some of the commonly available GPX files are names.gpx, oceanic_wpt.gpx, uk_ndb.gpx, uk_wpt.gpx, vors.gpx and wpointsnames.gpx. Using the Menu *Options, Chart, Overlay* only one of these files can be selected and overlaid at a time. Thus if one wanted to view an airport name and then waypoints local to that name one would have to select and view the names.gpx file overlay, then close it to open the uk_wpt.gpx overlay. Fortunately all of these files are in a fairly straight forward text format so extracts from the different files can be cut and pasted into a single file.

If users wish to make their own GPX Overlay then they can do so by downloading a program called EasyGPS from http://www.easygps.com.

Plane Symbols

When PlanePlotter is first installed it provides the user with a standard symbol to indicate the positions of aircraft when be displayed on a chart. The default symbol used to be a crude arrow shape but in the latest versions of PlanePlotter the shape of the standard symbol is defined in a special text file called *planesymbol.txt*. Users may notice that the PlanePlotter application directory contains a number of aircraft symbol bitmap images, one for each of the 36 ten degree points but these are associated solely with the Memory Map functions that few users use nowadays, and not with PlanePlotters own chart displays.

Users can create their own library of symbols by studying the existing *planesymbol.txt* file that contains a list of vertices to display a different polygonal symbol for the aircraft. Each vertex is a pair of x y integer coordinates as shown to the left. When PlanePlotter starts it imports the symbol from the *planesymbol.txt* file and use it instead of the default aircraft symbol. Note that the aircraft symbol size parameter already specified in *Options, Chart, Options* will still apply to the customised shape.

```
 0  8 n
 1  7 n
 1  3 n
 8 -4
 8 -6
 1 -2
 1 -6
 3 -8
 3 -9
 0 -8
-3 -9
-3 -8
-1 -6
-1 -2
-8 -6
-8 -4
-1  3 n
-1  7 n
 0  8 n
```

PlanePlotter colour codes the nose of the aircraft to indicate ascent or descent. In order to identify the vertices that constitute the nose, an "n" must be added after a space following the Y coordinate of each vertex that is part of the nose.

One can also make PlanePlotter display different symbols for different aircraft or different aircraft types. This is achieved by creating one or more text files with names *planesymbol1.txt*, *planesymbol2.txt*, planesymbol3.txt etc. Each file must follow the same convention as described above. The new symbols created in these files will be used when a User tag for an aircraft includes the sequence *$1*, *$2*, *$3* etc. For example when an aircraft is tagged with *$6* PlanePlotter will use the symbol contained in the file *planesymbol6.txt*. Tag numbers1-9 & A-F are supported however if no tag is used then the default symbol *planesymbol.txt* is used. An easy way of creating these is described next.

SymbolMaker2

There is now available a small utility program that enables users to create their own aircraft symbol from a silhouette image of an aircraft. All that is required to create a custom symbol is symbol is to run the SymbolMaker program from the Internet at *http://www.coaa.co.uk/SymbolMaker2.exe*. Silhouettes of aircraft are readily available form web site such as *http://www.au.af.mil/au/awc/awcgate/clip_af.htm*. They may be downloaded and saved to a local but should be inverted if necessary to create a black image on a white background.

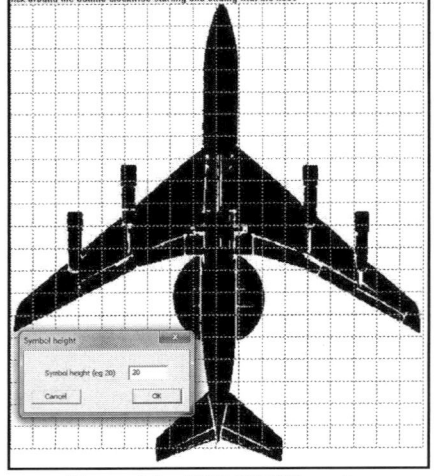

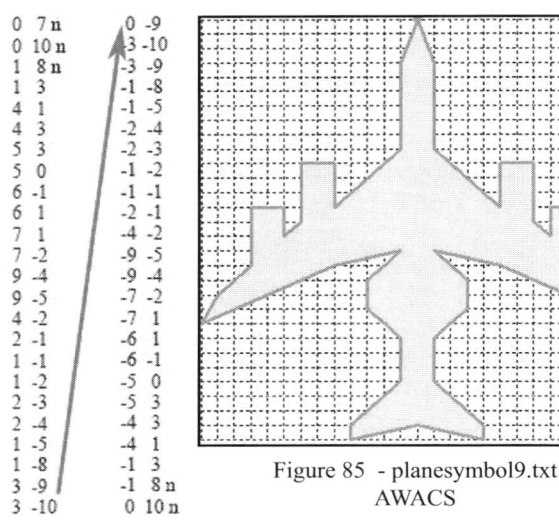

0 7 n	0 -9	
0 10 n	3 -10	
1 8 n	-3 -9	
1 3	-1 -8	
4 1	-1 -5	
4 3	-2 -4	
5 3	-2 -3	
5 0	-1 -2	
6 -1	-1 -1	
6 1	-2 -1	
7 1	-4 -2	
7 -2	-9 -5	
9 -4	-9 -4	
9 -5	-7 -2	
4 -2	-7 1	
2 -1	-6 1	
1 -1	-6 -1	
1 -2	-5 0	
2 -3	-5 3	
2 -4	-4 3	
1 -5	-4 1	
1 -8	-1 3	
3 -9	-1 8 n	
3 -10	0 10 n	

Figure 84 - AWACS Silhouette

Figure 85 - planesymbol9.txt AWACS

With SymbolMake2 running select *File, Silhouette* to open the downloaded image, in this case it's an AWAC silhouette. The next step is to specify the size of the symbol in pixels, 20 or 25 is a good starting point. Then starting at the nose of the aircraft and working clockwise click points on the grid to create the symbol outline finishing back at the nose. At this stage the symbol will be filled with a yellow colour. Finally click once more at the nose and two or three points more to define the nose area, this will change the nose area colour to blue. Finally save the plane symbol to a file called planesymbolX.txt where X is a numeral 1-9 or a letter A-H. The plane symbol text file contains a list of the vertices defining the symbol. A letter "n" after a pair of number indicates that point will change colour if the aircraft is ascending or descending. Some further examples are shown below. (Download *Planesymbol_files.zip* from *http://groups.yahoo.com/group/planeplotter/files*).

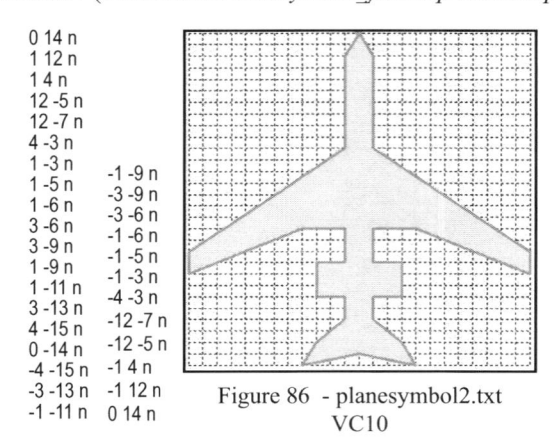

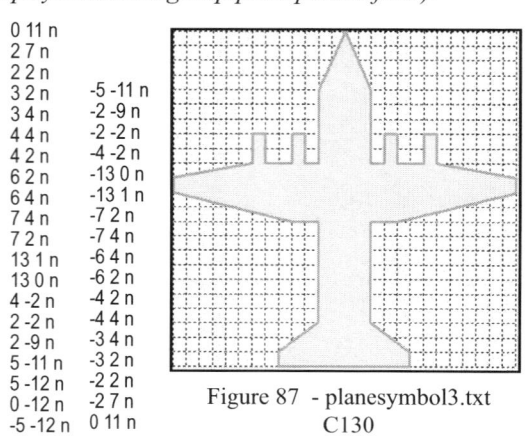

Figure 86 - planesymbol2.txt VC10

Figure 87 - planesymbol3.txt C130

Charts, Maps & Outlines

```
1 10 n
2 3 n
3 3 n
3 1 n
8 -6 n
8 -8 n
2 -7 n
3 -9 n
1 -9 n
1 -11 n     -2 -7 n
3 -13 n    -8 -8 n
3 -14 n    -8 -6 n
-3 -14 n   -3 1 n
-3 -13 n   -3 3 n
-1 -11 n   -2 3 n
-1 -9 n    -1 10 n
-3 -9 n    1 10 n
```

Figure 88 - planesymbol6.txt
Fast Jet

```
0 9 n     0 -11 n
1 9       -1 -10 n
2 9 n     -1 -9 n
2 6 n     -3 -9 n
9 11 n    -2 -7 n
10 10 n   -1 -7 n
3 3 n     -1 -2 n
4 0 n     -2 -1 n
10 -6 n   -9 -7 n
9 -7 n    -10 -6 n
2 -1 n    -4 0 n
1 -2 n    -3 3 n
1 -7 n    -10 10 n
2 -7 n    -9 11 n
3 -9 n    -2 6 n
1 -9 n    -2 9 n
1 -10 n   0 11 n
```

Figure 89 - planesymbol7.txt
Helo

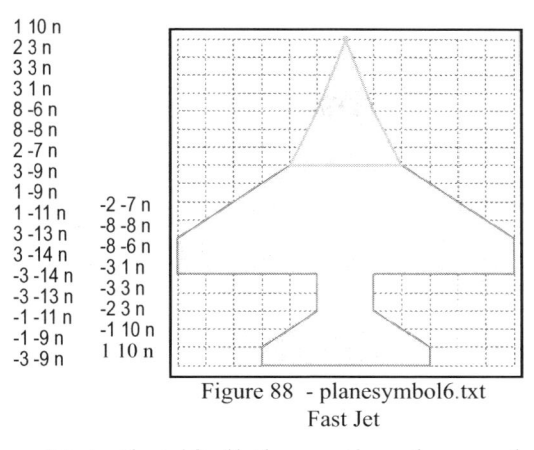

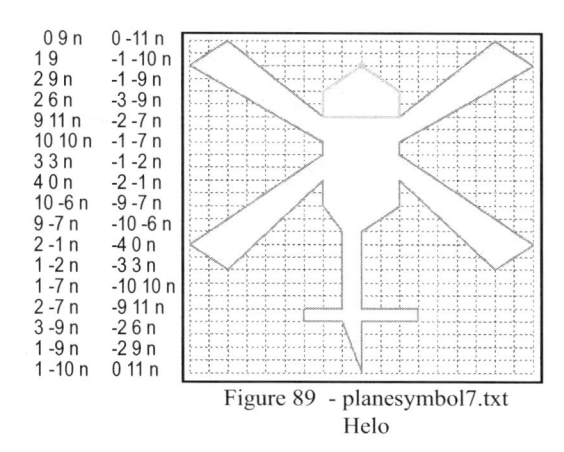

Note that if all the vertices have a letter "n" after them then the whole symbol will change colour as the aircraft ascends or descends and not just the colour of the nose.

Associating a symbol with an aircraft

Once the planesymbol*x*.txt files have been created anyone of them can be associated with an individual aircraft an aircraft type, an airline or a country. By *right clicking* an aircraft symbol displayed on a PlanePlotter chart Aircraft position report dialog will be opened as shown here. By clicking the Edit button a second dialog box will be displayed and the user may then enter a value in the *User tag*. By using the dollar sign and a number or letter such as $2 or $A PlanePlotter will associate the newly created symbol from in these cases either *planesymbol2.txt* or *planesymbolA.txt*.

Clicking OK will close the dialog and the aircraft symbol displayed will immediately be updated to the one specified in the text file. To ensure that association is retained as

Figure 90 - Aircraft position report dialog

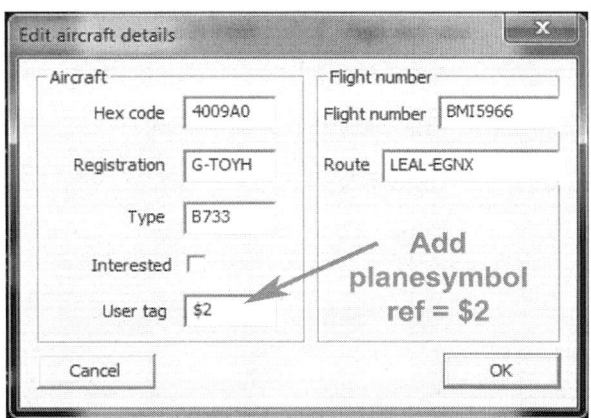

Figure 91 - Aircraft position report dialog

should then the PlanePlotter Menu *Options, Mode-S, Kinetic, Update registrations* box must be checked. Where the symbol needs to be associated with a number of aircraft this can be acheived by a bulk updating of the *User tags* in the basestation.sqb database. An explanation of how to do that is given in the following pages.

Editing the basestation.sqb database

Both the PlanePlotter databases basestation.sqb and flightroute.sqb use the SQLite relational database management system. Fortunately for PlanePlotter users there are a number of readily available database management tools that allow them to edit these databases without the need for a too-in-depth knowledge of SQLite.

The easiest tool to use is called *SQLite Expert Personal*. It can be downloaded from *http://www.sqliteexpert.com/SQLiteExpertPersSetup.exe* and as it is Freeware there is no cost nor expiration data.

After downloading, installing and then running the program the user will see a screen similar to that below. By then selecting *File, Open Database* the user can view the basestation.sqb database. First click on the *Aircraft* table indicator in the left panel and then click the *Data* button to display a table of all aircraft in the database. As the table contains 49 fields of data displayed as 49 columns it is necessary to scroll right and left to see all the data.

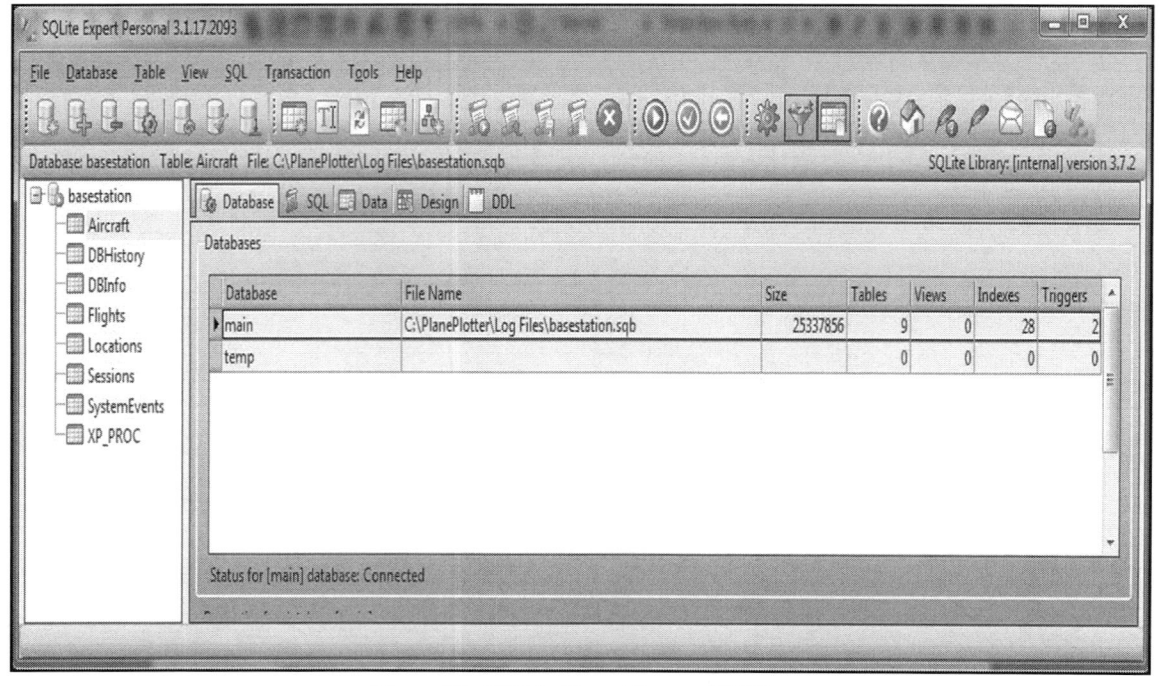

Figure 92 - SQLite Expert Personal Edition

Associating a User Tag with type of Aircraft

In this first example we want to use the aircraft symbol in the planesymbol7.txt file with any Airbus in the A32x series. The first step is click the *SQL* button that opens a panel where some SQL queries may be entered, these are effectively commands to manipulate the contents of a table in the database. In this case we enter two queries:

```
UPDATE Aircraft SET UserTag="$7" WHERE ICAOTypeCode LIKE "A3%"
SELECT * from Aircraft WHERE UserTag LIKE "$%";
```

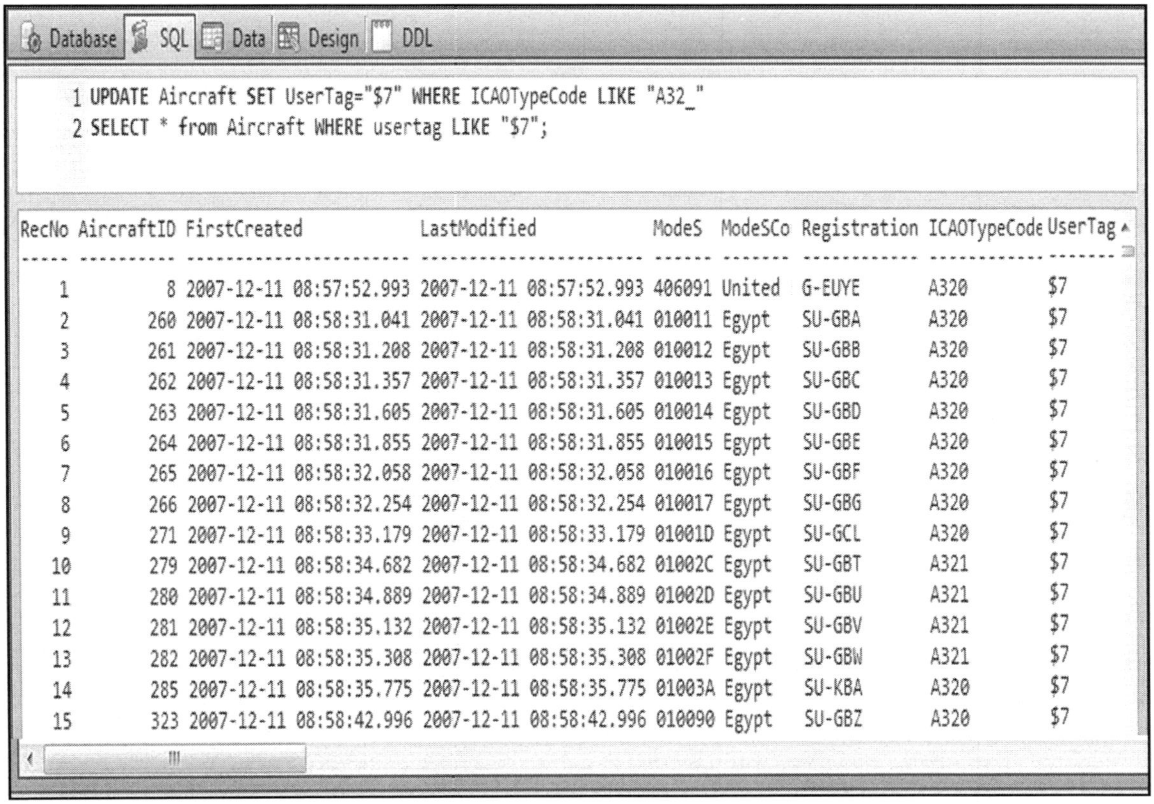

RecNo	AircraftID	FirstCreated	LastModified	ModeS	ModeSCo	Registration	ICAOTypeCode	UserTag
1	8	2007-12-11 08:57:52.993	2007-12-11 08:57:52.993	406091	United	G-EUYE	A320	$7
2	260	2007-12-11 08:58:31.041	2007-12-11 08:58:31.041	010011	Egypt	SU-GBA	A320	$7
3	261	2007-12-11 08:58:31.208	2007-12-11 08:58:31.208	010012	Egypt	SU-GBB	A320	$7
4	262	2007-12-11 08:58:31.357	2007-12-11 08:58:31.357	010013	Egypt	SU-GBC	A320	$7
5	263	2007-12-11 08:58:31.605	2007-12-11 08:58:31.605	010014	Egypt	SU-GBD	A320	$7
6	264	2007-12-11 08:58:31.855	2007-12-11 08:58:31.855	010015	Egypt	SU-GBE	A320	$7
7	265	2007-12-11 08:58:32.058	2007-12-11 08:58:32.058	010016	Egypt	SU-GBF	A320	$7
8	266	2007-12-11 08:58:32.254	2007-12-11 08:58:32.254	010017	Egypt	SU-GBG	A320	$7
9	271	2007-12-11 08:58:33.179	2007-12-11 08:58:33.179	01001D	Egypt	SU-GCL	A320	$7
10	279	2007-12-11 08:58:34.682	2007-12-11 08:58:34.682	01002C	Egypt	SU-GBT	A321	$7
11	280	2007-12-11 08:58:34.889	2007-12-11 08:58:34.889	01002D	Egypt	SU-GBU	A321	$7
12	281	2007-12-11 08:58:35.132	2007-12-11 08:58:35.132	01002E	Egypt	SU-GBV	A321	$7
13	282	2007-12-11 08:58:35.308	2007-12-11 08:58:35.308	01002F	Egypt	SU-GBW	A321	$7
14	285	2007-12-11 08:58:35.775	2007-12-11 08:58:35.775	01003A	Egypt	SU-KBA	A320	$7
15	323	2007-12-11 08:58:42.996	2007-12-11 08:58:42.996	010090	Egypt	SU-GBZ	A320	$7

Figure 93 - SQL Query and Result for ICAOTypeCode & UserTag

The first is a command to update every entry in the 'Aircraft' table 'UserTag' field where the 'ICAOTypeCode' is A320 or A321. (The use of a '%' after the 'A3' permits any following characters).

The second query then displays a list of all aircraft with the $7 UserTag. In this case we want to associate planesymbolB.txt file with any Estonian registered aircraft.

```
UPDATE Aircraft SET UserTag="$B" WHERE Registration LIKE "ES-%"
SELECT * from Aircraft WHERE Registration LIKE "ES-%";
```

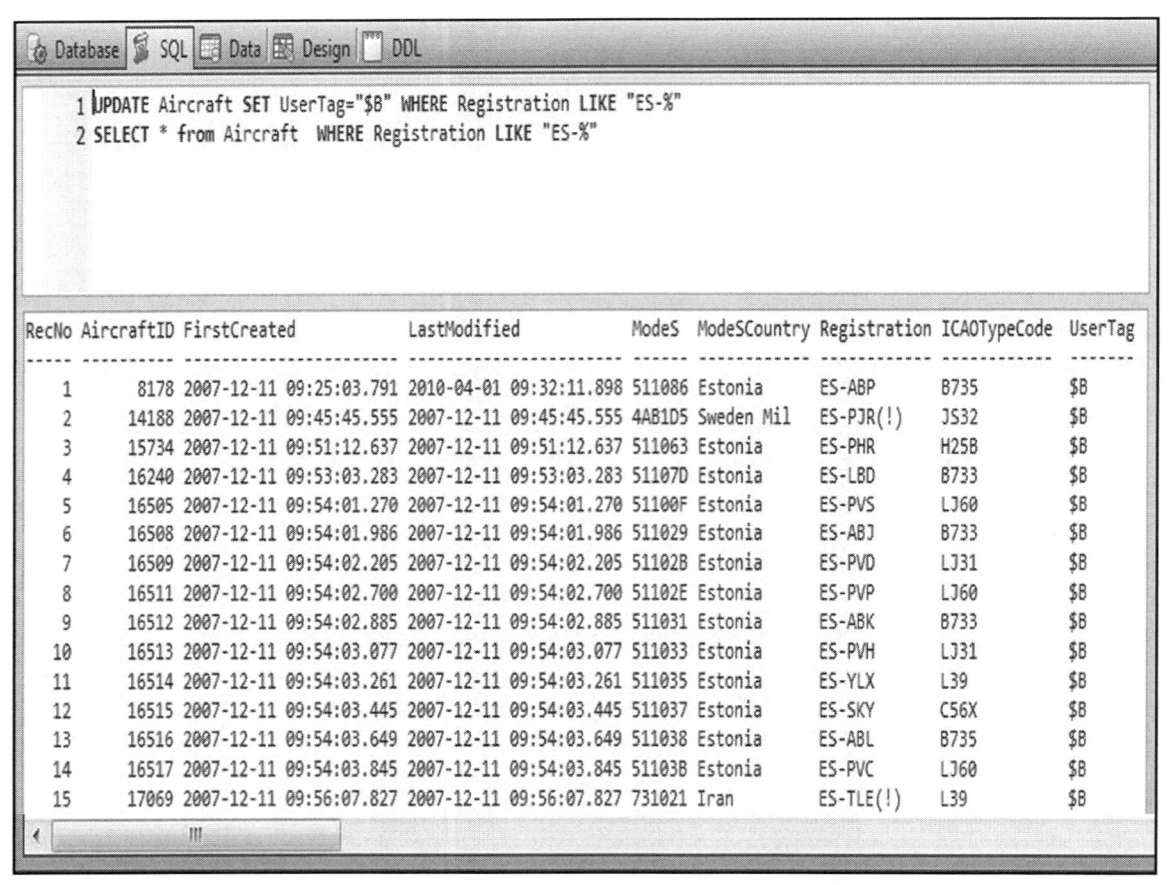

RecNo	AircraftID	FirstCreated	LastModified	ModeS	ModeSCountry	Registration	ICAOTypeCode	UserTag
1	8178	2007-12-11 09:25:03.791	2010-04-01 09:32:11.898	511086	Estonia	ES-ABP	B735	$B
2	14188	2007-12-11 09:45:45.555	2007-12-11 09:45:45.555	4AB1D5	Sweden Mil	ES-PJR(!)	JS32	$B
3	15734	2007-12-11 09:51:12.637	2007-12-11 09:51:12.637	511063	Estonia	ES-PHR	H25B	$B
4	16240	2007-12-11 09:53:03.283	2007-12-11 09:53:03.283	51107D	Estonia	ES-LBD	B733	$B
5	16505	2007-12-11 09:54:01.270	2007-12-11 09:54:01.270	51100F	Estonia	ES-PVS	LJ60	$B
6	16508	2007-12-11 09:54:01.986	2007-12-11 09:54:01.986	511029	Estonia	ES-ABJ	B733	$B
7	16509	2007-12-11 09:54:02.205	2007-12-11 09:54:02.205	51102B	Estonia	ES-PVD	LJ31	$B
8	16511	2007-12-11 09:54:02.700	2007-12-11 09:54:02.700	51102E	Estonia	ES-PVP	LJ60	$B
9	16512	2007-12-11 09:54:02.885	2007-12-11 09:54:02.885	511031	Estonia	ES-ABK	B733	$B
10	16513	2007-12-11 09:54:03.077	2007-12-11 09:54:03.077	511033	Estonia	ES-PVH	LJ31	$B
11	16514	2007-12-11 09:54:03.261	2007-12-11 09:54:03.261	511035	Estonia	ES-YLX	L39	$B
12	16515	2007-12-11 09:54:03.445	2007-12-11 09:54:03.445	511037	Estonia	ES-SKY	C56X	$B
13	16516	2007-12-11 09:54:03.649	2007-12-11 09:54:03.649	511038	Estonia	ES-ABL	B735	$B
14	16517	2007-12-11 09:54:03.845	2007-12-11 09:54:03.845	51103B	Estonia	ES-PVC	LJ60	$B
15	17069	2007-12-11 09:56:07.827	2007-12-11 09:56:07.827	731021	Iran	ES-TLE(!)	L39	$B

Figure 94 - SQL Query and Result for Estonian Registrations & UserTag

Here every entry in the 'Aircraft' table where the 'Registration' starts with ES- is given a UserTag of %B. Again the second query displays a list of the resulting update and changes.

The third example associates a plane symbol with an aircraft ModeSCountry indicator. Again there are two SQL queries.

```
UPDATE Aircraft SET UserTag="$F" WHERE ModeSCountry LIKE
"France"
SELECT * from Aircraft WHERE ModeSCountry LIKE "France";
```

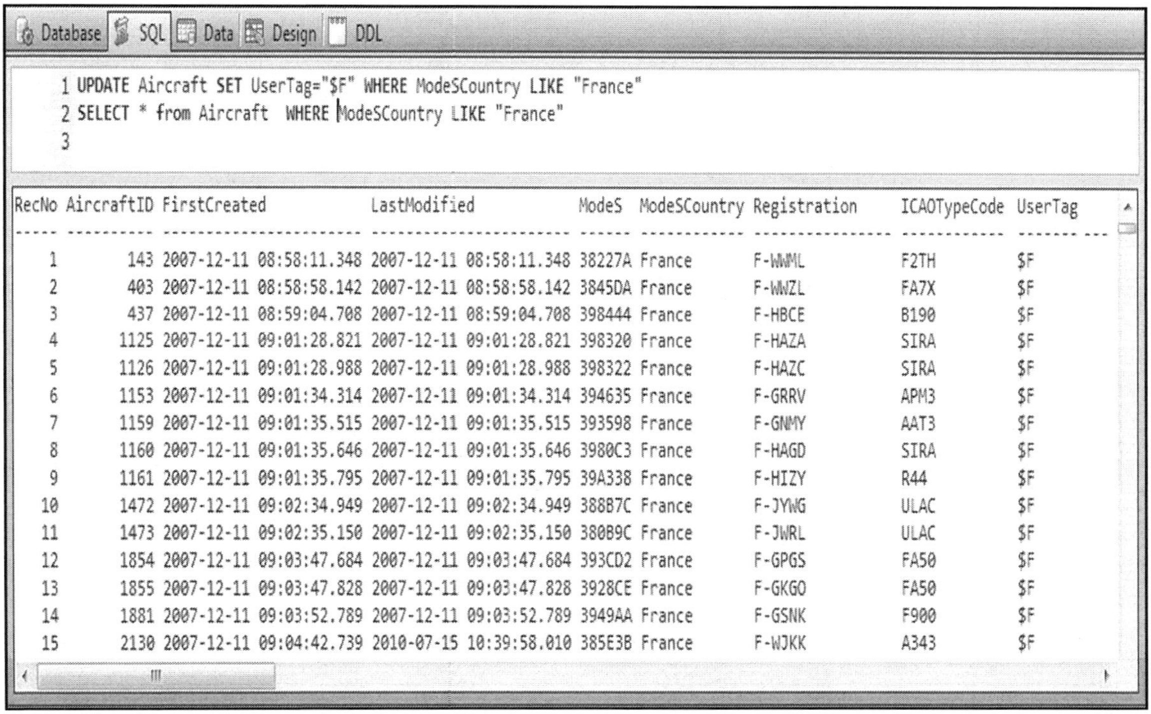

RecNo	AircraftID	FirstCreated	LastModified	ModeS	ModeSCountry	Registration	ICAOTypeCode	UserTag
1	143	2007-12-11 08:58:11.348	2007-12-11 08:58:11.348	38227A	France	F-WWML	F2TH	$F
2	403	2007-12-11 08:58:58.142	2007-12-11 08:58:58.142	3845DA	France	F-WWZL	FA7X	$F
3	437	2007-12-11 08:59:04.708	2007-12-11 08:59:04.708	398444	France	F-HBCE	B190	$F
4	1125	2007-12-11 09:01:28.821	2007-12-11 09:01:28.821	398320	France	F-HAZA	SIRA	$F
5	1126	2007-12-11 09:01:28.988	2007-12-11 09:01:28.988	398322	France	F-HAZC	SIRA	$F
6	1153	2007-12-11 09:01:34.314	2007-12-11 09:01:34.314	394635	France	F-GRRV	APM3	$F
7	1159	2007-12-11 09:01:35.515	2007-12-11 09:01:35.515	393598	France	F-GNMY	AAT3	$F
8	1160	2007-12-11 09:01:35.646	2007-12-11 09:01:35.646	3980C3	France	F-HAGD	SIRA	$F
9	1161	2007-12-11 09:01:35.795	2007-12-11 09:01:35.795	39A338	France	F-HIZY	R44	$F
10	1472	2007-12-11 09:02:34.949	2007-12-11 09:02:34.949	388B7C	France	F-JYWG	ULAC	$F
11	1473	2007-12-11 09:02:35.150	2007-12-11 09:02:35.150	380B9C	France	F-JWRL	ULAC	$F
12	1854	2007-12-11 09:03:47.684	2007-12-11 09:03:47.684	393CD2	France	F-GPGS	FA50	$F
13	1855	2007-12-11 09:03:47.828	2007-12-11 09:03:47.828	3928CE	France	F-GKGO	FA50	$F
14	1881	2007-12-11 09:03:52.789	2007-12-11 09:03:52.789	3949AA	France	F-GSNK	F900	$F
15	2130	2007-12-11 09:04:42.739	2010-07-15 10:39:58.010	385E3B	France	F-WJKK	A343	$F

Figure 95 - SQL Query and Result for Conuntry & UserTag

This time every entry in the 'Aircraft' table is given a UserTag of %F where the aircraft has a ModeSCountry of France. Once the update has been executed the second query displays a list of the resulting update and changes.

The queries only have to be entered as plain text in the Query panel. They can of course be copied and pasted from any source. The queries are initiated by either clicking the *F5 key* or selecting *SQL, Execute SQL* from the main menu bar. Where there is more than one query they can be initiated one at a time using *Shift-F5*.

Restoring Changed Data

It is wise to make a copy of any database before attempting to make any change to its contents. It is possible however to undo any changes that have been made using the queries described above by the use of a further simple query.

UPDATE Aircraft SET UserTag=Null WHERE UserTag LIKE "$%"

```
1 SELECT * from Aircraft
```

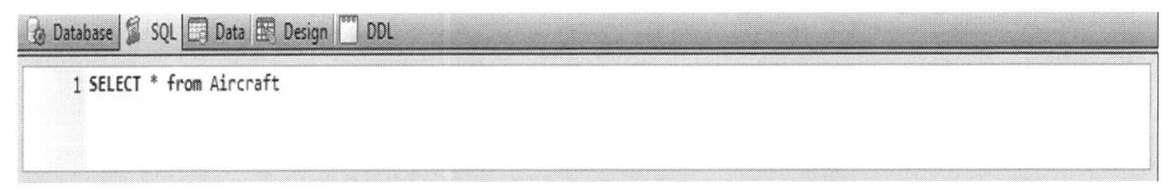

Figure 96 - SQL Query to display the complete database table

This query will remove all the UserTag entries having a $x entry. If other type of UserTag entries have been made then these will not be affected.

The list of aircraft displayed can be restored to the complete contents of the table at anytime with this query.

SELECT * from Aircraft WHERE Registration LIKE "ES-%";

```
1 UPDATE Aircraft SET UserTag=Null WHERE UserTag LIKE "$%"
```

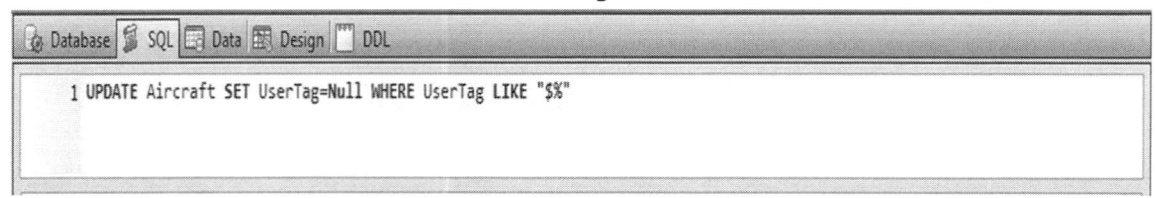

Figure 97 - SQL Query to remove Plane Symbol UserTags

A Note from COAA on the use of Plane Symbols

Although users can create and use different designs of plane symbols to their hearts content the original intention was that the symbols would be self selecting on the basis of the category code contained in the flight number Mode-S message. That code has four sets of eight entries with the first set 'A' being the most commonly used. A list of the ADS-B Emitter Category Set 'A' is as follows:

>   0 = No ADS-B Emitter Category Information
>   1 = Light (< 15 500 lbs.)
>   2 = Small (15 500 to 75 000 lbs.)
>   3 = Large (75 000 to 300 000 lbs.)
>   4 = High Vortex Large(aircraft such as B-757)
>   5 = Heavy (> 300 000 lbs.)
>   6 = High Performance (> 5 g acceleration and > 400kts)
>   7 = Rotorcraft

The overwhelming majority of aircraft encode 'A0' which mean that there is no category information available. The symbol selection is therefore currently based on a special coding ($x) in the User tag field for each aircraft. If users wish to use further updates and enhancements of the PlanePlotter program then they are encouraged to conform to the defined types 1-7 as far as possible and only to use the code space $8-$F for the other specific types of symbol.

It may be that in the future more aircraft start to encode the category field correctly and COAA may decide to use that field to control the symbol. Even if they then allow the User tag to take precedence problems may arise for the user as an aircraft with no User tag say with a code 'A1' will invoke the planesymbol1.txt design. If a user has decided to use say 'A1' for an SR61 then some puzzling plots will be displayed if a Cessna 152 should then signal 'A1'!

Figure 98 - Red Arrows at Farnborough 2010
Photo Courtesy and Copyright of Kevin Daws

Chapter 9
Sharing

PlanePlotter users can share data with other users by using the Internet to transfer data. Individual users can upload their locally received data to a sharing server from which other users can download it to provide a view of aircraft in other parts of the world.

The COAA Server

The most widely used server is the "COAA Server" and this is currently available without charge to all PlanePlotter registered users. To set up sharing using the COAA Server first select from the Menu bar *"Options, Sharing, Setup"* to open the dialog box shown in Figure 99. By checking or not checking the various boxes the user can determine what message data is to be uploaded and what downloaded.

As well as uploading the user own data to the COAA server for the PlanePlotter community the user may also upload their data to another sharing server by checking the *Enable secondary sharing* and then entering an IP address and script name for that server.

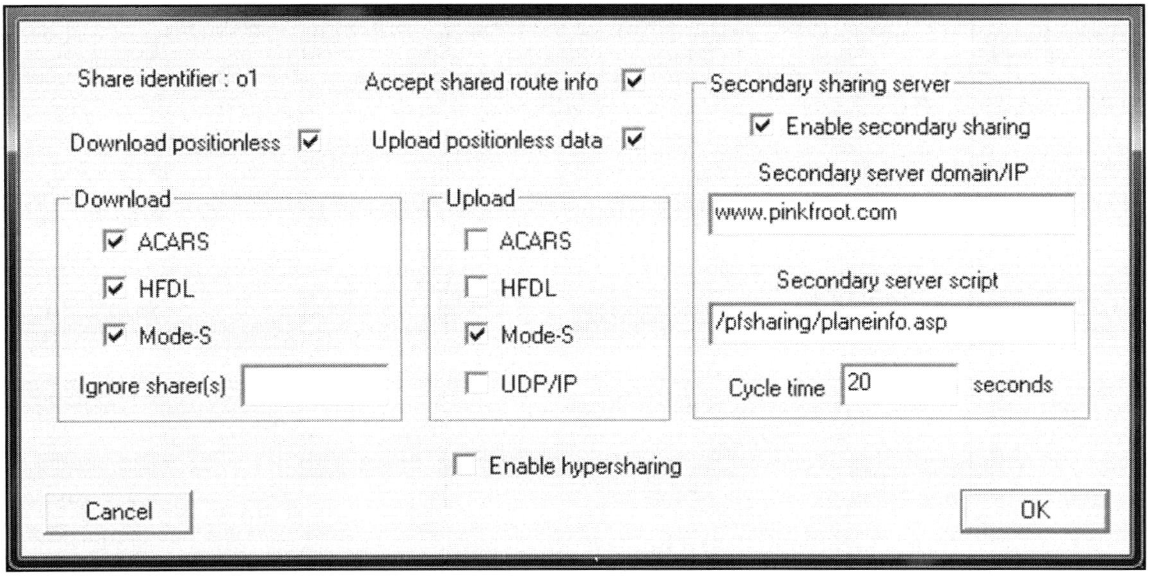

Figure 99 - Sharing

Start Sharing

To start sharing the user should ensure that the data is being processed by having first clicked on the Run button which should then be greyed out and then selecting *Options, Sharing, Enable* or clicking the *Share button*. This button will display one of three states. Each click of the button will change the state cycling through each one. The first state indicated by two arrows with a line through the button indicates that sharing is disabled. The second state is indicated by a single upwards facing arrow that indicates that the users local data will be uploaded to the sharing server. The third and final state is indicated by two arrows to show that the users local data will be uploaded to the sharing server and data from other users will be downloaded to the users computer.

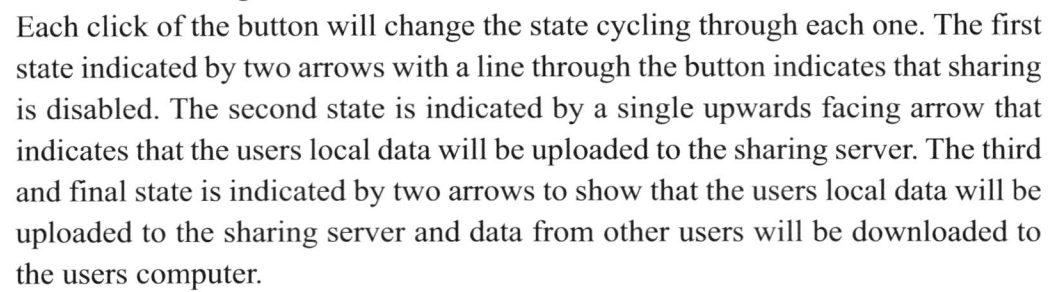

Once sharing is enabled the user should see in the lower left corner of the PlanePlotter screen messages *Sharing upload 1* followed by *Sharing download 1* followed by *Sharing download 1 processed*. If Secondary Sharing is enabled then the messages *Sharing upload 2* and *No data downloaded from server 2* may be seen.

Share Code

A share code is automatically allocated to a registered user the first time they enable sharing and it cannot be changed, the characters of a share code are randomly generated. The allocated share code can be found in the *About* box to the right of the registration number as seen in Figure 100.

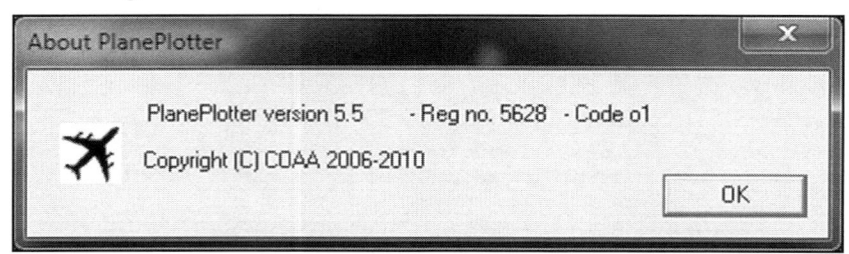

Figure 100 - About dialog

GPX Overlay of Sharers

PlanePlotter maintains a list of share codes that are available to registered users. The first is an automated list providing share code, latitude and longitude, and can be found at *http://www.coaa.co.uk/sharerlocations.php*. By clicking a link on that page a *sharerlocations.gpx* file will be created in the browser window and this should be saved as a GPX overlay in the PlanePlotter Chart files folder. When the overlay is enabled sharers will be displayed on the current chart as inverted triangular symbols with the two character share code above.

PlanePlotters Sharers Database

This can be found Yahoo PlanePlotters group at *http://groups.yahoo.com/group/planeplotter/database?method=reportRows&tbl=1*. Users can enter their own details and are able to enter some degree of personal information.

Sharer Location Map

There are two showing the locations of PlanePlotter sharers around the world. The first can be found at *http://www.coaa.co.uk/pp-user-charts.htm* and the second one at *http://scooterhound.com/WWWR/ships/sharelist.html*.

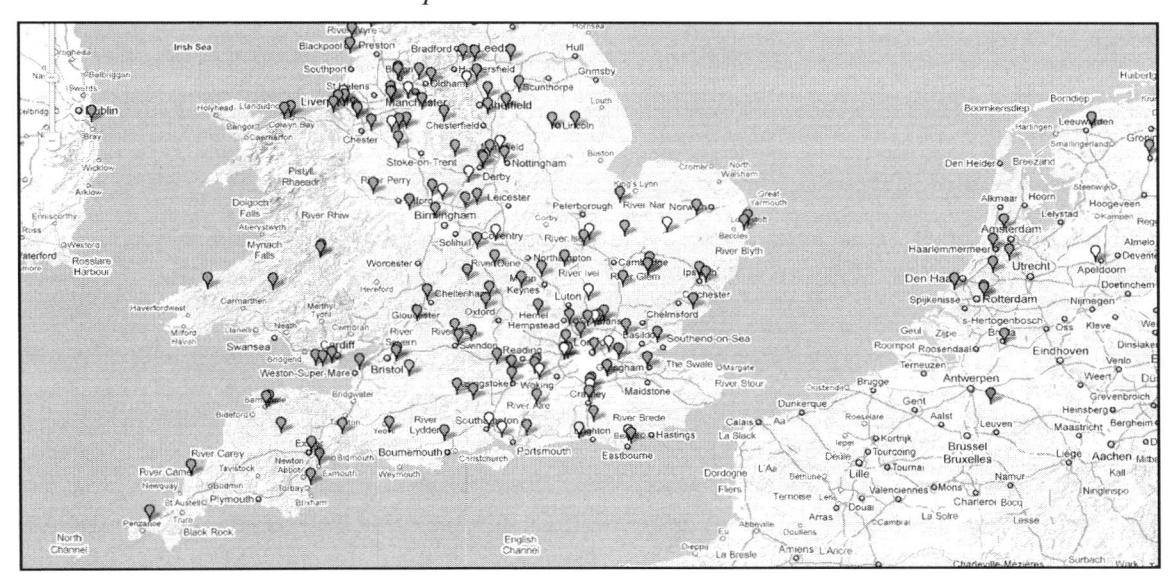

Figure 101 - Part of Curt Deegan's Map Showing Sharer Locations

Share ID on Labels

When a user clicks on a displayed aircraft the label will contain a share code for the station the message was received by. If the message was received by the users local station then the first character of the share code is replaced by an asterisk. It the message was received by another sharer and has been downloaded then the full two character code will be displayed as seen in Figures 102 & 103.

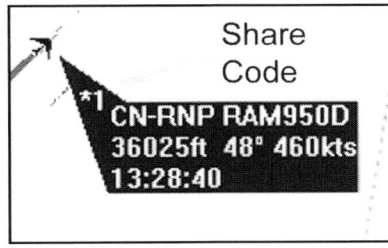

Figure 102 - Local Users Share Code

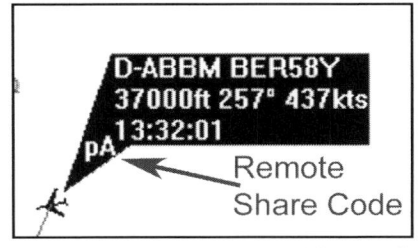

Figure 103 - Remote Users Share Code

Figure 104 - British Airways Boeing 747-436 at London Heathrow
Photo Courtesy and Copyright of Lee Shand

Chapter 10
Logs & Databases

There are a number of different logs and types of log available to PlanePlotter users. In the I/O options dialog the user can opt to have logs created of their Mode-S messages or of a designated aircraft by checking the appropriate boxes. One can also log ACARS messages in the Airmaster format. A fourth option enables support for the Memory-Map Navigator map plotting program which is a proprietary plotting program.

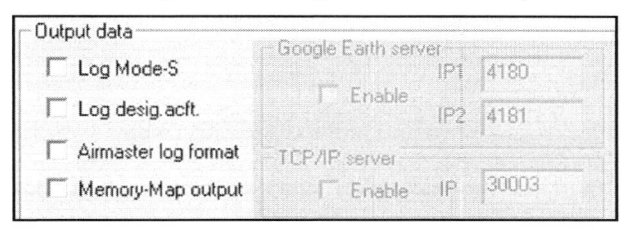

Figure 105 - I/O Output Data Logging Options

Log files will be saved in the nominated log files folder, typically *C:\PlanePlotter\Log files* with a filename of the format *planegadgetradar101002.bin*, where the first part of the name is the type of Mode-S receiver and the numerals are the date stamp. Note that the Mode-S and designated aircraft logs are only intended to be used to play back the logged event, they do not contain any tabular information and cannot be read by a text editor such as Notepad. To replay all the Mode-S aircraft or only the track of the designated aircraft alone user should select the Menu option *Review, Replay Mode-S* and the type of receiver in use.

Basic Log Files

If a user wishes to create a tabular text file type of log then there are a couple of visual basic scripts available from COAA to enable them to do so.

The first script will create a file called *pp_share_log.txt* containing a log of aircraft received every minute for 10 minutes. Part of such a log would look like this with the broken line indicating the end of each one minute block:

```
4B17A6,HB-IYW,SWR2104  ,39.9982,0.223938
4CA7A8,EI-EFR,RYR6386  ,40.40059,0.7904539
4CA647,EI-DYE,RYR7783  ,40.12007,0.394197
---------
4B17A6,HB-IYW,SWR2104  ,39.9319,0.1326474
4CA7A8,EI-EFR,RYR6386  ,40.34142,0.7064334
```

```
4CA647,EI-DYE,RYR7783  ,40.18976,0.4920127
400CD5,G-EZIV,         ,39.86215,9.433912E-02
---------
4B17A6,HB-IYW,SWR2104  ,39.90355,9.301758E-02
4CA7A8,EI-EFR,RYR6386  ,40.24436,0.5691043
4CA647,EI-DYE,RYR7783  ,40.27913,0.618042
400CD5,G-EZIV,EZY7MN   ,39.97252,0.1938477
---------
4B17A6,HB-IYW,SWR2104  ,39.90355,9.301758E-02
4CA7A8,EI-EFR,RYR6386  ,40.16518,0.4574932
4CA647,EI-DYE,RYR7783  ,40.39114,0.7769706
400FDC,G-EZBK,         ,39.59692,2.332703
---------
```

Logfile Script

```
1  ' *************************************************************
2  ' PlanePlotter script for logging active aircraft (including those
     received by sharing)
3  ' Create a file in the Log Files directory
4  Dim fso, f1, ts
5  Const ForWriting = 2
6  Set fso = CreateObject("Scripting.FileSystemObject")
7  fso.CreateTextFile ("C:\PlanePlotter\Log Files\pp_share_log.txt")
8  Set f1 = fso.GetFile("C:\PlanePlotter\Log Files\pp_share_log.txt")
9  Set ts = f1.OpenAsTextStream(ForWriting, True)
10 ' Tap in to PlanePlotter
11 Dim MyObject
12 Set MyObject = GetObject(,"PlanePlotter.Document")
13 ' Write all the aircraft data out to file every minute for 10 minutes
14 n = 10 ' this will loop just ten times - once per minute
15 while n > 0
16 i = 0
17  while i < MyObject.GetPlaneCount()
18    planeinfo = MyObject.GetPlaneData(i,0)
19    planeinfo = planeinfo + ","
20    planeinfo = planeinfo + MyObject.GetPlaneData(i,1)
21    planeinfo = planeinfo + ","
22    planeinfo = planeinfo + MyObject.GetPlaneData(i,2)
23    planeinfo = planeinfo + ","
24    plane3 = MyObject.GetPlaneData(i,3)
25    planeinfo = planeinfo + Cstr(plane3)
26    planeinfo = planeinfo + ","
27    plane4 = MyObject.GetPlaneData(i,4)
28    planeinfo = planeinfo + Cstr(plane4)
29    ts.WriteLine (planeinfo)
30    i = i + 1
31  wend
32  ts.WriteLine ("---------")
33  Wscript.Sleep 60000 ' 60 Seconds
34  n = n - 1
35 wend
36 ts.Close
```

Script - PPLOG.VBS courtesy of COAA

A second vbs script called *replayablepplog2.vbs* that will create a SBS1 style log file from shared and local data can be found in the Files area of Yahoo PlanePlotter groups. This script will create log files with names like *20101002_113244_pp_share_data.log* or *20101002_113244_pp_share_data.log*. An extract of the data from such a log is shown below.

```
"2010/10/02","09:03:45.000","1111111","020054","RAM650
","Unknown","0","34000","34000","40.65363","1.152344",
"0","0","506.2964","45.32008","0","0000"

"2010/10/02","09:07:36.000","1111111","405D0D","EZY591T
","Unknown","0","37000","37000","39.84082","5.433456E-03",
"0","0","416.125","227.3373","0","0000"

"2010/10/02","09:08:33.000","1111111","4CA73F","RYR6011
","Unknown","0","37000","37000","39.9391","0.1414795",
"0","0","418.116","226.8414","0","0000"

"2010/10/02","09:08:33.000","1111111","405F0C","EZY2734",
"Unknown","0","38025","38025","40.50957","0.9455733",
"0","0","474.2362","47.56377","0","0000"

"2010/10/02","09:08:36.000","1111111","34348E","ECNCM",
"Unknown","0","30000","30000","40.68777","0.3149206",
"0","0","383.2401","16.21236","0","0000"
```

Other Log Files

A number of other log files may be created by the PlanePlotter program itself and by some of PlanePlotters popular add-ons and will normally be saved in the Log files folder. Such files include AVI recordings, Mlat logs and alert messages.

Databases

When PlanePlotter is first installed it has no knowledge of aircraft registrations or of their flight details. When a Mode-S message is received from an aircraft it PlanePlotter may only find the aircrafts ICAO number. When the user clicks on an aircraft symbol in the display a label will be displayed similar to that shown in Figure 106 indicating that PlanePlotter is not aware of any registration for that particular aircraft. As well as not knowing the registration number PlanePlotter will also be unable to determine any flight details such as departure and destination airports.

Figure 106 - Aircraft Label

By right-clicking on the aircraft symbol an *Aircraft position report* dialog box will be displayed as shown in Figure 107. By then clicking the *Look up* button a search will be made on a web database to try and find the aircrafts registration and flight details. The mechanism for initiating this search is a small script called *lookup.vbs* that was placed in the main PlanePlotter program folder on installation.

PlanePlotter uses the Aircraft Registration database that can be found at *http://www.airframes.org/* which is an online database of international aircraft registry data and airline/operator data, including the unique 24bit ICAO24 aircraft addresses used in Mode-S and ADS-B. The database contains all kinds of airplanes including passenger and cargo airliners, business jets, helicopters, civil, military, and private aircraft.

Figure 107 - Aircraft Position Report

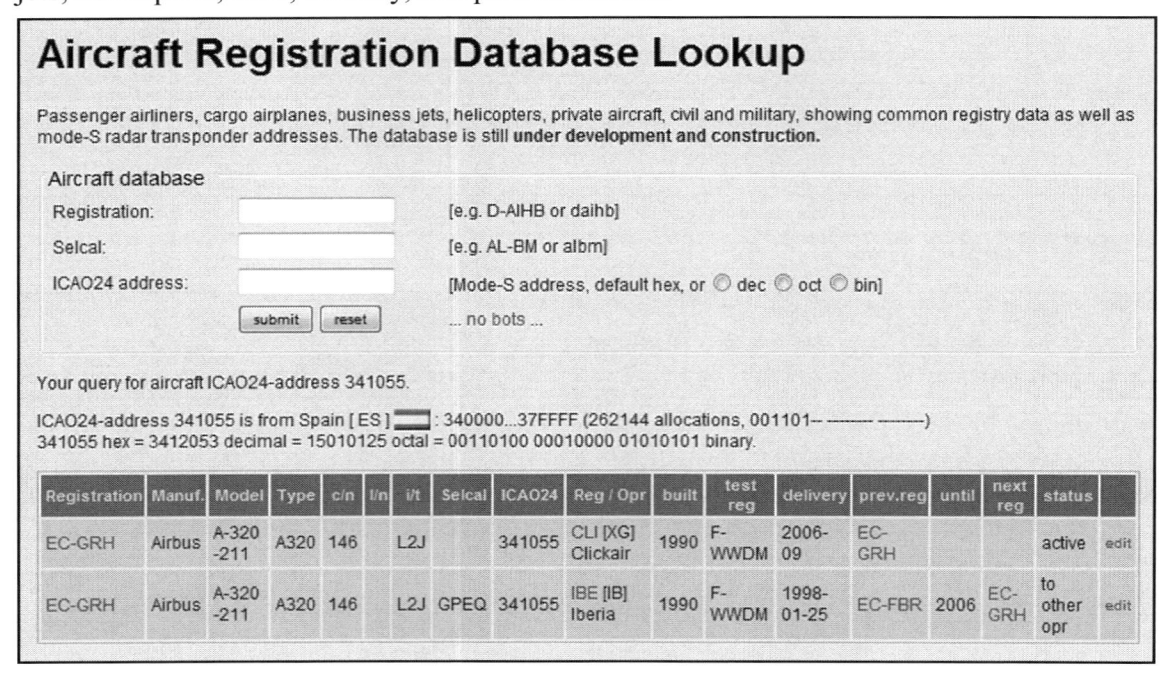

Figure 108 - Aircraft Registration Database

114

Logs & Databases

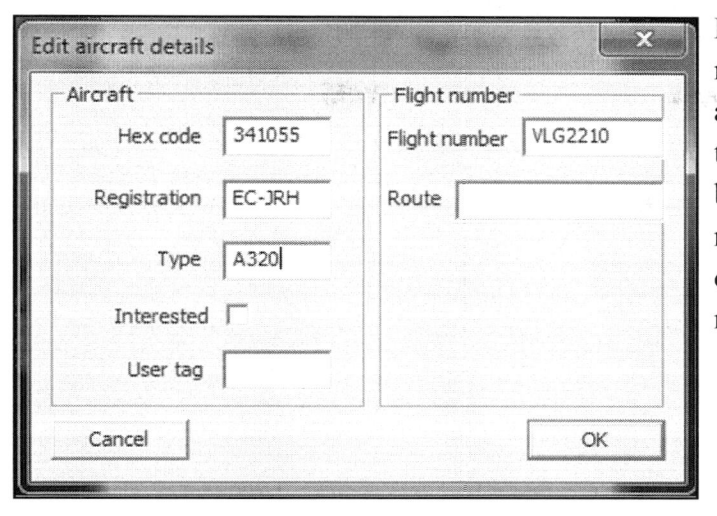

By clicking the *Edit* button the user may copy the aircraft registration and type from the database into the *Edit aircraft details* dialog box as shown in Figure 109. The next time that the aircraft label is displayed it will show the aircraft registration.

Figure 109 - Edit Aircraft Details   Figure 110 - Label with a/c Reg

The data entered into the aircraft details box is only retained by PlanePlotter whilst that aircraft is actively tracked. If the program is closed and that same aircraft is found at a later time the registration will not be known to PlanePlotter as no record has been kept of it.

This shortcoming can be overcome by installing some local databases on the users PC that can easily be accessed by PlanePlotter and the program already has the necessary hooks in place to use them.

PlanePlotter uses two databases that are called *basestation.sqb* and *flightroute.sqb* . These databases are in a common format known as SQLite which is a simple database that understands most of the standard SQL language.

Basestation.sqb

The first database *basestation.sqb* is not unique to PlanePlotter as it has its origins in the Kinetic SBS-1 BaseStation software. The database contains a large number of tables and fields most of which are not required by PlanePlotter. In fact PlanePlotter only uses the *ModeS, Registration* and *ICAOTypeCode* fields.

Users with a SBS-1 receiver this file will reside in the Kinetics application directory typically C:\Program files\Kinetic\Base station\ and this must be pointed to in the Menu *Options, Directories, SQB database (e.g. basestation.sqb)*.

For users that do not have the Kinetics box a copy of the *basestation.sqb* database that is specifically prepared for use with PlanePlotter can be downloaded from the *ManTMA* web site at *http://www.pp-sqb.mantma.co.uk/pp_support.html*. The database should be saved in

Figure 111 - Database Directory

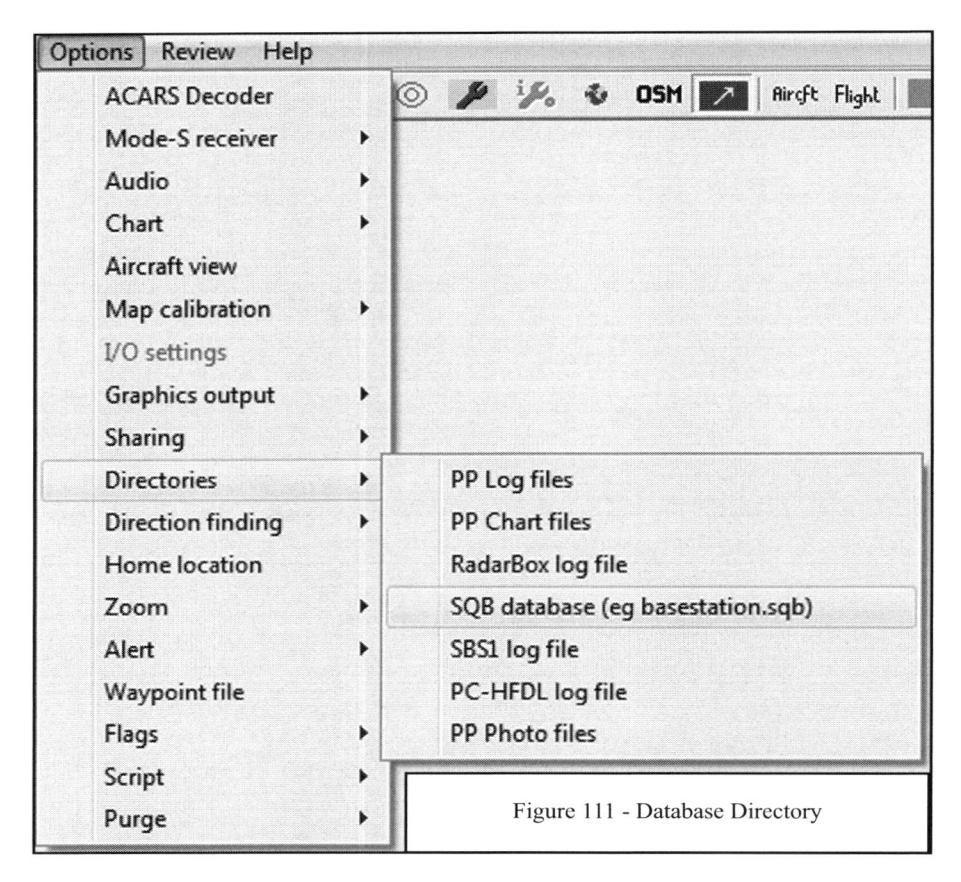

the *C:\PlanePlotter\Log files* folder with the name *basestation.sqb* and the Menu *Options, Directories, SQB database* pointed to *C:\PlanePlotter\Log files\basestation.sqb*. It is important to, point to the folder & filename and not just the folder.

Every time that PlanePlotter receives a message either from a local reception or from a sharing download the program examines the database to look for the registration and the aircraft type neither of which of course can be found in the Mode-S or ADS-B messages.

When simultaneously receiving ACARS messages and Mode-S or ADS-B messages if the same flight number is found in both an ACARS message and a Mode-S then as the ACARS messages contains the registration but Mode-S messages does not PlanePlotter will associate the registration with the Mode-S hex code and write it to the database.

Should the user invoke a *Look Up* on an aircraft as described earlier then the program will extract the registration and type from the response and write that data to the *basestation.sqb* database.

Note that the aircraft registration number data will only appear on PlanePlotter when a new flight is acquired. PlanePlotter will not update the on screen data of an existing flight because once a flight is being displayed PlanePlotters processing has moved past the stage where it looks for data in the database.

Flightroute.sqb

The second database *flightroute.sqb* is unique to PlanePlotter and stores only the relationship between flight numbers and routes together with a date/time stamp showing the time at which the route data was updated. as shown below.

```
Flight     Route          Updatetime
-------    ----------     ----------
NWA815     EDDF-KATL      1263982991
RYR1A      EIDW-EGNM      1257415992
CFE29G     EGLC-EGPF      1257415992
KLM1074    EGCC-EHAM      1257415992
EIN63T     EBBR-EIDW      1257415992
EXS907     EGCC-LLBG      1257415992
BAW48      KSEA-EGLL      1282385657
EIN20C     EIDW-EGCC      1257415992
KLM669     EHAM-KDFW      1257415992
AFR1669    EGCC-LFPG      1257415993
```

The *flightroute.sqb* database is stored in the PlanePlotter Log files folder and is automatically created the first time the program is used. It is possible to download a populated database from the Yahoo PlanePlotter Group.

When PlanePlotter receives a message from a newly detected aircraft it will get the aircraft registration number from the *basestation.sqb* database and check the *flightroute.sqb* database to see if the flight number and route are already known. If they are then they will be displayed. As PlanePlotter receives no route information in the Mode-S messages it cannot in itself add any information to the database. This is achieved using third party add-ons such as Curt Deegan's Findflight that uses the *SetRouteByHex* method.

SQLite Editors

There are a number of SQLite editors readily available on the Internet but one of the easiest to use is SQLite Expert that is available from *http://www.sqliteexpert.com/*. It should meet the needs of most users wanting to maintain or edit their own databases.

Figure 112 - Air France Boeing 777-328ER On its way to the U.S.A. from Paris F-GSQH passes overhead on a lovely winters day.
Photo Courtesy and Copyright of Lee Shand

Chapter 11
Multilateration

Basic Theory

First let us examine the basics of Multilateration or Mlat as it is commonly referred to. Firstly Multilateration is **not new**. Multilateration has been around since the 1940s when the Decca Navigator system was established in the United Kingdom after World War 2 and later used in many areas around the world. It was is a hyperbolic low frequency radio navigation system which operated by measuring the phase differences between continuous signals from a master and slave stations. The system used groups of at least three shore based transmitter stations called chains operating in the 70-130 kHz radio band. Each chain comprised of one Master and two or three Slave stations, usually located 80 to 110 km from the Master station. Secondly Multilateration is **not ADS-B**. It provides no cockpit information. It is a newer-technology alternative to today's secondary radar, offering the same effective range but providing better accuracy at significantly lower cost, and from much simpler equipment than large rotating radar antennas. It is a valuable complement to ADS-B in adding coverage of non-ADS-B traffic, for a substantially lower investment. Thirdly Multilateration should not be confused with trilateration or triangulation which use distances or absolute measurements of time of flight from three or more synchronized transmitter.

In aviation Multilateration is the process of locating aircraft based on the so called Time Difference Of Arrival (TDOA) of an aircrafts transponder signal to three or more strategically placed receiver stations. Mlat may be thought of as an inverted GPS system - in GPS there a number of moving satellite transmitting their individual time stamp signals to a stationery receiver that then calculates its position from those signals. Mlat triangulates a moving aircrafts position from a number of fixed ground stations at known locations receiving and time stamping an aircraft transponder squitter.

Let us take a moment to examine what a hyperbola curve is. It is an open curve formed by a plane that cuts the base of a right circular cone. In mathematics, a hyperboloid is a quadric, a type of surface in three dimensions as described by the equation over page

$$-\frac{x^2}{a^2} - \frac{y^2}{b^2} + \frac{z^2}{c^2} = 1$$

An easier way of describing a hyperboloid is visually as shown in Figure 113 below. Here an aircrafts transmitted squitter can be thought of as a cone of radiation. The cone is sliced where the it strikes the ground and the shape of the slice is a parabola, a specific type of a hyperboloid.

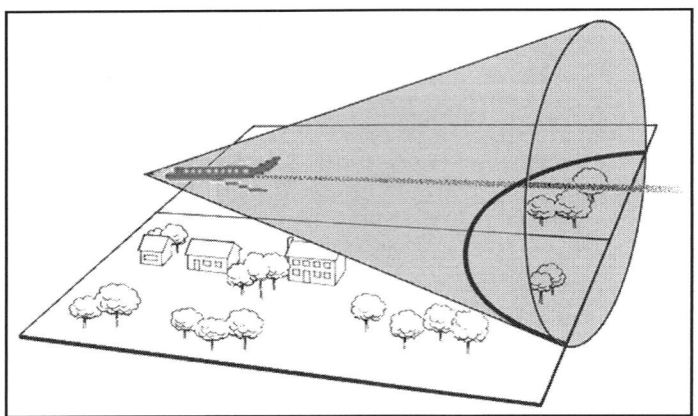

Figure 113 - Hyperbola Curve

Where there are two receiving ground stations the squitter emitted from the aircraft will arrive at slightly different times, the TDOA being due to the different distances of each receiver from the platform. And for given locations of the two receivers a whole set of aircraft locations would give the same measurement of TDOA. Thus with two known receiver locations and a known TDOA the locus of possible aircraft locations can be derived a three-dimensional surface characterised as one half of a hyperboloid for which

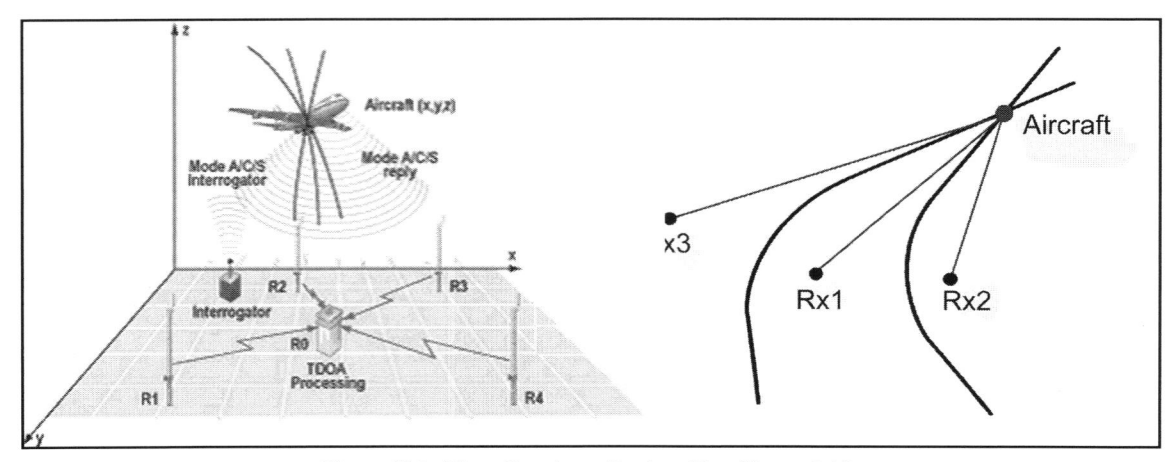

Figure 114 - Three Receivers Produce Two Hyperoloids

Multilateration

any two points on said surface will have the same differential distance from said receivers, i.e. a signal transmitted from any point on the surface will have the same TDOA measured by the receivers as a signal transmitted from any other point on the surface.

Therefore in practice the TDOA corresponding to a moving transmitter when measured will produce a corresponding hyperbolic surface with the transmitter is located somewhere on the that surface. The results is a new hyperbola curve and probably one that is not as symmetrical as the original. If there are three receiving stations then two curves will be calculated and the intersection of the two will be the location of the transmitter.

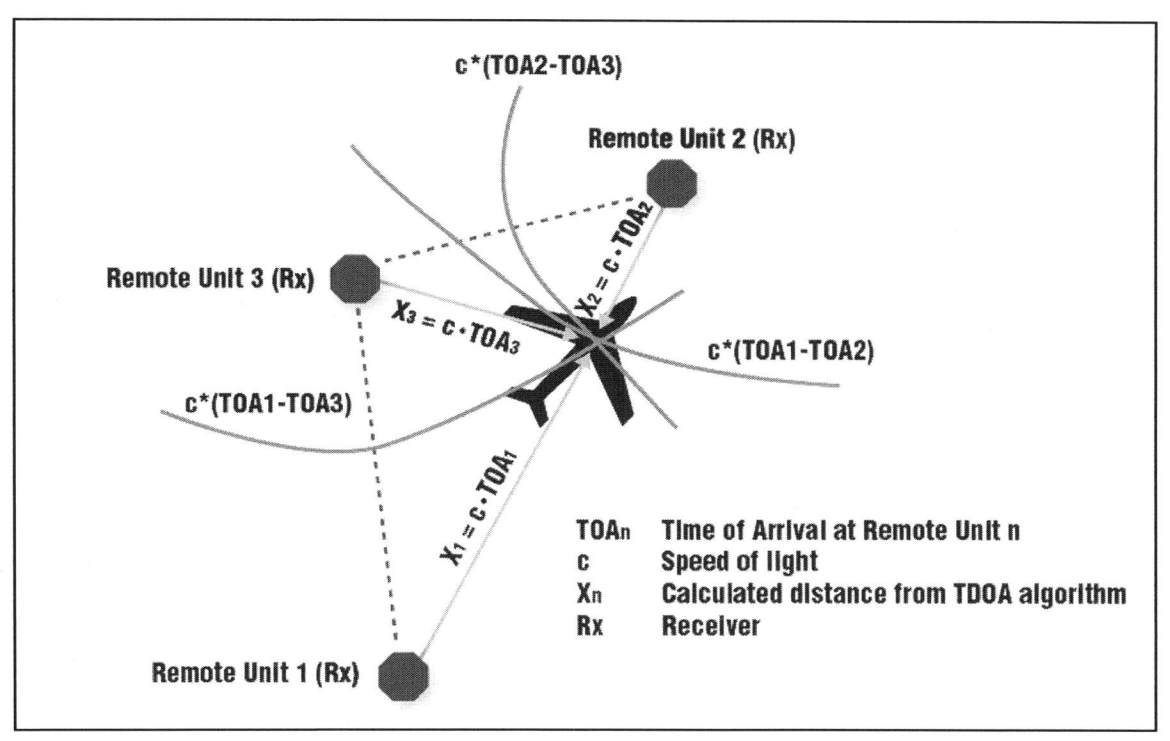

Figure 115 - The Principle of the TDOA Technique

If a fourth receiver is now introduced, a third TDOA measurement is available and the intersection of the resulting third hyperboloid with the curve already found with the other three receivers defines a unique point in space. This then fully locates the aircraft in 3D. Errors in the measurement of the time of arrival of pulses mean that enhanced accuracy can be obtained with more than four receivers and in general N receivers will provide N - 1 hyperboloids. When there are N > 4 receivers the N - 1 hyperboloids should intersect on a single point but in reality the surfaces rarely intersect because of various errors. These can be reduced by optimising the results using the least squares method or an extended Kalman filter.

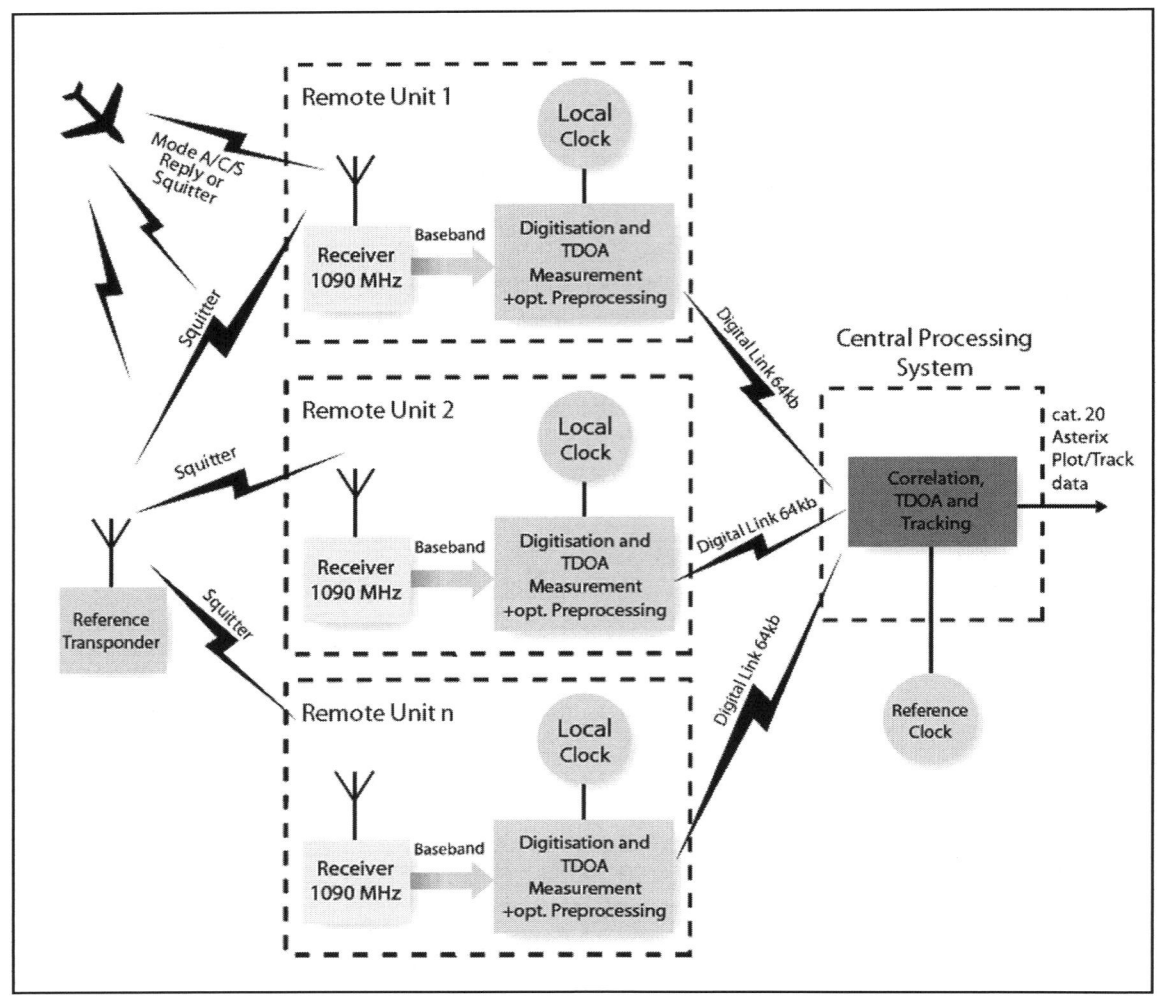

Figure 116 - Schematic of a commercial Wide Area Multilateration used at Innsbruck

PlanePlotters Multilateration

PlanePlotter must be able to receive *raw data* from the users receiver to enable Multilateration processing. At the time of writing only the Kinetics SBS-1er and PlaneGadget Radar PGR receivers are able to do so. Users who are able to report raw data are termed *Ground Stations* and must configure their PCs and PlanePlotter installation in a specific manner. Users who are able to make Mlat position requests are termed *Master Users* and have to approved and enabled to obtain that status.

As explained earlier the more Ground Stations there are the greater the number of hyperboloid curves. If the Ground Stations are well distributed and plentiful, the intersection will be unique. With fewer Ground Stations then fewer curves will be calculated and there may be multiple intersections with the location of the position less aircraft being ambiguous.

As Mlat works on the Time Difference Of Arrival of signals this places some critical timing constraints on the system thus it is essential that the users PC's clock is accurately fixed to an atomic standard (see Appendix B). It is also essential that each Ground Station reports its Home location accurately (see Chapter 6).

PlanePlotters Multilateration Process

When a user wants to ascertain the location of a position less aircraft is done by PlanePlotter he initiates a request that is immediately passed to the sharing server. All Ground Stations are in regular contact via the standard 60 second sharing refresh cycle so the he request for raw data form other Ground Stations is forwarded by the sharing server to every Ground Station. Each Ground Station then responds by sending the requested data directly to the requesting Master User and not to the sharing server but by

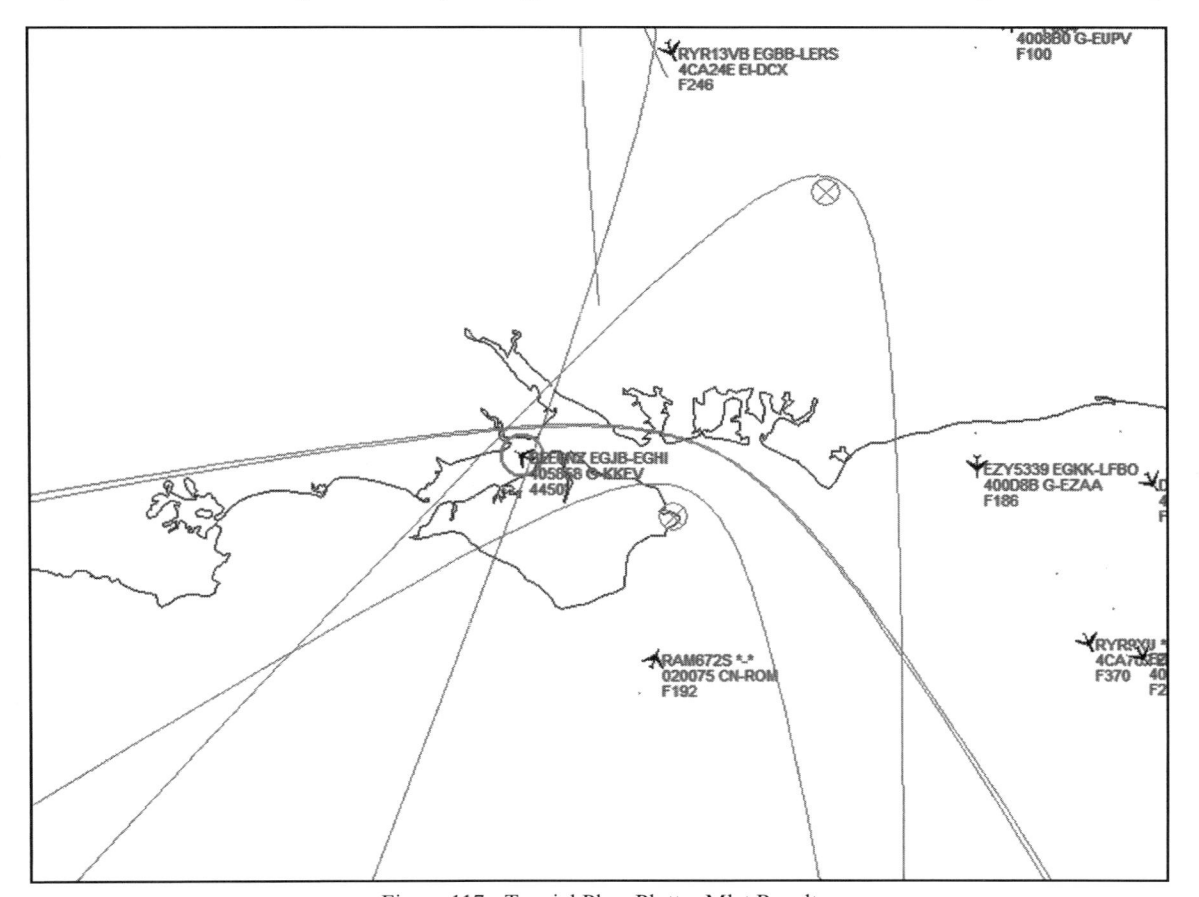

Figure 117 - Typcial PlanePlotter Mlat Result

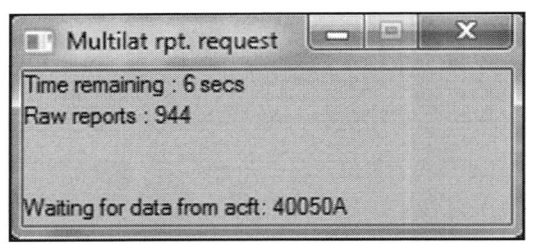

Figure 118 - PlanePlotter Mlat Information Boxes

the UDP/IP protocol. The sharing cycle for different users occurs at difference instances thus the requesting Master User will receive data from the participating ground stations at intervals over the 60 second period following the request being made. During this time, an information box appears on the Master User's screen showing the number of raw data messages that have been received from any Ground Stations. At the end of this period the requesting Ground Station then analyses the data received to calculate any curves that can be plotted from the composite raw data. Theses curves are then plotted on the Master User's chart. The information box displays the results of the analysis for a few seconds after the process is completed. PlanePlotter then calculates an estimated position of the aircraft from the intersection of the curves and plots it on the PlanePlotter chart together with the curves.

To fully understand PlanePlotters Mlat it is worth emphasising that there are three distinct and separate functions within PlanePlotter that interact with the Internet. These three functions are Sharing, Ground Station (GS) and Master User (MU). PlanePlotters sharing system uses standard HTTP posts communicating with sharing servers to relay uploads and downloads of shared data messages. The GS function sends UDP packets of raw data directly to a specific user but where that user has requested the data through a message to the sharing server. The raw data is collected from the local receiver and processed by PlanePlotter and transmitted directly in a brief burst each time the raw data request is requested. The MU function accepts UDP packets of raw data directly from other GSs. The data is transmitted by those GSs in response to a request embedded in the MU's sharing upload. The Sharing server immediately forwards the request to the other GSs using a UDP message.

Thus each GS must have PlanePlotter enabled to send UDP/IP datagrams to any IP address on the Internet. Also each MU must be able to receive UDP/IP datagrams from sources on the Internet. This facility must be enabled in the user's firewall and also their router. The first time a request is made the firewall may issue a warning to which the user should respond to always allow the connection.

PlanePlotters Multilateration Preparation

When a router is used to connect a PC to the Internet it must be configured to send incoming UDP datagrams addressed to port 9742 to the local IP address of the user's PC. The first step is to determine the users own LAN TCP/IP address. PlanePlotter provides a simple means of doing by selecting Menu *Test, Networking, Check own LAN address*.

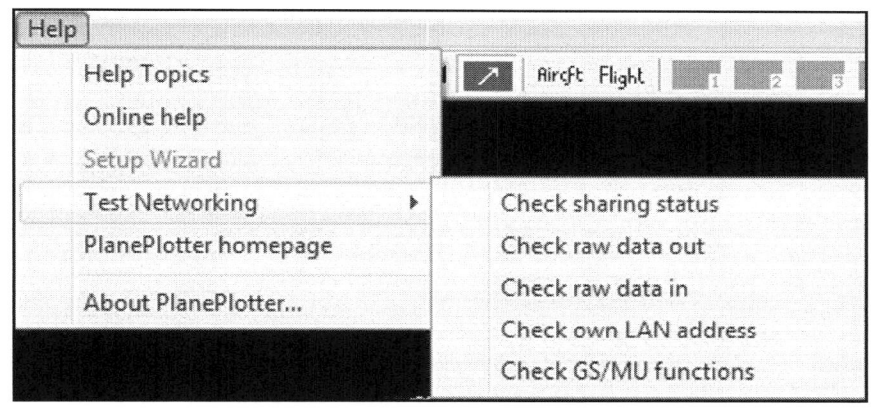

Figure 119 - PlanePlotter Test Networking Options

This will open command screen view showing the relevant information which in the illustration below is 192.168.0.32. A note should be made of this and retained.

Figure 120 - PCs LAN Address.

As there are many different makes of Router it is not possible here to provide details for all them. The router must be told to pass any packets for UDP port 9742 to port 9742 at the local IP address of the PC running PlanePlotter. The user should login to their router as Administrator and look the appropriate advanced setting area that deals with virtual servers. A Virtual Server allows the user to direct incoming traffic from WAN side of the Internet connection to the Internal server

An example of a typical set-up is shown below. Note that external port number of 9742 is linked to the local LAN server IP address 192.168.0.32.

NAT -- Virtual Servers Setup

Virtual Server allows you to direct incoming traffic from WAN side (identified by Protocol and External port) to the Internal server with private IP address on the LAN side. The Internal port is required only if the external port needs to be converted to a different port number used by the server on the LAN side. A maximum 32 entries can be configured.

Server Name	External Port Start	External Port End	Protocol	Internal Port Start	Internal Port End	Server IP Address	Remove
PP	9742	9742	UDP	9742	9742	192.168.0.32	☐

Figure 121 - Typical Router Virtual Server Setup.

The router may also need to have its DoS detection adjusted so that it does not interpret the sudden flood of raw data as a Denial of Service (DoS) attack. The router will usually be transparent to outgoing packets but may need to have a parameter *Maximum TCP/UDP sessions* increased to allow PlanePlotter to send possibly 200 or more UDP packets in quick succession.

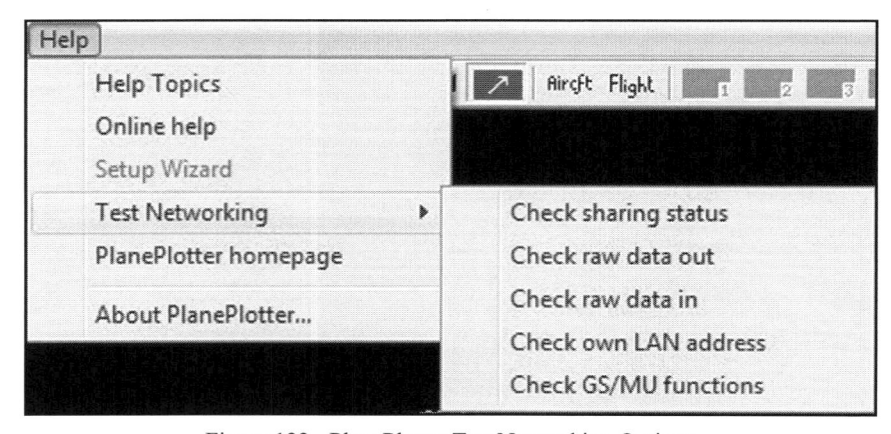

Figure 122 - PlanePlotter Test Networking Options

Once the router has been set up then the connections should be verified and the various *Test Networking* functions tested in the following order not down the menu options list.

Multilateration

Check Sharing Status

The first test that should be conducted is to test the sharing status of the user and to confirm that they are a Master User. The message shown in Figure 123 is received by the users PC and displayed on their web browser.

PlanePlotter Network Test

PlanePlotter installation PP999999

Time last accessed (UTC)	User ID	Clock err (secs)	Last acft up	Last acft down	Tot acft up	Raw data en.	Raw data valid	Master User
2010-10-04 11:16:15	o1	2	0	0	112988	yes	yes	yes

Figure 123 - PlanePlotter Test Networking Sharing Status

Check Raw Data In

The second test that confirms that the user can receive a UDP message directly into their PC. Again a message is displayed in the user's web browser but an additional message id shown in a pop up dialog box in the PlanePlotter display itself.

PlanePlotter Network Test

You are currently a PlanePlotter Master User.
A test UDP message is now being sent to your instance of PlanePlotter with serial number PP999999.
A report should pop up in PlanePlotter immediately.
If it does not, check your PlanePlotter I/O settings and your router port forwarding of data for UDP port 9742 and then try the test again.
Test ended.

Figure 124 - PlanePlotter Test Networking Check Raw Data In Browser Message

PlanePlotter

The following UDP test message has been successfully received from the sharing server [188.64.185.55:51374]

Test UDP datagram inbound to port 9742 on PlanePlotter PP999999.

Your router settings appear to be good for Multilateration.

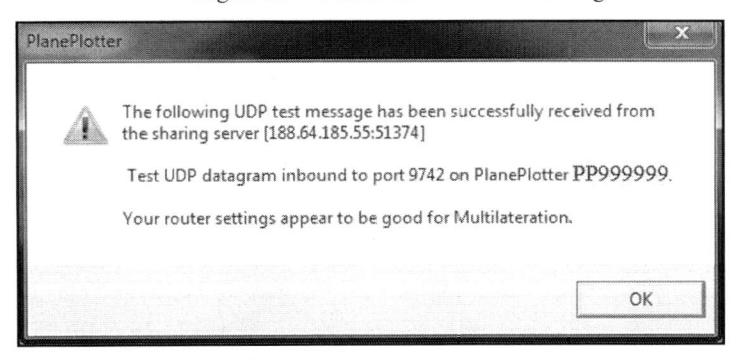

OK

Figure 125 - PlanePlotter Test Networking Check Raw Data In Dialog Message

Check Raw Data Out

The third test that confirms that the user is a master user and that raw data is being received from the Mode-S receiver and is responding to a UDP message from the server for that data to be transmitted. Again a message is displayed in the user's web browser.

PlanePlotter Network Test

You are currently a PlanePlotter Master User.
A test UDP request is now being sent to your instance of PlanePlotter with serial number PP999999.
You should see a response below indicating the number of messages currently in your raw data buffer.

Response from your PlanePlotter
PP694564 raw buffer content = 4029

The response suggests that your router settings are good for Mlat.
Your raw data buffer contains data which suggests that you have successfully performed the DLL substitution for raw data access.
You may wish to exercise your browser refresh button, to make sure that the buffer content is changing.
An unchanging value would indicate that something has interrupted raw data.
In case you wondered, the value is a pointer into a circular buffer. The value increments modulo 32768.
Any non zero value indicates that raw data has been accumulated but the actual value is not significant.
Test ended.

Please remember to close this page after use.

Figure 126 - PlanePlotter Test Networking Check Raw Data Out

Check GS/MU Functions

The final and most important test and at its satisfactory conclusion will confirm that the installation can function as a PlanePlotter Mlat GS/MU. Prior to starting the test PlanePlotter should be receiving and processing messages from its Mode-S receiver and the previous tests have completed satisfactorily. The test may take a minute or two to complete and it may be necessary to repeat a number of times.

The test first reports that it has found an actively sharing PlanePlotter installation on the specific LAN address. IP 88.9.161.241. It then associates the serial number of the installation with a sharer code and reports the sharer code, that the user is both a Master User and a Ground Station. It then confirms that the user has enabled the raw data setting in I/O settings before starting a number of tests to poll for data. During these polls it check that there is data in the raw buffer content before then testing that data can be transferred. Finally it requests a specific number of raw data reports to confirm how may are sent.

Multilateration

If the correct amount of raw data is not received a request will be made to run the test again. If the correct amount of raw data is still not received then could indicate a problem with the router so the user will be advised to check their router DoS/SPI settings. Sometimes the problem may lie not with the router setting but with the Internet Service Provider throttling back the amount of UDP traffic permitted.

PlanePlotter Ground Station/Master User Test

Make sure that PlanePlotter is running, processing and sharing before you perform this test!

There is 1 actively sharing PlanePlotter installation on IP 88.9.88.9.

Sharer o1

    is currently a Master User.
    is currently a Ground Station
    has "raw raw" enabled in Options I/O settings according to the information submitted in sharing.
Starting 1st simple poll test.....
    responded to 1st simple poll test
Starting additional simple poll test.....
    also has "raw data" enabled in Options I/O settings according to the information obtained by direct polling.
Starting 2nd simple poll test.....
    responded to 2nd simple poll test (PP999999 raw buffer content = 4302)
    raw data is available in PlanePlotter.
Starting plane data transfer test....
    responded to a request for data with 204 bytes.
Starting raw data transfer test....
    responded to a request for 100 raw data reports with 100 reports.

    Please remember to close this page after use.

Figure 127 - PlanePlotter Test Networking Check GS/MU Functions

Initiating a PlanePlotter Mlat Computation

When a user wishes to start a PlanePlotter Mlat computation they must first open the *Aircraft View* window. It may help to first select *Aircraft with Mlat possible* from the *Aircraft list display options* in the View Menu.

PlanePlotter User Guide

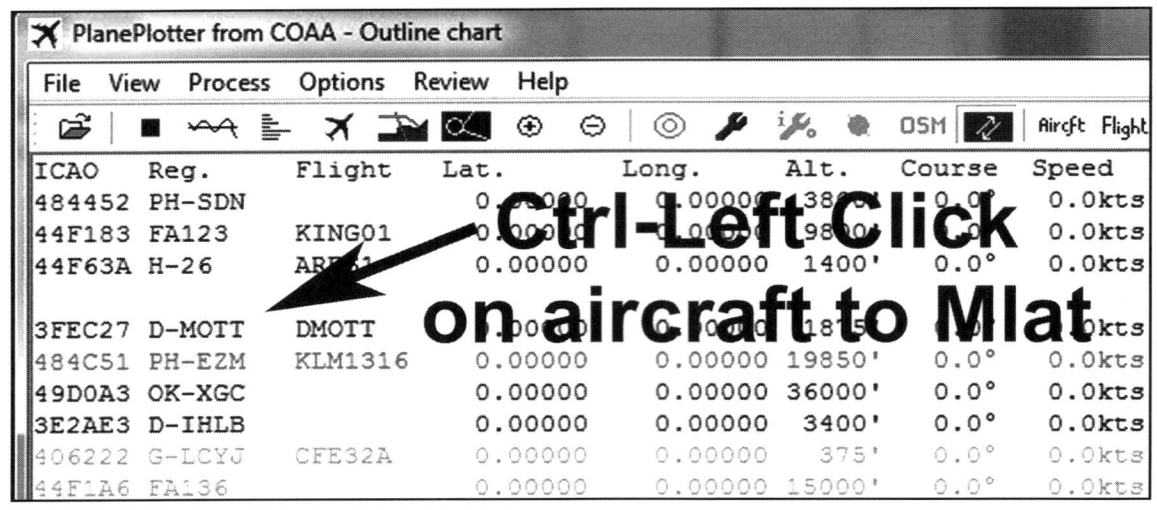

```
✈ PlanePlotter from COAA - Outline chart
  File  View  Process  Options  Review  Help
  ICAO    Reg.    Flight   Lat.        Long.       Alt.    Course   Speed
  484452  PH-SDN           0.Ctrl-Left Click       0.0°    0.0kts
  44F183  FA123   KING01   0.00000  0.00000  9800'  0.0°    0.0kts
  44F63A  H-26    AR               0.00000  1400'  0.0°    0.0kts
  3FEC27  D-MOTT  DMOTT    on aircraft to Mlat kts
  484C51  PH-EZM  KLM1316  0.00000  0.00000  19850' 0.0°    0.0kts
  49D0A3  OK-XGC           0.00000  0.00000  36000' 0.0°    0.0kts
  3E2AE3  D-IHLB           0.00000  0.00000  3400'  0.0°    0.0kts
  406222  G-LCYJ  CFE32A   0.00000  0.00000  375'   0.0°    0.0kts
  44F1A6  FA136            0.00000  0.00000  15000' 0.0°    0.0kts
```

Figure 128 - PlanePlotter View Window

The user should then select the aircraft for Mlat by a *Ctrl-Left Click* on the line of data for that aircraft. This will then send a request to the sharing server and at the same time open an Mlat request box. The sharing server will then send a message to all downloading Ground Stations via the standard 60 second sharing refresh cycle. Ground Stations then will then respond by sending data on the nominated aircraft directly back to the by the UDP/IP protocol. This process is repeated for a period of 10 seconds at the end of which the Mlat request box will have been updated to show whether the Mlat fix was successful or not. It will also show the number of raw reports received and how many ground station responded. Finally it will show how many hyperboloid curves have been calculated. The chart display will then be illuminated and the computed hyperboloids plotted on the Master User's chart. PlanePlotter then calculates an estimated position of the aircraft from the intersection of the curves and plots it on the PlanePlotter chart together with the curves.

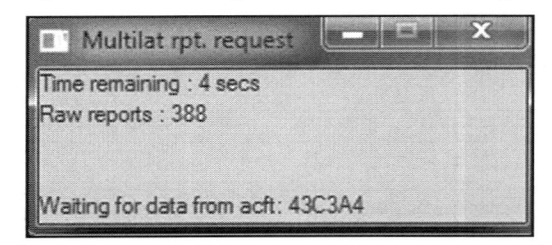

Figure 129 - Mlat Interim Request Box

Figure 130 - Mlat Final Request Box

It is worth reminding users that for PlanePlotters Multilateration to work problem then the clock on their PC must be accurately synchronised to an atomic time standard and that the setting of their home location is precise.

Multilateration

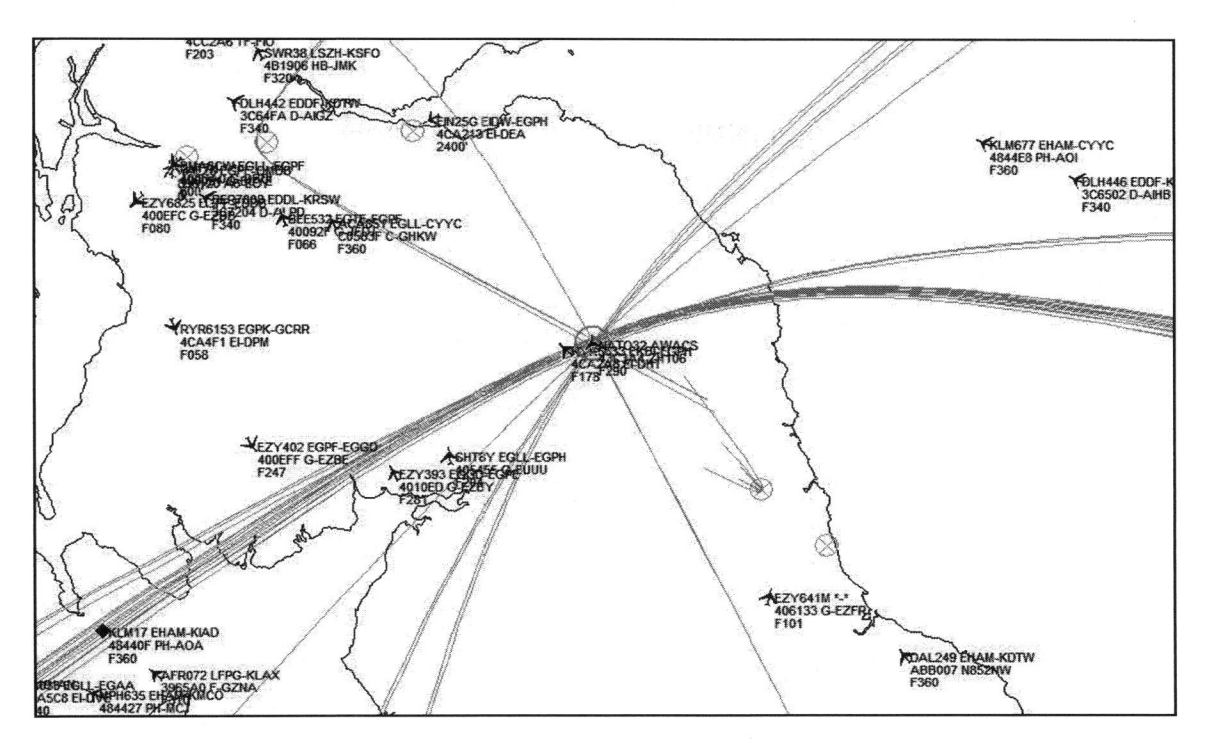

Figures 131 & 132 - Examples of Typical PlanePlotter Multilateration Computations

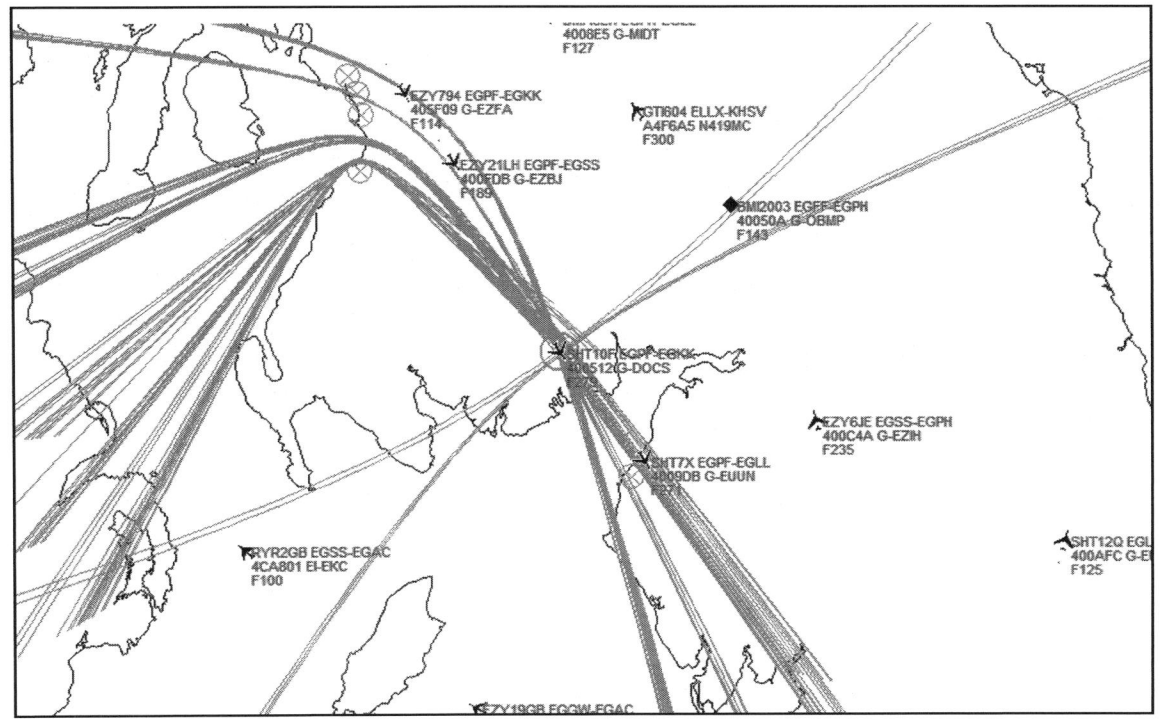

Figure 133 - Sunset at Kraków John Paul II Balice International Airport
Photo Courtesy & Copyright of Grzegorz Rafalski

Chapter 12
Google Earth, Google Maps & Other Add-Ons

As well as displaying the positions, movements and information about aircraft on PlanePlotters own charts and screen displays PlanePlotter can transfer the data to other programs such as Google Earth and Maps to create custom web page displays.

Additionally there are a number of add-on computer programs known as scripts that have been created and placed in the public domain by other PlanePlotter users and enthusiasts that users can readily incorporate into their own systems to enhance the capability of this excellent program.

Google Earth

PlanePlotter can act as a Google Earth server by checking the box in the I/O settings as described on Pages 78 & 79. If the Google Earth program is installed on the users computer then PlanePlotter can provide a birds eye view of the aircraft are currently being

Figure 134 - Typical Google Earth View

133

processed by the program on a Google Earth view. Alternative a view from the cockpit of a designated aircraft can be displayed in Google Earth as it flies on its journey.

Note that that PlanePlotter uses two KML files called *google_aircraft.kml* & *google_cockpit.kml* that are created in the installation of the PlanePlotter and placed in the main program application folder, these should not be removed.

Before starting the Google Earth program PlanePlotter must be running and processing aircraft messages. To select the cockpit view right-click on an aircraft and check the *Flight deck view* box. Alternatively open Google Earth menu bar and select *File, Open* to then open either *google_cockpit.kml* or C:\PlanePlotter*google_aircraft.kml*.

Google Maps

PP2GM (PlanePlotter-to-Google Maps) is an add-on to PlanePlotter that enables users to create a copy of their chart display that can be displayed on a browser web page. To do this the user does not need to know anything about web page design or web hosting as the only requirement is for a web browser to be installed on the computer and for the computer to be connected to the Internet.

Figure 135 - Typical Google Map showing aircarft pop-up.
See http://www.kenandglen.com/aisDual.html

Google Earth, Google Maps and other Add-Ons

PP2GM is comprised of two specific parts, the first is a small program that resides on the user's PC and creates an XML file from the current PlanePlotter messages. The program then uploads this file to the users web site to be used by the second part of PP2GM which is a custom web page. The small program is a visual basic script named *PP2GM.vbs* which when started opens a graphical interface as shown in Figure 136. All that is needed to start the collection of data and uploading of the XML is to complete all the entries in the GUI and then press the *Execute* button.

To upload the XML file PP2GM uses the *MoveItFreely* FTP program. This program has improved error recovery that will eliminate many problems that might result from ISP or FTP server access problems as well as local PC issues. The program is self-installing can be downloaded without charge from *http://www.stdnet.com/products/?category_number=6&subcategory_number=1.*

When the PP2GM web page is opened in a web browser it automatically read the data in the uploaded XML file and displays a Goggle Map with aircraft symbols. Clicking a symbol will produce a pop-up containing information about the aircraft and an image of it if available.

A Zip file containing the PP2GM script and instructions on how to implement it can be downloaded from the Files section of the PlanePlotter Forum at *http://groups.yahoo.com/group/planeplotter/files/*. The downloaded Zip file contains 7 files that should be extracted to the PPTools folder *C:\PlanePlotter\PPTools*. The user is advised to read *PP2GM.txt* before using the script as it contains a lot of very useful information on how PP2GM should be installed and used.

Figure 136 - PP2GM Control Panel

FindFlight

As discussed in Chapter 10 when PlanePlotter first receives a message from an aircraft neither the aircrafts registration nor its flight details are known. PlanePlotter then has to interrogate its databases to see if the aircraft is already known. If it is not known then a manual lookup using external sources has to be implemented.

FindFlight is another excellent utility from Curt Deegan that solves this problem for PlanePlotter Users. The program once started will run in the background and examine each aircraft message as it is received. When FindFlight processes a missing RegNum it sends the data to PlanePlotter and it is added to the Basestation.sqb database. FindFlight does not update the PlanePlotter display as the RegNum data will only appear on PlanePlotter when a new flight is acquired. PlanePlotter will not update the on-screen data of an existing flight as once a flight is displayed the program is past the stage where it looks for data in the SQB database. Therefore until the flight has been acquired by PlanePlotter FindFlight is unaware of it so the update of the SQB file is after the fact as far as what is displayed on a PlanePlotter chart.

The process may be summarised as:

1. PP acquires a flight and displays .NO-REG because it's not in the SQB.
2. FF interrogates PP and finds a new flight with no RegNum and proceeds to look it up.
3. FF tells PP the RegNum and PP updates the SQB.
4. The next time PP sees that flight, it goes back to step 1, only now it finds a RegNum.
5. FF finds a new flight in step 2. but it has a RegNum so nothing needs to be looked up.

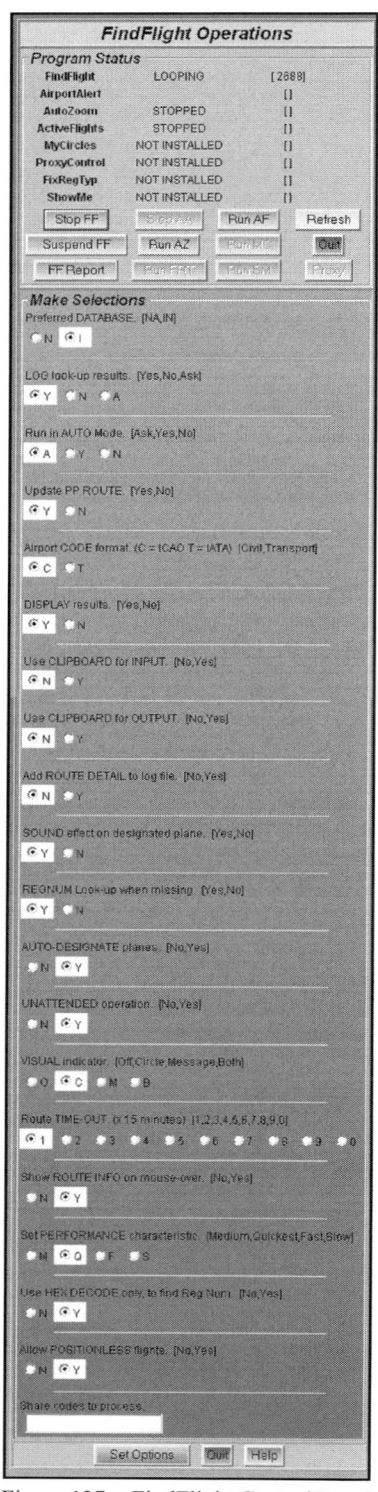

Figure 137 - FindFlight Control Panel

A Zip file containing the FindFlight script and instructions on how to implement it can be downloaded from the Files section of the PlanePlotter Forum at *http://groups.yahoo.com/group/planeplotter/files/*. The downloaded Zip file contains 17 files that should be extracted to the PPTools folder *C:\PlanePlotter\PPTools*. The user is advised to read *Findflight.txt* before using the script as it essential to understand how the program should be installed. It is most important to read and understand the section on creating shortcuts as without doing so the program may not function correctly.

The program is started by using a file called *FFset.hta* which opens a GUI as shown in Figure 137. Only the top portion of the GUI will be displayed when it is first open so the user should expand the window to show the full content. Once the option have been set the program is started by clicking the *Start FF* button. As the GUI does not automatically update its status it may be necessary to occasionally click the *Refresh* button. With FindFlight running the GUI may be minimised or closed as it will continue to function in the background until PlanePlotter itself is closed.

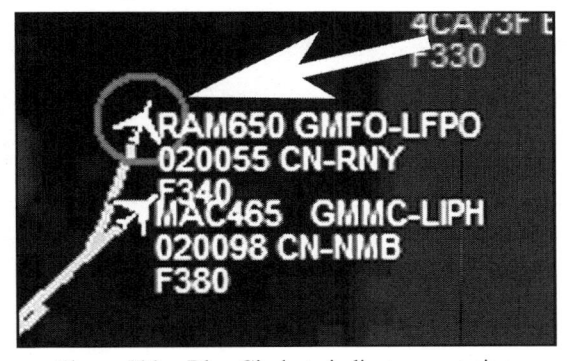

Figure 138 - Blue Circle to indicate processing

Each time a new aircraft is detected FindFlight will place a blue circle around the aircraft symbol and sound a beep to indicate that it is searching some external databases to obtain the aircrafts registration and flight routing information.

Once this is found the data will be written into the local databases. If the user then moves the cursor to hover about an aircraft symbol on the PlanePlotter chart then an information window will pop up as can be seen in Figure 139.

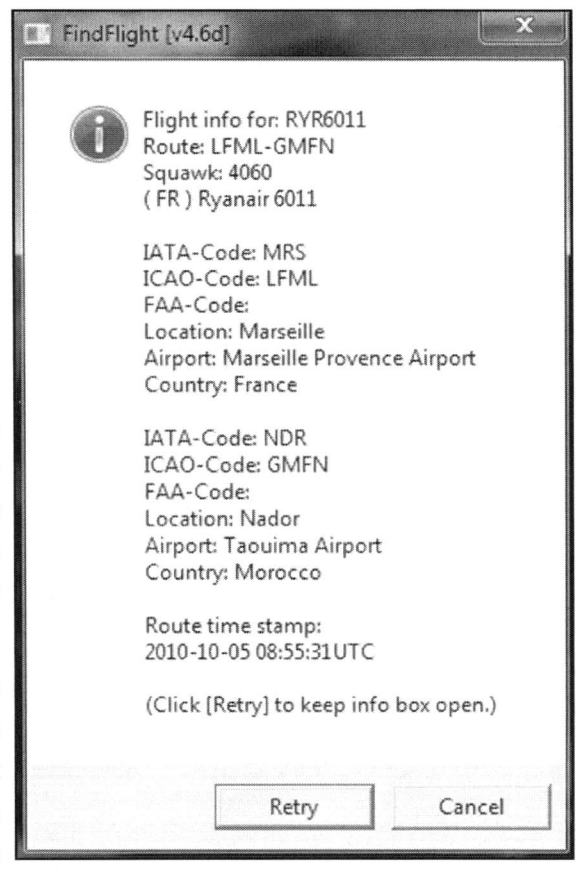

Figure 139 - FindFlight Information Window

137

FlightDisplay

```
FRL = LHR:EGLL:Heathrow
Airport:London:UK:::51.47139:-0.452778:PHL:KPHL:Philadelphia International
Airport:Philadelphia:USA:Pennsylvania:PA:39.86805555:-75.24861111~Not Known

FlightStats =LHR:EGLL:Heathrow
Airport:London:UK:::51.47139:-0.452778:PHL:KPHL:Philadelphia International
Airport:Philadelphia:USA:Pennsylvania:PA:39.86805555:-75.24861111~7h 55m

Flytcom=LHR:EGLL:Heathrow
Airport:London:UK:::51.47139:-0.452778:PHL:KPHL:Philadelphia International
Airport:Philadelphia:USA:Pennsylvania:PA:39.86805555:-75.24861111~Not Known

Flightaware=LHR:EGLL:Heathrow
Airport:London:UK:::51.47139:-0.452778:PHL:KPHL:Philadelphia International
Airport:Philadelphia:USA:Pennsylvania:PA:39.86805555:-75.24861111~Not Known

Libhomeradar=LHR:EGLL:Heathrow
Airport:London:UK:::51.47139:-0.452778:PHL:KPHL:Philadelphia International
Airport:Philadelphia:USA:Pennsylvania:PA:39.86805555:-75.24861111
```

Figure 140 - FlighDisplay Information Window

An alternative to FindFlight is an add-on called FlightDisplay that can be downloaded from *http://sites.google.com/site/0flightdisplay/Download*. Unlike FindFlight that only searches through a single nominated database FlightDisplay searches through several and displays the results from each one as shown in Figure 140.

ZoneMe

ZoneMe is another of Curt Deegan's PlanePlotter add-on application that provides a three-dimensional alert zone capability. It can be used on its own or in conjunction with the PlanePlotters own alert function but when altitude and/or elevation needs to be included in the alert consideration. ZoneMe has provision for Range Limits, Aircraft Type and Share Code Exclusions in the alert determination. PlanePlotter Quick Chart switching and Zooming can be assigned to ZoneMe alert zone definitions & alert messages logged. The latest version of ZoneMe incorporates Elevation as a factor in defining an alert zone. By including Elevation along with the other 3-D bounds of an alert zone plane spotters can have alerts generated for aircraft as they enter a zone and rise sufficiently above the horizon to be viewable. This function works in conjunction with the PP My Sky View. The files to install ZoneMe can be downloaded from *http://groups.yahoo.com/group/planeplotter/files/*

PPDataGrid

This add-on creates an interactive data grid that runs alongside Plane Plotter. By clicking on either the data grid rows or Plane Plotter screen highlights aircraft in both applications. PPDataGrid enables sorting of columns and filtering of aircraft data. Columns may be dragged to a specific position or particular columns chosen for display.

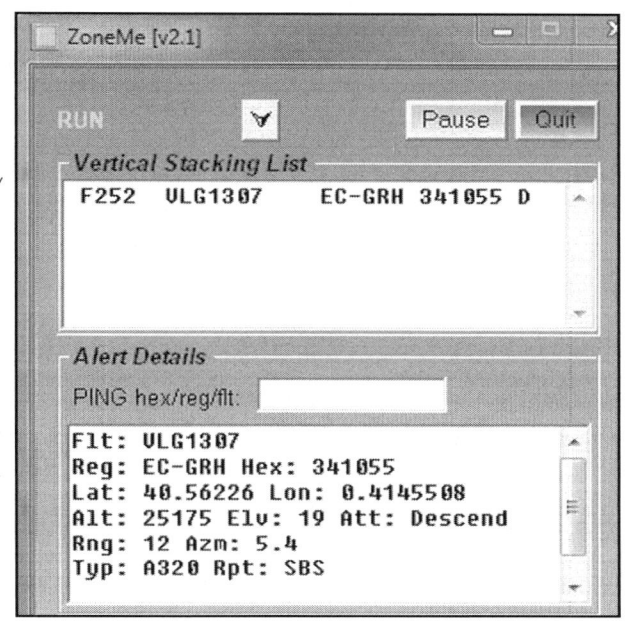

Figure 141 - ZonMe Control Panel

Double clicking on an aircraft entry will zoom in on that aircraft in the Plane Plotter view. The add-on also enables additional aircraft images, operator logos, country flags and climb & descent indicators. The add-on may be downloaded and installed from *http://www.thehangar.co.uk/downloads/ppdatagrid/publish.htm*.

MyCircles

MyCircles is a tool for determining PlanePlotter sharer range circle scale factors. It includes multiple share code support, circle labels, dynamic analysis, scanning for designated planes, sharer Locate function, sharer status display, aircraft selection by share code, sharer info display, enhanced GPX waypoint support, experimental tracking function, squawk code translation, airport location coverage circles, current/last squawk and altitude change display. The files to install MyCircles can also be downloaded from *http://groups.yahoo.com/group/planeplotter/files/*.

MapIt

This is a useful utility that accepts coordinates in various formats, converts and plots them to a Google Map displayed on the users web browser. It can also be used show coordinates of map positions and compute range and bearing. The files to install MyCircles are available for download from the Yahoo Planeplotter Group at *http://groups.yahoo.com/group/planeplotter/files/*.

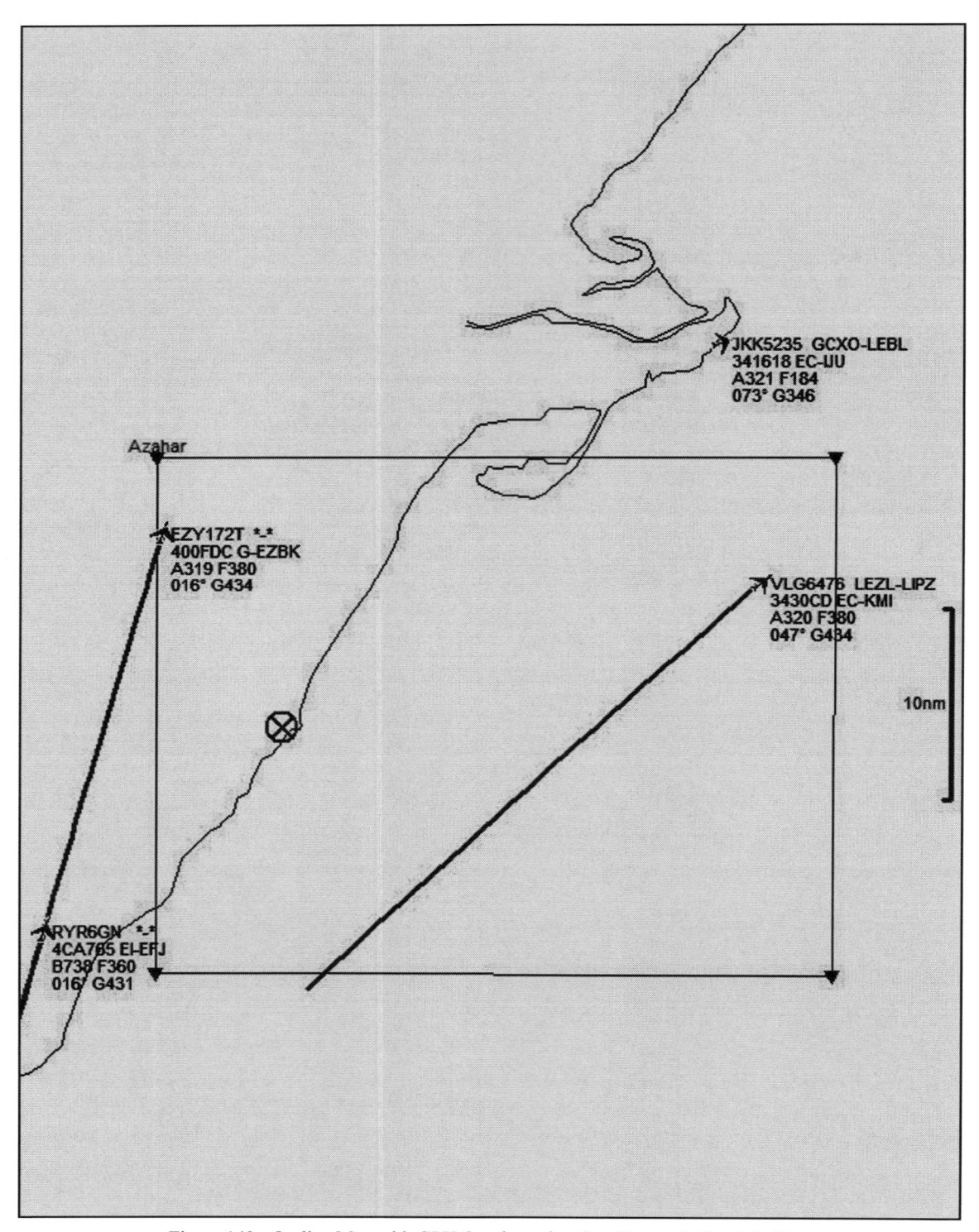

Figure 142 - Outline Map with GPX Overlay using Curt Deegan's ZoneMe Alert

Chapter 13

Understanding Alerts

Although some mention about Alerts has been made in Chapters 6 and 7 it is worthwhile to discuss the subject in further detail. PlanePlotter provides some very interesting and useful features in the way it allows users to raise and log alerts. A PlanePlotter Alert is a message to the user that an aircraft with particular credentials is now being processed by PlanePlotter and displayed on the currently viewed chart.

Figure 143 - An Alert Zone

PlanePlotter User Guide

PlanePlotter Alert Zone

A user can create or define a number of different Alert zones. Each Alert zone is a user defined polygon that is outlined in red on the current chart. To create an Alert zone the user should first select Menu *Options, Alert, Alert Zone, Add* and then click on a number of points or vertices to create the outline of the zone. When the zone is defined then Menu *Options, Alert, Alert Zone, Add* should be selected once more to disable adding any further vertices.

The Alert zone polygon may be saved to a file by selecting Menu *Options, Alert, Alert Zone, Save*. This option enables the user to save a number of different Alert zones although only one can be active at a time. Note that if PlanePlotter is closed the displayed Alert zone is still defined and will be automatically reloaded when the program next runs.

A previously saved Alert Zone can be re-established by selecting Menu *Options, Alert, Alert Zone, Load*.

The current Alert zone can be removed completely by selecting Menu *Options, Alert, Alert Zone, Clear*.

Chart Options and Settings

Clicking the *Blue Spanner* toolbar button will open the Chart Settings dialog box. The bottom left section of this as shown in Figure 144 is concerned with PlanePlotters Alert settings.

The user chooses what parameters should raise an Alert by checking one or more of the boxes, selecting from the *Interested/Not Interested* selector or by clicking the *..match* button. When one or more options are checked they are treated as a logical *OR*, that is for example if both the *aircraft in zone* box is checked and the selector is set to *Interested* then either an aircraft entering the Alert zone or an aircraft with its Interested flag set will raise an Alert message.

The message will be displayed on top of any open window and continue to be displayed as long as the Alert condition is satisfied.

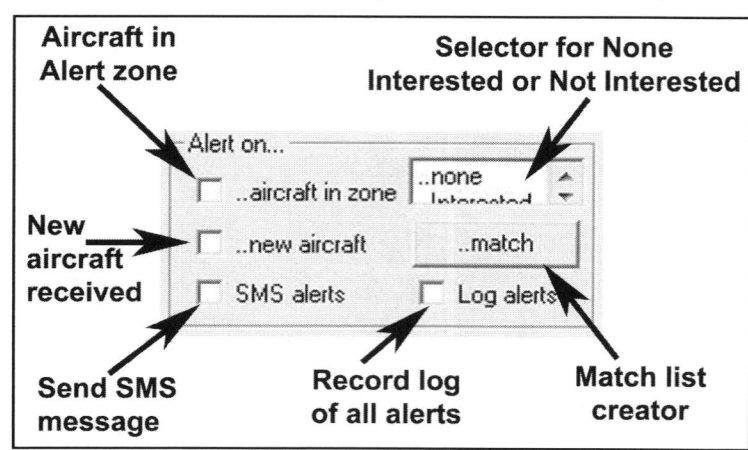

Figure 144 - Restore Point with Quick Charts

Zone Alert

When the *..aircraft in zone* box is checked then any aircraft entering the Alert zone will raise an Alert. The alert message will continue to be displayed for as long as the aircraft remains in the zone. Where there are a number of aircraft within the Alert zone the Alert message will list them all.

Figure 145 - Zone Alert

Figure 146 - Restore Point with Quick Charts

Interested

With the selector set to *..Interested* then any aircraft on the chart with an *Interested flag* set will raise an Alert. The flag can be set for individual aircraft by right clicking an aircraft symbol, clicking the *Edit* button and then checking the *Interested* box in Edit

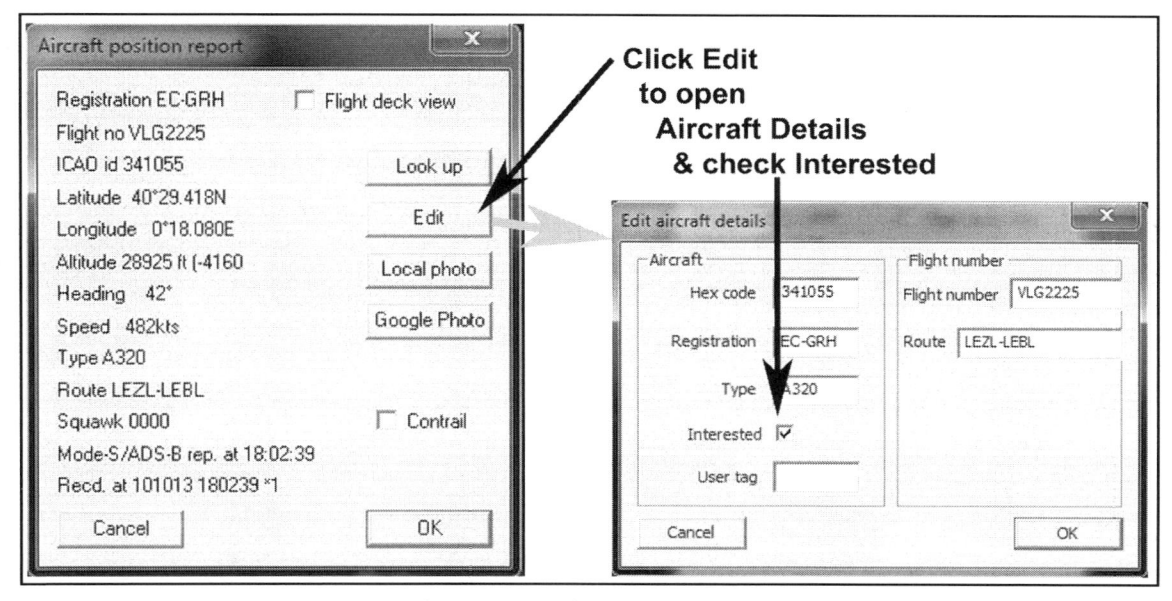

Figure 147 - Setting the Aircraft Flag

Figure 148 - Alert from an Interested aircraft

143

Aircraft Details dialog box. This flag will remain set in the basestation.sqb database until reset.

As well as setting an individual Interested flag a whole group of aircraft could be flagged by setting their flags in the basestation.sqb database using a SQL command as discussed in Chapter 8. For example one could set the flag for each of the new Boeing Dreamliners so that an Alert is raised whenever they should appear on the users chart by using the SQL:-

UPDATE Aircraft SET Interested="1" WHERE ICAOTypeCode LIKE "B772";
SELECT * from Aircraft WHERE ICAOTypeCode LIKE "B772";

Not Interested

With the selector set to ..*Not Interested* then aircraft on the chart that do not have the *Interested flag* set will raise an Alert. It is envisaged that normally the user will flag only a few aircraft as *Interested* so choosing this option will raise alerts on a multitude of aircraft!

None

No Alerts will be raised with the selector set to ..*None* whether or not aircraft displayed on the chart have an *Interested flag* set or not set, providing of course that neither the ..*aircraft in zone* nor ..*new aircraft* boxes are checked and also there are no entries in the *match* list

New Aircraft

With the ..*new aircraft* box checked then any aircraft not previously displayed on the chart will raise an Alert. The alert message is transitional and only remains on the screen for a very short period. An audible alert is also sounded in case the users misses the screen message.

Figure 149 - Zone Alert

Figure 150 - New Aircraft Alert Message

Understanding Alerts

SMS Alerts

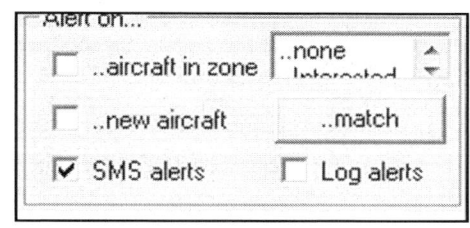

Figure 151 - SMS Alerts

PlanePlotter provides a novel feature to registered users that allows SMS Text Message Alerts to be transmitted.

A user must first complete a web form similar to that shown below in Figure 153.

On completion of this form COAA will enable a user account that enables the user to check/purchase message credits. When you SMS alerts are enabled in PlanePlotter, the system will send a text messages for every alert generated. Therefore it is important to only configure those alerts SMS notification is required of or the text messages credits may be used up too quickly!

Figure 153 - SMS Registration Form

Log Alerts

Figure 154 - Log Alerts

With the *..Log alerts* box checked then a log file will be created recording all the Alerts raised.

A new file will be created every day with a filename of the form *pp_alertYYMMDD.log* e.g. pp_alert110110 for the 10th January 2011.

A sample extract of a log file is shown below. Alerts inside a zone are shown as such, Interested are preceded with *Int* and Not Interested with *!Int*.

BAW894 G-EUUT 401240 : BAW894 G-EUUT 401240 is inside the Alert Zone at 10-10-14 10:34:58 UTC
Int : TOM2HD G-OOPH 4006C5: Msg fm: G-OOPH TOM2HD 4006C5 A321 EGCN-GCLP at 10-10-14 10:36 UTC
Int : TOM2HD G-OOPH 4006C5: Msg fm: G-OOPH TOM2HD 4006C5 A321 EGCN-GCLP at 10-10-14 10:36 UTC
!Int : RYR4673 EI-DPA 4CA4E5 : Msg fm: EI-DPA RYR4673 4CA4E5 B738 LIME-LEMG at 10-10-14 10:27 UTC
!Int : RYR5447 EI-EKK 4CA80B : Msg fm: EI-EKK RYR5447 4CA80B B738 LFML-LEMD at 10-10-14 10:30 UTC

Match List

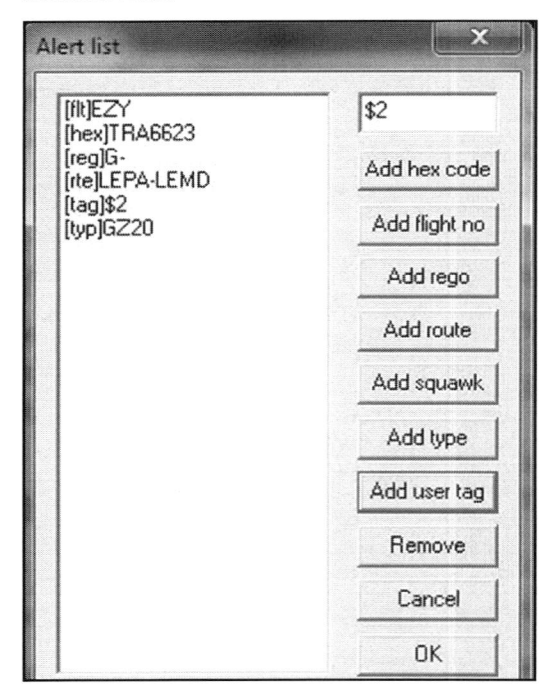

Figure 155 - Alert List

The Match list function is perhaps the best feature of all in PlanePlotters Alerts. When the *..match* button is clicked a new dialog box is displayed where a whole range of alert criteria can be set-up.

Hex Code - by entering a hex code in the top right box and then clicking the *Add hex code* button an entry will be placed in the list like [hex]TRA6623.

Flight No - by entering a flight number or part of a flight number in the top right box and then clicking the *Add fight no* button an entry will be placed in the list like [flt]EZY. That will raise an Alert on any EasyJet flight.

Registration - by entering an registration or part of hex code in the top right box and then clicking the *Add rego* button an entry will be placed in the list like [reg]G-. Doing so will raise an Alert on any British registered aircraft.

Route - by entering a flight route in the top right box and then clicking the *Add route* button an entry will be placed in the list like [rte]LEPA-LEMD. Doing so will raise an Alert on an aircraft flying that particular route. The route identifier must be entered in upper case characters.

Squawk - by entering a four digit squawk code in the top right box and then clicking the *Add squawk* button an entry will be placed in the list like [skq]7265.

Type - by entering an ICAO aircraft code in the top right box and then clicking the *Add type* button an entry will be placed in the list like [typ]GZ20. Abbreviated codes such as A3 will raise Alerts on A319, A320's etc.

User Tag - by entering a User Tag button an entry will be placed in the list like [tag]$2. Aircraft with one of the new plane symbols will cause an Alert.

The user may of course enter a combination of any of the above or add multiple instances of each one.

ZoneMe

Whilst PlanePlotters inbuilt Alert functions will meet the needs of most users it is worthwhile mentioning a third party add on.

Curt Deegan's ZoneMeis a PlanePlotter add-on application that provides a three dimensional alert zone capability. ZoneMe can be used instead of, or in conjunction with the built-in PP alert features, when altitude and/or elevation needs to be included in the alert consideration. ZoneMe also incorporates Range Limits, Aircraft Type, and Share Code Exclusions in the alert determination. PlanePlotter Quick Chart switching and Zooming can all be assigned to ZoneMe alert zone definitions.

PlanePlotters Alert functions provide a two dimensional, polygonal alert zone whereas ZoneMe2 offers a three dimensional cuboid alert zone for when altitude must be included in alert zone considerations. A ZoneMe 3D alert zone has a 2D footprint of a rectangle versus the polygonal shape of the PlanePlotter 2D alert zone feature. In this regard ZoneMe may be regarded as less flexible. However ZoneMe has the added benefit of the third dimension of altitude and a ZoneMe alert zone is bounded by north and south latitudes, east and west longitudes and upper and lower altitude.

Figure 156 - Air Tahiti Nui Airbus A340-313X F-OJTN at 32000 Feet on a cold Boxing Day morning, Photo Courtesy & Copyright of Lee Shand

Chapter 14

Tips and Tricks

PlanePlotter is a comprehensive software program that can meet at many levels the needs of the aviation enthusiast. Within a few minutes of installing the software a user can be tracking and identifying aircraft not only in their local area but around the world. The program has a multitude of features may of which are not obvious to start off with. For those willing to spend time learning all the intricacies then they will find that there is no other program to match PlanePlotter. To make life easier and reduce the learning time the user will find in this chapter of the book some useful tips and tricks.

The ABCD Buttons

This group of buttons provide a powerful way of saving different configurations of PlanePlotter set-up and layout. Users should not be afraid to use and should ignore any adverse comments about their use. In the simplest of terms they enable a 'backup' to be taken of how PlanePlotter is being used at a particular time.

The user is recommended once they have a PlanePlotter set-up they are happy with to save the configuration using the Menu *File, Save Restore point, A*. The what ever changes are made from then on that configuration can be recalled using Menu *File, Load Restore point, A*.

Now a user may wish to have a number of different charts where they would like to change the view with a single click of the mouse. Once a chart has been loaded then it

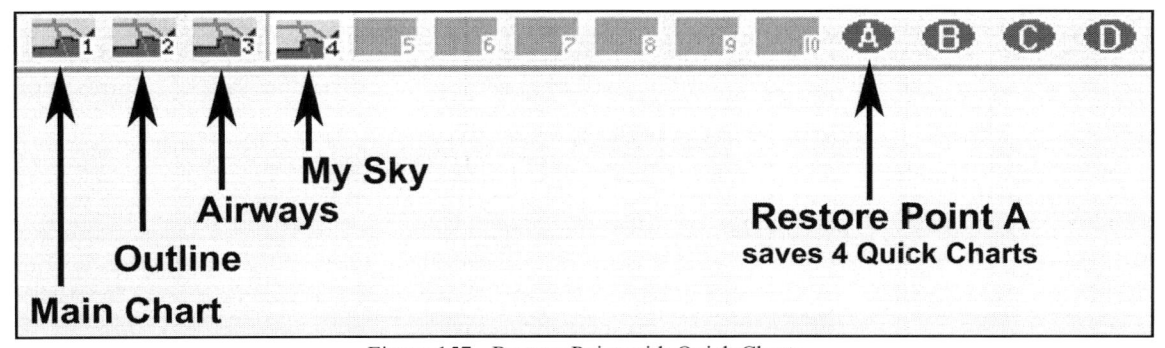

Figure 157 - Restore Point with Quick Charts

can be assigned a *Quick Chart* button. An example here could be one geographical chart for normal viewing, an *OSM* map showing an enlarged area, an *Outline map* showing an alert zone. Each could be assigned a *Quick Chart* button, one could also have a *My Sky* view assigned to another button. The user may also like a second outline map with some airway and airport outlines superimposed on it and could be assigned to yet another *Quick Chart*. Once those assignments have been made then a further *File, Save restore point, A* will retain them in that backup.

Option settings should as *Sharing Upload* only, defining a script or setting a list of parameters to matched for an alert will all be saved in a *Restore Point*.

When a user is experimenting with different settings or view it is recommend that they save their work in *Restore Point D*. This is a handy little trick as it saves their work in case of a PC failure or power cut but always allow them to restore their default known good configuration in *Restore Point A*.

My Sky

The vertical elevation of PlanePlotters My Sky view is dependent upon the aspect ratio of the window. If the window is made narrower then the elevation available is increased

Aircraft Photographs

When the user right clicks an aircraft symbol an *Aircraft Position Report* dialog is displayed. The user may view a photograph of the aircraft by selecting either *Local Photo* or *Google Photo*. The latter will open a Google search page showing a number of different images.

The user can create their own library of photos. Each photo should be named with the registration letters of the aircraft. The name is not case sensitive and any hyphen separating the country letters from the aircraft specific letters is ignored. Photos should normally be stored in the C:\Planeplotter\Photo files\ folder and the PP Photo Files folder directory should be pointed at that location.

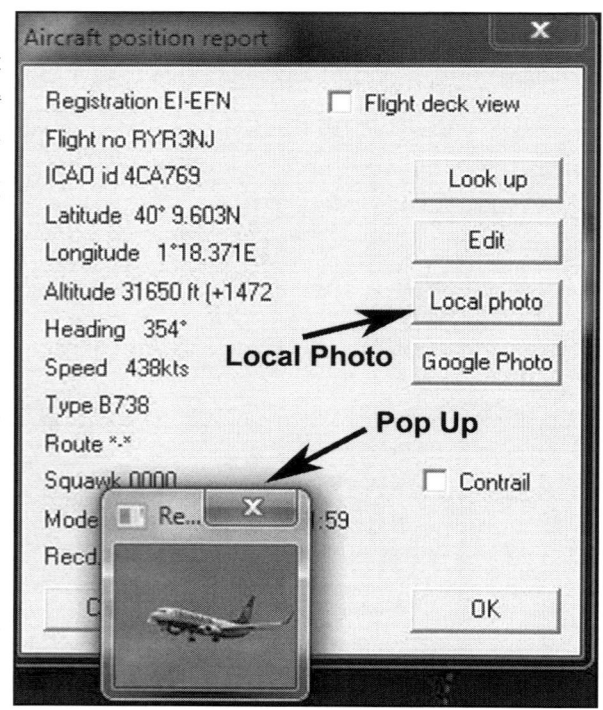

Figure 158 - Restore Point with Quick Charts

Aircraft type codes and symbol colours

PlanePlotter allocates a type code and colour to each symbol shown on a chart. These are:

    Type 0   ACARS position - red
    Type 1   ACARS ADS 'next estimated position' report - blue
    Type 2   ACARS AMDAR position - green
    Type 3   SBS or RB log file data position - yellow (real-time) or orange (delayed)
    Type 4   HFDL position - cyan
    Type 5   ACARS flight-plan waypoint position - magenta
    Type 6   SBS-1 TCP position delayed - yellow or orange
    Type 7   RadarBox position delayed - yellow or orange

PlanePlotter also changes the colour of the symbols nose a blue nose indicates that the aircraft is ascending whilst a brown nose shows it is descending. If the nose is the same colour as the major portion of the symbol the aircraft is in level flight.

Users can of course create their own symbols and this has been dealt with in quite some detail earlier in Chapter 8.

Missing Plots

If any starting PlanePlotter there are no aircraft shown on the screen then the user should first check that the I/O options have been set correctly. Then that the *Start* processing button has been pressed and the symbol changed from a round green circle to a black square dot.

The user should also check that PC's computers clock is correctly synchronised to their time zone. See Appendix B for details of the NTP timekeeping utility.

Fuzzy Screens

Sometimes zooming into a chart will make it appear fuzzy. Should that happen then use the *OSM* button for the area and after each change of zoom level click the *OSM* button again. This is another example of where the *Quick Chart* feature can be used.

Cleaning a Messy Screen

Sometimes the screen can become very cluttered with old aircraft symbols and tracks and contrails. The easiest way to clean up is the screen is to click *Restore A* button provided that a restore point has been created as discussed earlier.

Do not be tempted to us the Purge option unless you are really sure that you want to clear out all the recorded data.

More Maps for use with PlanePlotter

Users can find a comprehensive range of both calibrated and vector maps at the ManTMA PP Support site. *http://www.mantma.co.uk/pp_support.html*

Outline Colours

When PlanePlotter is first used any Vector maps or outlines will use the PlanePlotter default colours. However these colours can be changed to suit an individual user. This is easily done by using a *Ctrl-Left Click* on any part of an outline which will open a colour palette where a new colour can be allocated to the particular outline. If the user wishes to retain the new colours then they should be saved in a *Restore Point*.

GPX overlays and Vector Maps

GPX overlay files will work with vector maps. Once a vector map has been loaded then a GPX overlay say containing waypoints can also be defined and switched on.

Overcoming the 'No-Reg' Label

With a new PlanePlotter installation a user may find that the first aircraft acquired display a No-Reg label. This is because the message details received from the aircraft do not contain the aircrafts registration letters but do contain hex-code identifier. PlanePlotter needs to convert that to the registration and it does this by looking up the identifier in a database. That database has to be installed by the user as it is not included as part of the PlanePlotter installation package. Further details can be found in Chapter 10.

Updating PlanePlotter to a later version

COAA is constantly striving to add new features to the PlanePlotter program and from time to time will release new versions of the software. To update a newer version is quite straight forward. All that is needed is to download the update and then install to the same location as the original copy was installed to. All the original configurations will be retained. Sometimes when the updated version is started for the first time the PCs firewall may request that access permissions ca be granted.

Using PlanePlotter on more than one Computer

The registration payment is a lifetime subscription and the program may be installed on a second PC. However each PC requires a separate number, this is obtained by going to the Registration Section in the PlanePlotter website.

Using the original email address when first registered the user should enter the new Personal number and await the new registration number to be emailed to them.

Chapter 15

User Support

User Forums

There are a number of web forums providing support to the PlanePlotter user community. On of these is the well-established *Yahoo PlanePlotter User Group* that that can be found at *http://groups.yahoo.com/group/planeplotter*.

Others include *RadarSpotters* at *http://radarspotters.eu/forum/index.php*, *PlanePlotters* at *http://www.planeplotters.co*, *FlightRadar24* at *http://forum.flightradar24.com/index.php* and the *ManTMA Overflights Group* at *http://www.mantma.co.uk/pp_support.html*.

There is also an online Wiki at *http://planeplotter.pbworks.com*.

Email Support

Additional support can be obtained from COAA the authors of the PlanePlotter software program by email to *info@coaa.co.uk* or to the author of this book at *pp_support@kenandglen.com*.

Other Software

There are a number of useful documents and software add-ons that can be used in conjunction with PlanePlotter together with a number of calibrated charts for different locations around the world. These can be found in the Files section of the ShipPlotter User Group at *http://groups.yahoo.com/group/planeplotter/files*.

This book

Copies of the scripts and google.html as well as electronic updates to the book will be provided free of charge to bona fide purchasers of the book. Comments or suggestions on this book should be addressed by email to the author at *pp_support@kenandglen.com* or by visiting his web pages at *http://www.kenandglen.com*.

Appendix A
Radar Frequency Bands

The spectrum of electro magnetic waves used in radar systems cover the range of a few hundred hertz up to 200Ghz. Traditionally frequency bands were classified by the letters L, S, C and X that covered 1Ghz to 12Ghz. Developments in microwave technology led to higher frequencies being developed and the letters Ku, K, Ka, V & W added. There was no particular logic in the use of these letters and some say they were used to keep the frequencies secret although they are defined as an IEEE Standard.

Figure A1 IEEE & NATO Radar Frequenct Band Comparison

With the development of Electronic Warfare (EW) the frequency bands have been given a new nomenclature using the alphabetical sequence A through to M. This is now defined as a NATO standard. Figure A1 shows a comparison between the two systems whilst Tables A1 and A2 detail the frequencies and wavelengths of each band.

Radar Bands (IEEE)

Band Designator	Frequency (GHz)	Wavelength (centimeters)
L band	1 to 2	30.0 to 15.0
S band	2 to 4	15 to 7.5
C band	4 to 8	7.5 to 3.8
X band	8 to 12	3.8 to 2.5
Ku band	12 to 18	2.5 to 1.7
K band	18 to 27	1.7 to 1.1
Ka band	27 to 40	1.1 to 0.75
V band	40 to 75	0.75 to 0.40
W band	75 to 110	0.40 to 0.27

Table A1 IEEE Radar Frequency Bands

Electronic Warfare Bands

Band	Frequency Range(MHz)	Channel Width(MHz)
A	0 to 250	15
B	250 to 500	25
C	500 to 1,000	50
D	1,000 to 2,000	100
E	2,000 to 3,000	100
F	3,000 to 4,000	100
G	4,000 to 6,000	200
H	6,000 to 8,000	200
I	8,000 to 10,000	200
J	10,000 to 20,000	1000
K	20,000 to 40,000	2000
L	40,000 to 60,000	4000
M	60,000 to 100,000	4000

Table A2 NATO EW Frequency Bands

Appendix B
Setting the PC Clock Accurately

For users sharing their data and especially for users participating in the Multilateration project it is essential that the PCs clock is accurate. It is not sufficient just to use the standard Windows Internet tome synchronising time server as PlanePlotter requires much more accurate time standard.

Fortunately for PlanePlotter users there is a readily available precise time source that can be easily downloaded and installed on the PC and will set and keep the computers clock exact without any further intervention. This is a Network Time Protocol (NTP) utility that can be downloaded from http://www.meinberg.de/english/sw/ntp.htm.

Installation and Set Up of the NTP Utility on Windows

From the Meinberg web site download the set-up file 'ntp-4.2.4p8@lennon-o-win32-setup.exe' and save it to a know location on the PCs hard disk. Then run the setup.exe which will then display the License Agreement screen shown in Figure B1. After clicking the 'I Agree' button a screen will be displayed as shown in Figure B2 prompting for a folder location. It is recommended to accept the default setting as once the utility is installed there should be no need to interfere with the software.

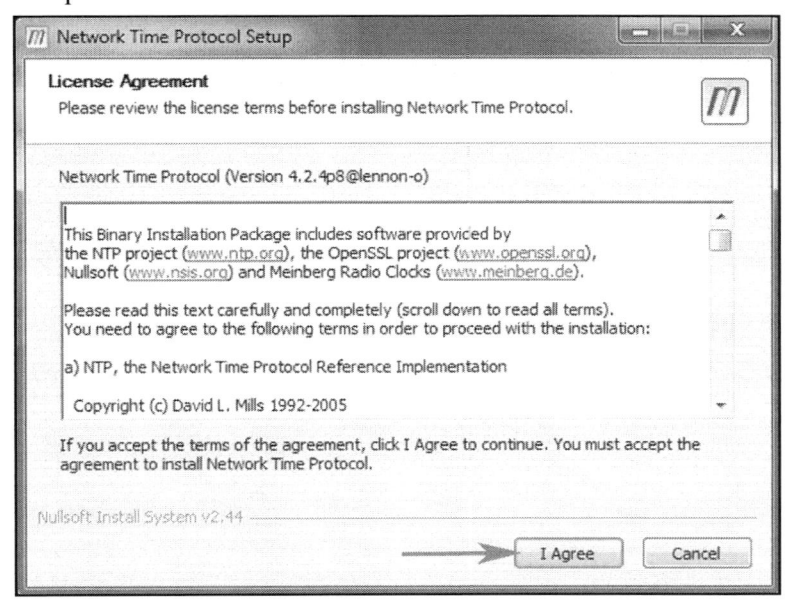

Figure B1 - NTP Licence Agreement

Having accepted the default folder location click on the 'Next button' to continue.

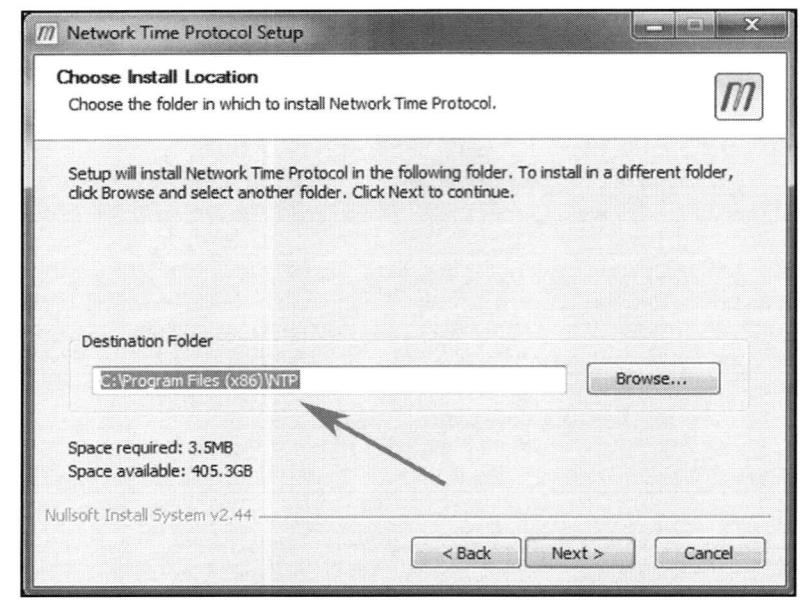

Figure B2 - NTP Installation Folder Location

It is important to leave all the components selected as shown in Figure B3 below, do **not** change them, before clicking the ´Next button to proceed to the next step.

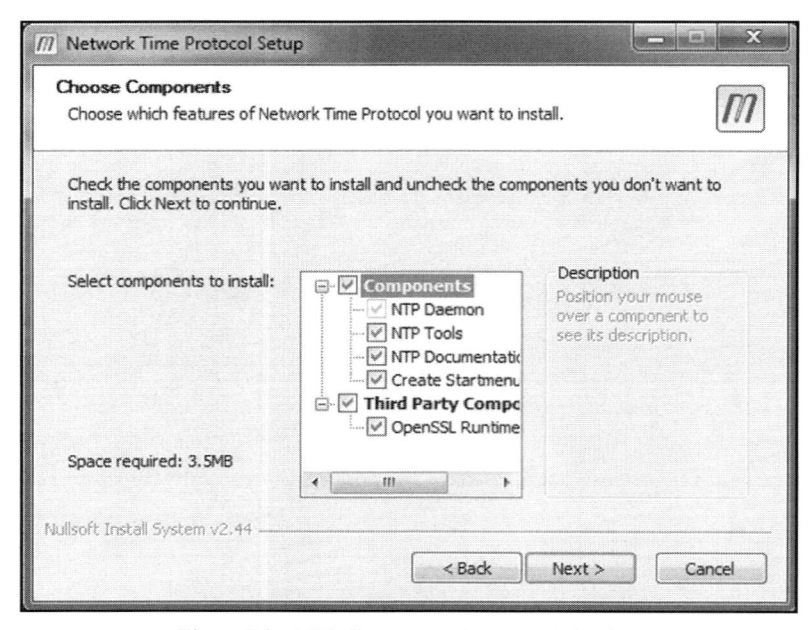

Figure B3 - NTP Component Features Selection

The NTP installation program will now create an custom configuration. The user should select a country or region from the drop-down list as shown in Figure B4 below. The other settings should **not** be changed.

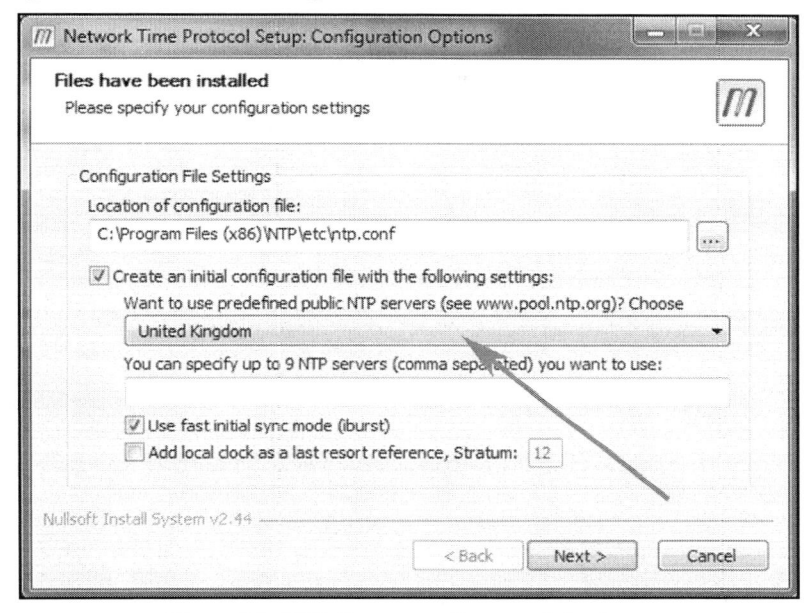

Figure B4 - NTP Component Features Selection

The next screen shown in Figure B5 allows the professional user to edit the NTP configuration. The PlanePlotter user is recommended **not** to change anything and should just click the 'No button' to proceed to the next stage.

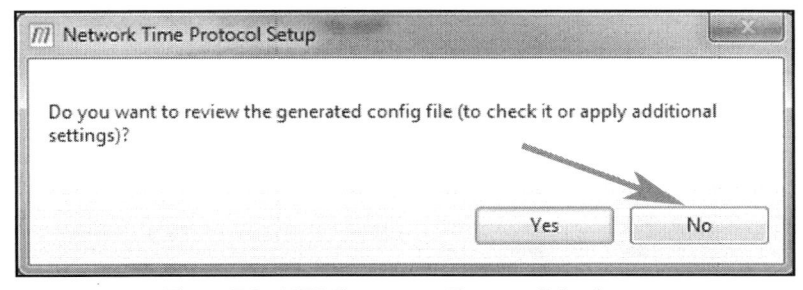

Figure B5 - NTP Component Features Selection

NTP will now create a user account for the software and will run whenever the PC is switched on to ensure a high degree of timekeeping stability. The next screen shown in Figure B6 allows the installer program to create a special user account for the PlanePlotter user. Ensure that box is checked and leave the other settings unchanged.

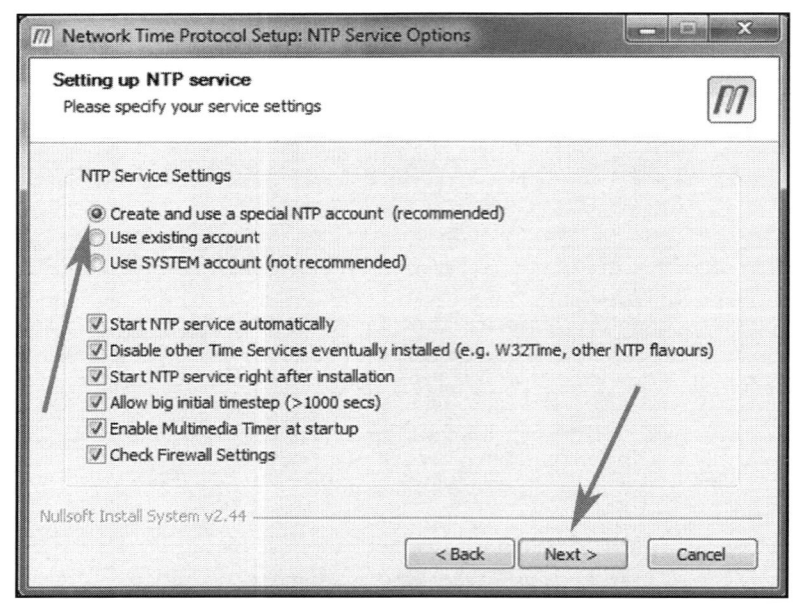

Figure B6 - NTP Server Settings

Next enter a password for the newly created account as shown in Figure B7. Your standard PC password should be sufficient.

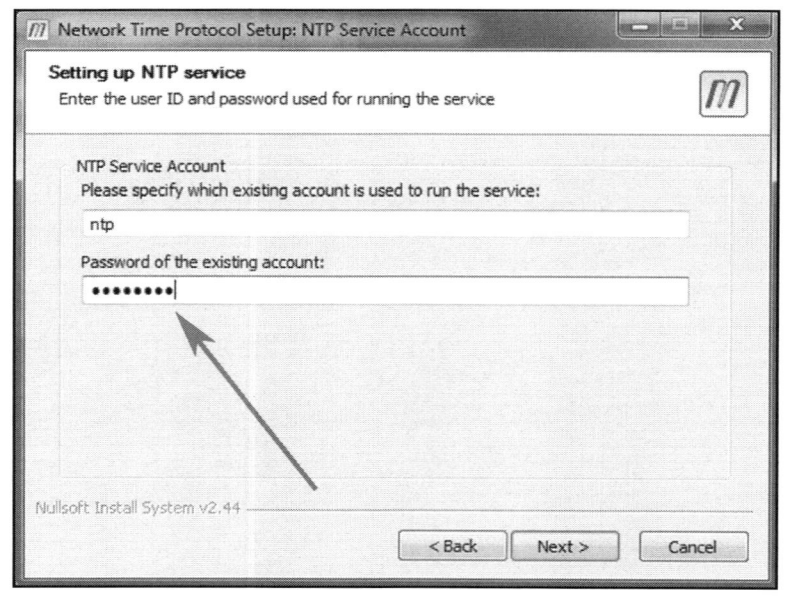

Figure B7 - NTP Server Settings

After clicking the 'Next button' the software installation will complete and the NTP service will be started.

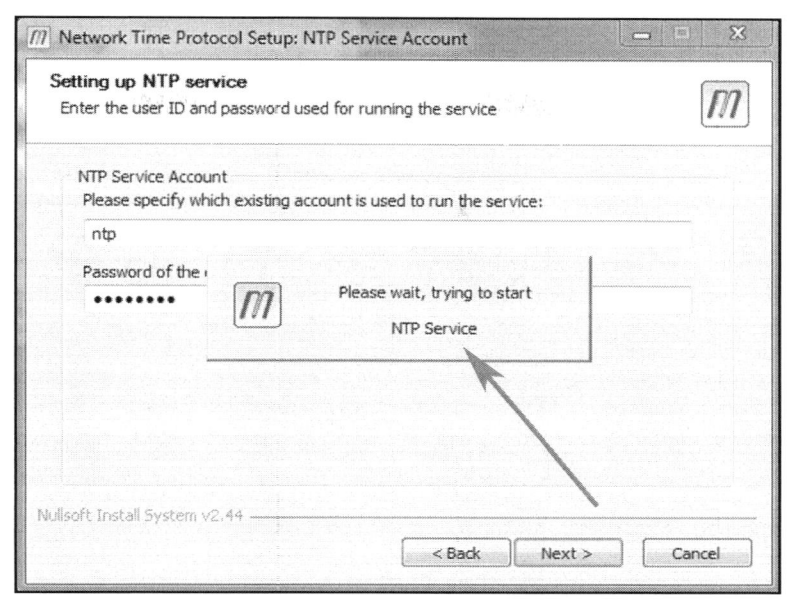

Figure B8 - NTP Installation Complete and Service Starting,

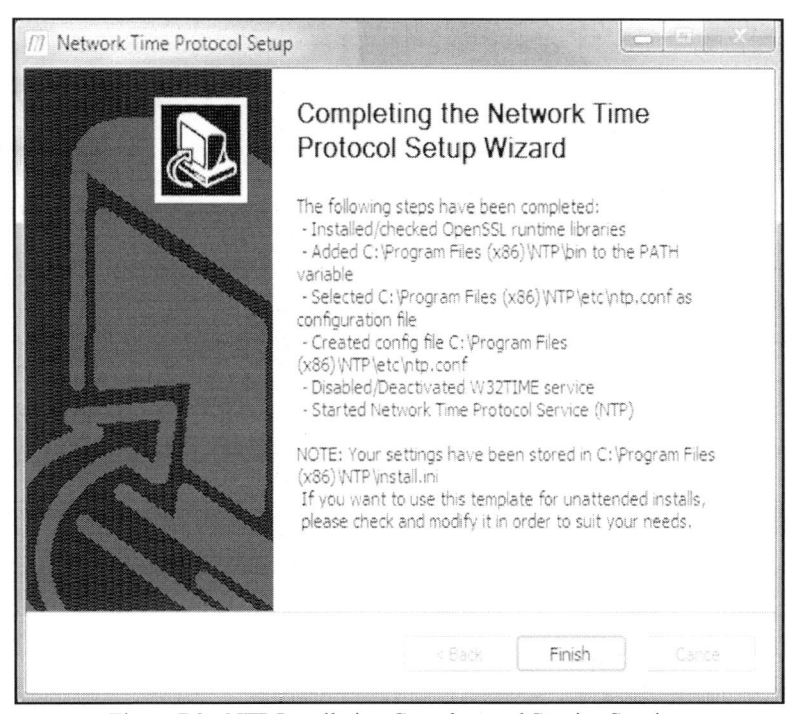

Figure B9 - NTP Installation Complete and Service Starting,

At this stage the user should uncheck Internet server synchronising in the Windows Control Panel, Date and Time box as NTP is now performing that function.

To Check NTP is working correctly

Once the installation is complete the user should check that the timekeeping utility is working correctly. This is done at the DOS or command line level. Select Programs, Accessories, Command Prompt and then enter the command "ntpq -p" as shown in Figure B10.

Figure B10 - Entering the "ntpq -p" command

Once the test is complete then results similar to those shown in Figure B12 should be produced. To exit the test enter the command "exit."

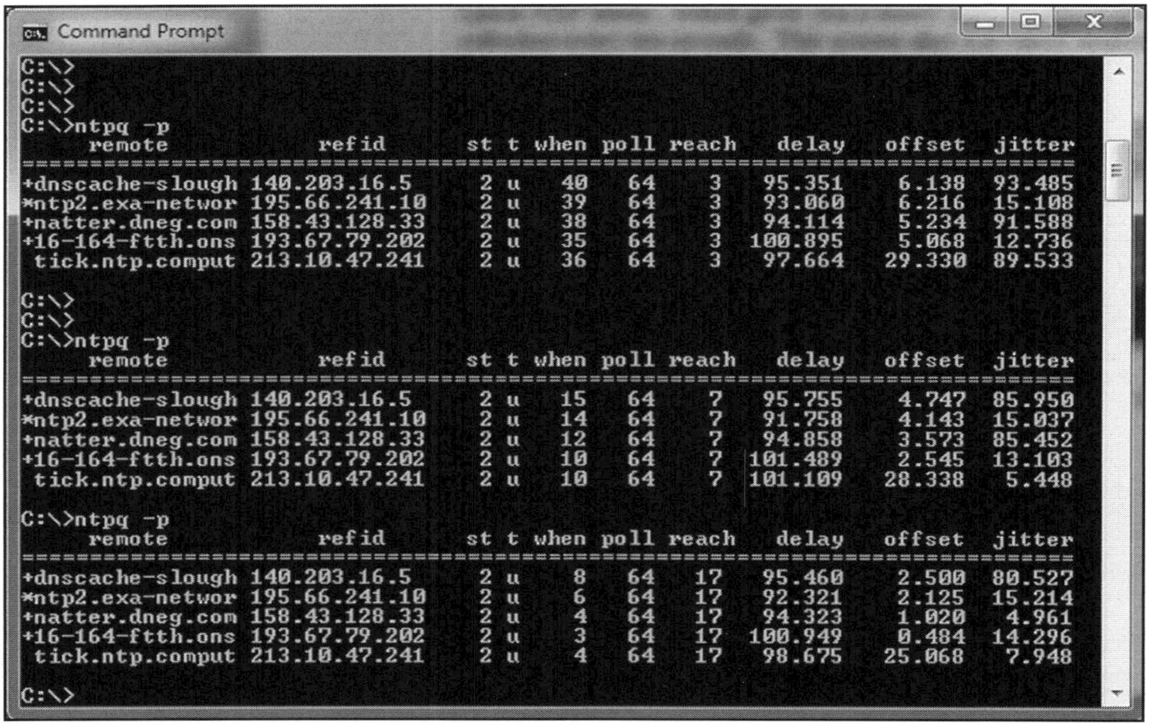

Figure B11 - ntpq -p test result

Appendix C
PlaneGadget Radar Installation

Users should read the installation manual that comes with the PlaneGadget Radar receiver. If they do not have a copy of the manual then they should follow the instructions here.

The PlaneGadget Radar is supplier complete with an installation CD containing the software drivers for the Windows XP, Windows Vista, Windows 7 and Windows 2000 operating system. The Windows Vista and Windows 7 drivers are only for the 32 bit versions of those operating systems as the 64 bit versions are not supported by PlaneGadget. However at the end of this Appendix there is note of how to install the PlaneGadget Radar software on those 64 bit versions.

Installation

First install the PlaneGadgets Radar CD into the PCs DVD or CD drive. After having connected the antenna to the receiver the USB cable should be plugged first into the

Figure C1 - Windows Found New Hardware Wizard

receiver and then into the PC. The PC should then report having found some new hardware and start the 'Found New Hardware Wizard' as shown in Figure C1 & C2.

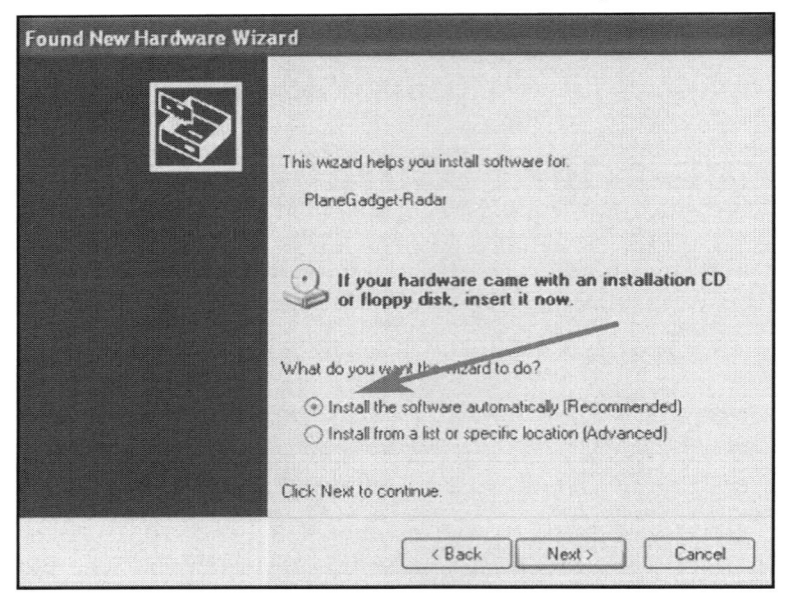

Figure C2 - Windows Found New Hardware Wizard

Having checked the 'Install the software automatically and clicked the 'Next button' the Wizard will then search the CD for the correct drivers as illustrated in Figure C3.

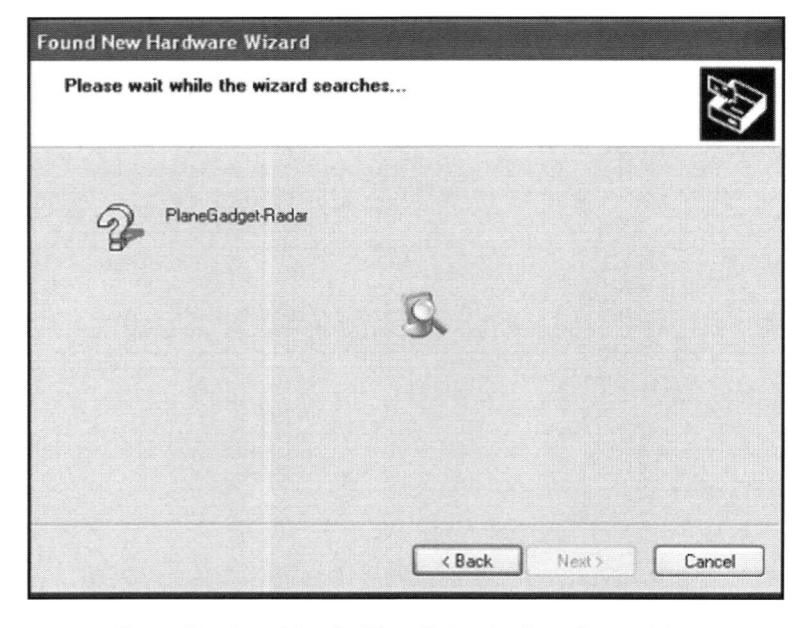

Figure C3 - Searching for PlaneGadget Radar software drivers

Once a driver has been found a message may be displayed as shown in Figure C4 stating that the Windows Logo test has failed. This message should be ignored and the installation process continued by clicking the 'Continue Anyway button.'

Figure C4 - Windows Logo Test Message

The installation process will then continue and determine which Comms Port the PlaneGadget Receiver is connected to.

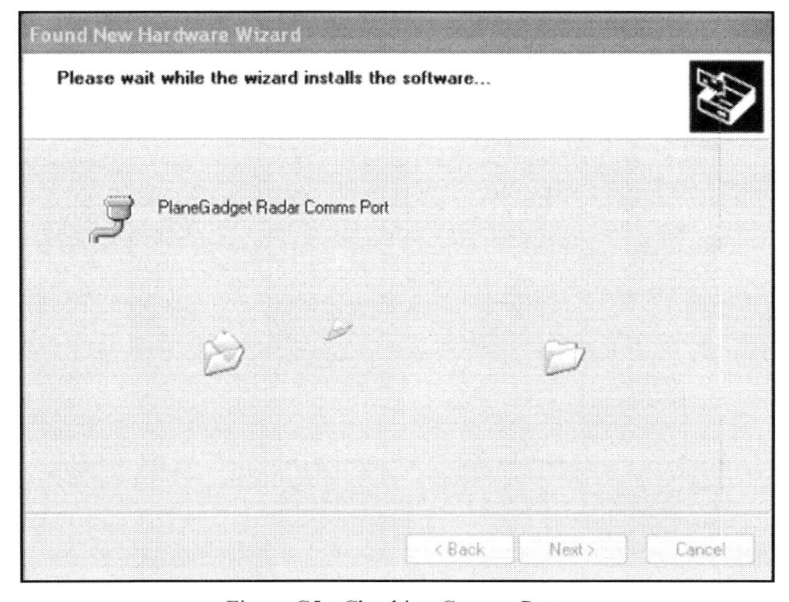

Figure C5 - Checking Comms Ports

Once the connection to the Comms Port has been verified the installation process is complete and the 'Finish button' may be clicked to close the Wizard.

Figure C6 - Found New Hardware Wizard Completion

Checking Comms Port

The user should make a note of which Comm Port the PlaneGadget Radar is using by examining the Hardware Device Manager in the Windows Control Panel as shown below in Figure C7.

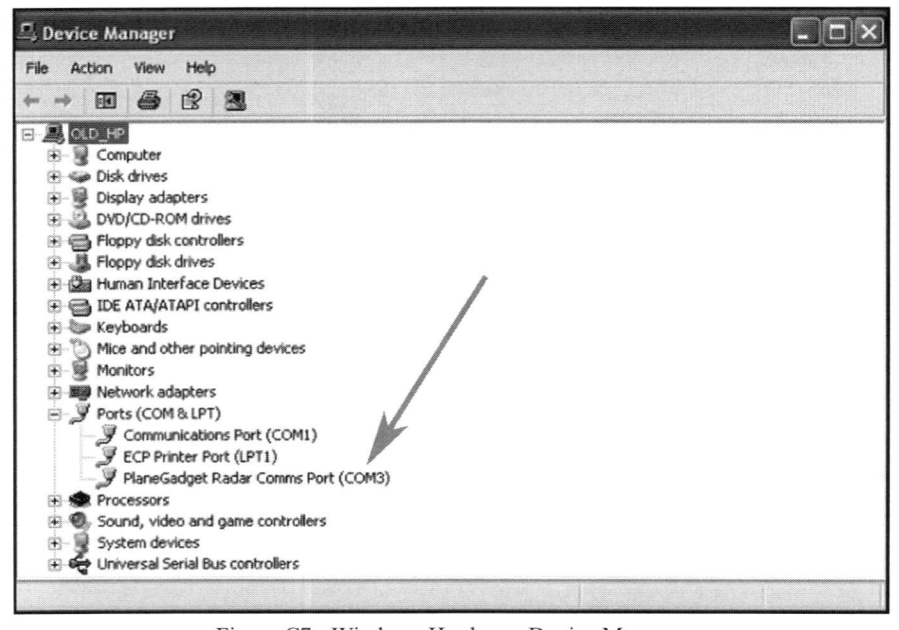

Figure C7 - Windows Hardware Device Manager

Windows Vista 64 Bit and Windows 7 64 Bit.

If the user tries to install the PlaneGadget Radar onto a 64 bit version of the Windows Vista or Windows 7 operating system then an error message will be displayed to the effect that the PlaneGadget Radar software is not supported on those operating system.

At this stage let the install fail, then disconnect the PlaneGadget Radar by removing the USB cable. Then the PC and reconnect the PlaneGadget Radar ignoring any prompts to install software and ignoring any 'Found New Hardware' messages. Then open the Windows Control Panel and select

  - Hardware and Sound,

  - - View Devices and Printers

  - - - Device Manager

  - - - - Ports.

Then find and right click on the 'PlaneGadget Radar Comms Port', then select 'Update Device Driver Software.'

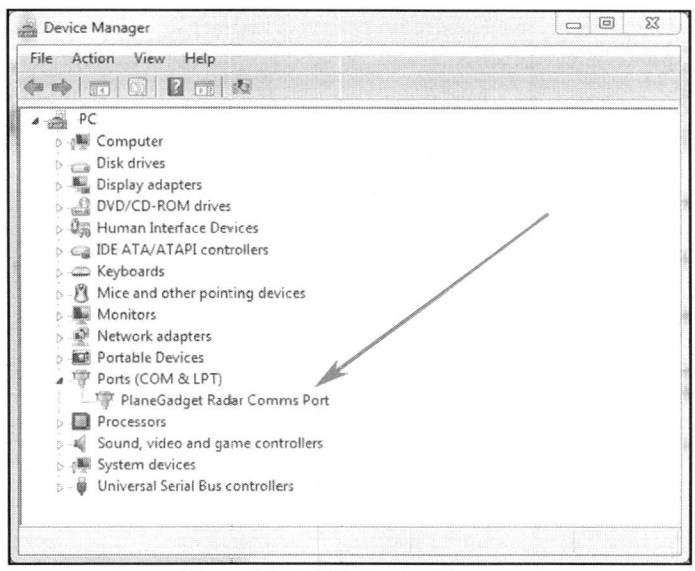

Figure C8 - 64 Bit Device Manager

Then select the CD drive as the source for the software driver which then allow the PlaneGadget Radar software to be properly installed for the 64 bit operating system.

Once the installation has completed make a note of the assigned Comm Port number as with a regular installation this will be needed when setting up PlanePlotter to use the PlaneGadget Radar receiver.

Appendix D
ACARS

The ACARS radio transmission is an amplitude-modulated (AM) signal to make it consistent with the historical use of AM voice mode on the aircraft bands in the early days of radio. The transmission is in the VHF band around 130 MHz and uses a 2400 bps NRZI-coded coherent audio frequency MSK (Minimum Shift Keying—a particular form of FSK) on AM to make use of standard aircraft AM communications equipment. A list of the frequencies commonly used for ACARS can be found in Appendix A.

The demodulated audio signal fits perfectly in the audio spectrum of an AM receiver (300-3000 Hz); thus, almost any make and model of air band receiver or scanner covering the VHF 130Mhz band is suitable for receiving ACARS transmissions.

In most cases the earphone output of your receiver can be used to connect the receiver to a standalone ACARS Decoder, to a PC, or to a Laptop Computer running one of the many suites of ACARS Decoder software.

It is essential that the air-band receiver or scanner should have a Squelch control such that the Squelch is turned off. It will also be necessary to adjust the volume control of the receiver so that the decoder is not overloaded. As a brief sync tone is sent before each transmission, many scanners' squelch will not open up fast enough, thus another reason to keep the squelch turned off. Suit some receivers and scanner are listed below.

N.B Squelch is normally used to remove the background noise and static when receiving voice transmissions. But as the ACARS signal is in effect a short burst of noise typically under one second in length, it necessary to be able to hear it.

Make	Model
Signal	R532/R53
Uniden	USC230, UBC-30XLT, USC230, UBC3500XLT
Yaesu	FT-817
Yupiteru	VT-125/150 , VT-225, MVT-3100, MVT-7100, MVT-7200

Table 1 - Typical Scanners used for ACARS

Appendix D

ACARS Message Frame Format

Each message frame consists of at least 50, and up to a maximum of 272 characters or bytes. Each character uses a 7-bit ACSII code with an additional eighth parity bit. This results in a total message transmission duration of between 0.17 and 0.91 seconds.

The message frame format is rigidly defined to include synchronization, address, acknowledgment, mode and error checking characters, in addition to the actual message text. Imbedded message-label characters indicate the type of message. The message format is shown in Table 2 below.

Parameter	Value	No of Characters
Pre-key	Tx warm-up/Rx AGC adjustment	16 characters
Bit Sync	Establish bit synchronisation	2 characters „+„ , „*„
Character Sync	Establish character synchronisation	2 characters SYN, SYN (16h)
Start of Heading	Start of Heading	1 character SOH (01h)
Mode	Ground system interface configuration	1 character
Address	Aircraft registration number	7 characters
Ack/Nak	Acknowledge/Non-Acknowledge flag	1 character
Label	Type of message	2 characters
Block Identifier	Message block number	1 character
STX	STX (02h) - if no text ETX (03h)	1 character
Flight No	Airline flight number	6 characters
Text	printable characters only	maximum 220 characters
Suffix	ETX or ETB (17h)	1 character
BCS	Block Check Sequence	16 bits
BCS Suffix	DLE (7fh)	1 character

Table 2 - ACARS Message Frame Format

The sixteen pre-key characters are all binary 1 values, thus resulting in the 0.05 second 2400 Hz beep that can be heard at the start of every message.

The Block Check Sequence field contains the value of an error-detection polynomial that can be used to determine if the entire message was received free of errors. Standard 7-bit ASCII is used, bit 8 is an odd parity bit used for the text field of the ACARS message, and the content may be free text or a mixture of formatted and free text.

ACARS communications are divided into Category A and Category B. An aircraft uses a Category A transmission to broadcast its messages to all listening ground stations. This is denoted by an ASCII *2* (Hex32) in the Mode field of the downlink message. When using Category B, an aircraft transmits its message to a single ground station. This is denoted by an ASCII character in the range @ to *]* (Hex40 - Hex5D) in the Mode field of the downlink message.

All ground stations support Category A, but may uplink ' to *}* (Hex27 - Hex7D) in the Mode field. The ground station may use either *2* or the range ' to *}* (Hex27 - Hex7D) in the Mode field.

At the receiving end, a Block Check calculation is made and compared to the calculation appended to the packet by the transmitting station. If the downlink messages contain errors, no response will be given and the transmitting station will retransmit the packet a number of times until a positive acknowledgement is received and the message can be deleted from storage, or the aircrew be alerted to its nontransmission.

If an uplink message is found in error, the airborne equipment will generate a negative acknowledgement (NAK) which triggers an uplink retransmission. Retransmission is also triggered by timeout.

Positive acknowledgement from the aircraft consists of the transmission of the Uplink Block Identifier of the correctly received block. Positive acknowledgement from the ground station consists of a similar transmission of the Downlink Block Identifier. Acknowledgements are placed in the Technical Acknowledgement field. The general-response-message label is **_DLE** (Hex5F Hex7F). Messages with this label contain no information except acknowledgements and are used for link maintenance.

A typical raw undecoded message might look like this:

```
□□□□□□□□□□□□□□□<SYN><SYN><SOH>2..KC693<NAK>Q05<STX>S68!2\42H;C<
FS<ACK><BEL>#########+*#<NAK>*T#^U*#*<DC2>*#*T#U+L#MjU#UM#)5R#+E
#T*J#5U#+jU#<SYN>*U#U##################################vw<S
OH>2.<SO>II0?(0_ □C<ETX>@<DLE>
```

Decoding ACARS Messages

The raw message shown at the end of the last chapter could be decoded and translated into plane English as follows, "KC693 operating as MC4208 was first contacted at 19:18 on 21/06/2010 using Ground Station 2 with broadcast message number S68A was conducting Q0 type link test."

Appendix D

Fortunately, users don't have to work through the process of translating such raw messages as there are a readily available number of decoders that will do the job for them.

First, though, it is necessary to look at the structure of decoded messages and discuss some of the standard-message formats.

ACARS Downlink Messages

Consider this decoded message:

```
(#5)15-05-2010 15:33:36 M=06 ADDR= F-ZKXF TA=Q ML=Q0 B=9 MSN=D09D FID=AF9876
```

which can be displayed more simply in a table like that below.

Decoded	Interpretation
(#5)	Decoder generated message number
15-05-2010 18:43:32	Decoder generated timestamp (optional)
M=06	Mode Category A = A,
ADDR=F-KZXF	Aircraft address
TA=Q	Technical acknowledgement
ML=Q0	Message Label (message type)
B=9	Uplink/Downlink Block Identifier
MSN=D09D	Message Sequence Number
FID=AF9876	Flight Identifier

Table 3 - Typical ACARS Decoded Message Format

In this case, record #5 decoded at 15:33:36 contains a message from a French aircraft with registration F-ZKXF using logical channel 06 to transmit and acknowledge uplink block Q and a link test (Q0) with block identifier 9 and message sequence number 0635 (here the time in minutes and seconds after the hour is used—other formats are also in use). The flight is Air France AF9876.

A method commonly used by many software decoders for displaying the decoded details shown below.

ACARS mode: 06 Aircraft reg: F-ZKXF
Message label: Q0 Block id: 9 Msg. no: D09D
Flight id: AF9876

Some more examples of the more important or frequently seen ACARS messages:

```
M=06 ADDR= F-KZXF TA=NAK ML=_? B=3 MSN=2810 FID=AF9876
```

Decoded	Interpretation
(#7)	Decoder generated message number
16-05-2010 09:33:12	Decoder generated timestamp (optional)
M=06	Mode Category A = A,
ADDR=F-KZXF	Aircraft address
TA=NAK	Technical acknowledgement
ML=_?	Message Label (message type)
B=3	Uplink/Downlink Block Identifier
MSN=2810	Message Sequence Number
FID=AF9876	Flight Identifier

Using logical channel 06 an unsolicited (TA=NAK) general response _? without information is transmitted as block 3 from aircraft F-KZXF on flight AF9876 with sequence number 2810. General responses are mainly used for block acknowledgement purposes.

```
M=06 ADDR= ?????? TA=NAK ML=SQ B= 00XSSRH
```

Decoded	Interpretation
(#8)	Decoder generated message number
16-05-2010 15:13:52	Decoder generated timestamp (optional)
M=06	Mode Category A = A,
ADDR=??????	Aircraft address
TA=NAK	Technical acknowledgement
ML=SQ	Message Label (message type)

This message is called a "squitter." That is an ID and Uplink test message that is transmitted at regular intervals from ground stations. This particular one is a squitter (SQ) version 0 (00) from a SITA (XS) ground station in Singapore (SRH). The ? denotes the ASCII NUL character (00h) used for broadcast. A block identifier is not used.

Although users can enjoy the fun and challenge of decoding ACARS messages the PlanePlotter program does it for them.

Appendix E

ICAO / IATA Airport Codes

ICAO	IATA	Airport & Country	ICAO	IATA	Airport & Country
AGGH	HIR	Honiara Intl, Solomon Islands	CCA6	YWM	Williams Harbour, Canada
AGGM	MUA	Munda, Solomon Islands	CCB4	YDI	Davis Inlet, Canada
ANAU	INU	Nauru Intl, Nauru	CCD4	YSO	Postville, Canada
AYGA	GKA	Goroka, Papua New Guinea	CCE4	YBI	Black Tickle, Canada
AYLA	LAE	Nadzab, Papua New Guinea	CCH4	YHG	Charlottetown, Canada
AYMD	MAG	Madang, Papua New Guinea	CCK4	YFX	Fox Harbour, Canada
AYMH	HGU	Mount Hagen, Papua New Guinea	CCP4	YHA	Port Hope Simpson, Canada
AYPY	POM	Port Moresby, Papua New Guinea	CCZ2	YRG	Rigolet, Canada
AYRB	RAB	Rabaul, Papua New Guinea	CEB3	YCK	Colville Lake, Canada
AYWK	WWK	Wewak Intl, Papua New Guinea	CEB5	ZFW	Fairview, Canada
BGAM	AGM	Angmagssalik Airport, Greenland	CEC4	YJA	Jasper, Canada
BGBW	UAK	Narssarssuaq Airport, Greenland	CEM3	YLE	Wha Ti, Canada
BGCH	JCH	Christianshab Airport, Greenland	CEQ5	YGC	Grande Cache, Canada
BGCO	CNP	Constable Pynt Airport, Greenland	CER3	YDC	Industrial, Canada
BGDU	DUN	Dundas Airport, Greenland	CEV4	YRA	Rae Lakes, Canada
BGEM	JEG	Auisiait Airport, Greenland	CFC4	XMP	Macmillan Pass, Canada
BGFH	JFR	Frederikshab Airport, Greenland	CFJ2	YFJ	Wekweti, Canada
BGGH	GOH	Nuuk Airport, Greenland	CJL2	YDJ	Hatchet Lake, Canada
BGGN	JGO	Godhavn Airport, Greenland	CJL8	YDU	Kasba Lake, Canada
BGHB	JHS	Holsteinsborg Airport, Greenland	CJS3	XCL	Cluff Lake, Canada
BGJN	JAV	Jakobshavn Airport, Greenland	CJV7	SUR	Summer Beaver, Canada
BGKA	QPW	Kangaatsiaq Airport, Greenland	CJY3	YTT	Tisdale, Canada
BGKK	KUS	Kulusuk Airport, Greenland	CKB6	YAX	Wapakeka, Canada
BGNN	JNN	Nanortalik Airport, Greenland	CKL3	WNN	Wunnumin Lake, Canada
BGNS	JNS	Narssaq Airport, Greenland	CNE3	XBE	Bearskin Lake, Canada
BGQA	NAQ	Qaanaaq Airport, Greenland	CNK4	YPD	Georgian Bay, Canada
BGSC	OBY	Scoresbysund Airport, Greenland	CNL3	XBR	Brockville, Canada
BGSF	SFJ	Kangerlussuaq Airport, Greenland	CNM5	KIF	Kingfisher Lake, Canada
BGTL	THU	Thule Air Base, Greenland	CNS7	YKD	Kincardine, Canada
BGUM	UMD	Uummannaq Airport, Greenland	CNT3	YOG	Ogoki Post, Canada
BGUP	JUV	Upernavik Airport, Greenland	CNZ3	XCM	Chatham, Canada
BIAR	AEY	Akureyri Airport, Iceland	CPV8	KEW	Keewaywin, Canada
BIBL	BLO	Blonduos Airport, Iceland	CSH4	YLS	Lebel, Canada
BIEG	EGS	Egilsstaoir Airport, Iceland	CSR8	SSQ	La Sarre, Canada
BIHN	HFN	Hornafjorour Airport, Iceland	CYAC	YAC	Cat Lake, Canada
BIIS	IFJ	Isafjorour Airport, Iceland	CYAG	YAG	Ft Frances Mun, Canada
BIKF	KEF	Keflavik Intl, Iceland	CYAL	YAL	Alert Bay, Canada
BIKR	SAK	Alexander Airport, Iceland	CYAM	YAM	Sault Ste Marie, Canada
BINF	NOR	Norofjorour Airport, Iceland	CYAQ	XKS	Kasabonika, Canada
BIPA	PFJ	Patreksfjorour Airport, Iceland	CYAS	YKG	Kangirsuk, Canada
BIRG	RFN	Raufarhofn Airport, Iceland	CYAT	YAT	Attawapiskat, Canada
BIRK	RKV	Reykjavik Airport, Iceland	CYAW	YAW	Halifax, Canada
BISI	SIJ	Siglufjordur Airport, Iceland	CYAY	YAY	St Anthony, Canada
BIST	SYK	Stykkisholmur Airport, Iceland	CYAZ	YAZ	Tofino, Canada
BITE	TEY	Pingeyri Airport, Iceland	CYBB	YBB	Pelly Bay, Canada
BITH	THO	Sauoanes Airport, Iceland	CYBC	YBC	Baie, Canada
BIVM	VEY	Vestmannaeyjar Airport, Iceland	CYBD	QBC	Bella Coola, Canada
CAB5	YBM	Bronson Creek, Canada	CYBE	YBE	Uranium City, Canada
CAD5	YMB	Merritt, Canada	CYBF	YBY	Bonnyville, Canada
CAH3	YCA	Courtenay, Canada	CYBG	YBG	Bagotville, Canada
CAJ4	YAA	Anahim Lake, Canada	CYBK	YBK	Baker Lake, Canada
CAM3	DUQ	Duncan, Canada	CYBL	YBL	Campbell River, Canada
CAP3	YHS	Sechelt, Canada	CYBQ	XTL	Tadoule Lake, Canada
CAT5	YMP	Port Mc Neill, Canada	CYBR	YBR	Brandon Muni, Canada
CBM5	YTX	Telegraph Creek, Canada	CYBT	YBT	Brochet, Canada
CBX5	TNS	Cantung, Canada	CYBV	YBV	Berens River, Canada
CBX7	TUX	Tumbler Ridge, Canada	CYCA	YRF	Cartwright Nws, Canada

ICAO	IATA	Airport & Country	ICAO	IATA	Airport & Country
CYCB	YCB	Cambridge Bay, Canada	CYHE	YHE	Hope, Canada
CYCC	YCC	Cornwall Regl, Canada	CYHF	YHF	Hearst Mun, Canada
CYCD	YCD	Nanaimo, Canada	CYHH	YNS	Nemiscau, Canada
CYCE	YCE	Huron, Canada	CYHI	YHI	Holman, Canada
CYCG	YCG	Castlegar, Canada	CYHK	YHK	Gjoa Haven, Canada
CYCH	YCH	Miramichi, Canada	CYHM	YHM	Hamilton, Canada
CYCL	YCL	Charlo, Canada	CYHN	YHN	Hornepayne Mun, Canada
CYCN	YCN	Cochrane, Canada	CYHO	YHO	Hopedale, Canada
CYCO	YCO	Kugluktuk, Canada	CYHR	YHR	Chevery, Canada
CYCQ	YCQ	Chetwynd, Canada	CYHT	YHT	Haines Junction, Canada
CYCR	YCR	Sinclair Meml, Canada	CYHU	YHU	Montreal, Canada
CYCS	YCS	Chesterfield Inlet, Canada	CYHY	YHY	Hay River, Canada
CYCT	YCT	Coronation, Canada	CYHZ	YHZ	Halifax Intl, Canada
CYCW	YCW	Chilliwack, Canada	CYIB	YIB	Atikokan Mun, Canada
CYCY	YCY	Clyde River, Canada	CYID	YDG	Digby, Canada
CYCZ	YCZ	Fairmont Hot Springs, Canada	CYIF	YIF	St, Canada
CYDA	YDA	Dawson, Canada	CYIK	YIK	Ivujivik, Canada
CYDB	YDB	Burwash, Canada	CYIO	YIO	Pond Inlet, Canada
CYDF	YDF	Deer Lake, Canada	CYIV	YIV	Island Lake, Canada
CYDL	YDL	Dease Lake, Canada	CYJF	YJF	Ft Liard, Canada
CYDM	XRR	Ross River, Canada	CYJN	YJN	St, Canada
CYDN	YDN	Barker, Canada	CYJQ	ZEL	Bella Bella, Canada
CYDO	YDO	Dolbeau, Canada	CYJT	YJT	Stephenville, Canada
CYDP	YDP	Nain, Canada	CYKA	YKA	Kamloops, Canada
CYDQ	YDQ	Dawson Creek, Canada	CYKC	YKC	Collins Bay, Canada
CYEG	YEG	Edmonton Intl, Canada	CYKD	LAK	Aklavik, Canada
CYEK	YEK	Arviat, Canada	CYKF	YKF	Kitchener/Waterloo Regl, Canada
CYEL	YEL	Elliot Lake Mun, Canada	CYKG	YWB	Kangiqsujuaq Wakeham Bay, Canada
CYEM	YEM	Manitoulin East Muni., Canada	CYKJ	YKJ	Key Lake, Canada
CYEN	YEN	Estevan, Canada	CYKL	YKL	Schefferville, Canada
CYER	YER	Ft Severn, Canada	CYKO	AKV	Akulivik, Canada
CYET	YET	Edson, Canada	CYKQ	YKQ	Waskaganish, Canada
CYEU	YEU	Eureka, Canada	CYKX	YKX	Kirkland Lake, Canada
CYEV	YEV	Zubko, Canada	CYKY	YKY	Kindersley, Canada
CYEY	YEY	Amos Mun, Canada	CYKZ	YKZ	Toronto, Canada
CYFA	YFA	Ft Albany, Canada	CYLA	YPJ	Aupaluk, Canada
CYFB	YFB	Iqaluit, Canada	CYLB	YLB	Lac La Biche, Canada
CYFC	YFC	Fredericton, Canada	CYLD	YLD	Chapleau, Canada
CYFH	YFH	Ft Hope, Canada	CYLH	YLH	Lansdowne House, Canada
CYFO	YFO	Flin Flon, Canada	CYLJ	YLJ	Meadow Lake, Canada
CYFR	YFR	Ft Resolution, Canada	CYLL	YLL	Lloydminster, Canada
CYFS	YFS	Ft Simpson, Canada	CYLQ	YLQ	La Tuque, Canada
CYFT	YMN	Makkovik, Canada	CYLR	YLR	Leaf Rapids, Canada
CYGB	YGB	Gillies Bay, Canada	CYLT	YLT	Alert, Canada
CYGH	YGH	Ft Good Hope, Canada	CYLU	XGR	Kangiqsualujjuaq, Canada
CYGK	YGK	Kingston, Canada	CYLW	YLW	Kelowna, Canada
CYGL	YGL	La Grande Riviere, Canada	CYMA	YMA	Mayo, Canada
CYGM	YGM	Gimli Industrial Park, Canada	CYME	YME	Matane, Canada
CYGO	YGO	Gods Lake Narrows, Canada	CYMG	YMG	Manitouwadge, Canada
CYGP	YGP	Gaspe, Canada	CYMH	YMH	Marys Harbour, Canada
CYGQ	YGQ	Greenstone Regl, Canada	CYMJ	YMJ	Moose Jaw, Canada
CYGR	YGR	Iles, Canada	CYML	YML	Charlevoix, Canada
CYGT	YGT	Igloolik, Canada	CYMM	YMM	Ft Mc Murray, Canada
CYGV	YGV	Havre St, Canada	CYMO	YMO	Moosonee, Canada
CYGW	YGW	Kuujjuarapik, Canada	CYMT	YMT	Chapais, Canada
CYGX	YGX	Gillam, Canada	CYMU	YUD	Umiujaq, Canada
CYGZ	YGZ	Grise Fiord, Canada	CYMW	YMW	Maniwaki, Canada
CYHA	YQC	Quaqtaq, Canada	CYMX	YMX	Montreal Intl, Canada
CYHB	YHB	Hudson Bay, Canada	CYNA	YNA	Natashquan, Canada
CYHD	YHD	Dryden Regl, Canada	CYNC	YNC	Wemindji, Canada

Appendix E

ICAO	IATA	Airport & Country	ICAO	IATA	Airport & Country
CYND	YND	Ottawa/Gatineau, Canada	CYSB	YSB	Sudbury, Canada
CYNE	YNE	Norway House, Canada	CYSC	YSC	Sherbrooke, Canada
CYNH	YNH	Hudson's Hope, Canada	CYSD	YSD	Suffield, Canada
CYNL	YNL	Points North Landing, Canada	CYSE	YSE	Squamish, Canada
CYNM	YNM	Matagami, Canada	CYSF	YSF	Stony Rapids, Canada
CYOC	YOC	Old Crow, Canada	CYSH	YSH	Montague, Canada
CYOD	YOD	Cold Lake, Canada	CYSJ	YSJ	Saint John, Canada
CYOH	YOH	Oxford House, Canada	CYSK	YSK	Sanikiluaq, Canada
CYOJ	YOJ	High Level, Canada	CYSL	YSL	St, Canada
CYOO	YOO	Oshawa, Canada	CYSM	YSM	Ft Smith, Canada
CYOP	YOP	Rainbow Lake, Canada	CYSN	YCM	St Catharines, Canada
CYOS	YOS	Billy Bishop Reg'l, Canada	CYSP	YSP	Marathon, Canada
CYOW	YOW	Ottawa, Canada	CYSR	YSR	Nanisivik, Canada
CYPA	YPA	Prince Albert, Canada	CYST	YST	St Theresa Point, Canada
CYPC	YPC	Paulatuk, Canada	CYSU	YSU	Summerside, Canada
CYPD	YPS	Port Hawkesbury, Canada	CYSY	YSY	Sachs Harbour, Canada
CYPE	YPE	Peace River, Canada	CYTA	YTA	Pembroke, Canada
CYPG	YPG	Southport, Canada	CYTE	YTE	Cape Dorset, Canada
CYPH	YPH	Inukjuak, Canada	CYTF	YTF	Alma, Canada
CYPL	YPL	Pickle Lake, Canada	CYTH	YTH	Thompson, Canada
CYPM	YPM	Pikangikum, Canada	CYTJ	YTJ	Terrace Bay, Canada
CYPN	YPN	Port Menier, Canada	CYTL	YTL	Big Trout Lake, Canada
CYPO	YPO	Peawanuck, Canada	CYTQ	YTQ	Tasiujaq, Canada
CYPQ	YPQ	Peterborough, Canada	CYTR	YTR	Trenton, Canada
CYPR	YPR	Prince Rupert, Canada	CYTS	YTS	Timmins, Canada
CYPW	YPW	Powell River, Canada	CYTZ	YTZ	Toronto, Canada
CYPX	YPX	Puvirnituq, Canada	CYUB	YUB	Tuktoyaktuk, Canada
CYPY	YPY	Ft Chipewyan, Canada	CYUL	YUL	Montreal Intl, Canada
CYPZ	YPZ	Burns Lake, Canada	CYUT	YUT	Repulse Bay, Canada
CYQA	YQA	Muskoka, Canada	CYUX	YUX	Hall Beach, Canada
CYQB	YQB	Quebec, Canada	CYUY	YUY	Rouyn, Canada
CYQD	YQD	The Pas, Canada	CYVB	YVB	Bonaventure, Canada
CYQF	YQF	Red Deer Industrial, Canada	CYVC	YVC	La Ronge, Canada
CYQG	YQG	Windsor, Canada	CYVG	YVG	Vermilion, Canada
CYQH	YQH	Watson Lake, Canada	CYVK	YVE	Vernon, Canada
CYQI	YQI	Yarmouth, Canada	CYVM	YVM	Qikiqtarjuaq, Canada
CYQK	YQK	Kenora, Canada	CYVO	YVO	Val, Canada
CYQL	YQL	Lethbridge, Canada	CYVP	YVP	Kuujjuaq, Canada
CYQM	YQM	Moncton, Canada	CYVQ	YVQ	Norman Wells, Canada
CYQN	YQN	Nakina, Canada	CYVR	YVR	Vancouver Intl, Canada
CYQQ	YQQ	Comox, Canada	CYVT	YVT	Buffalo Narrows, Canada
CYQR	YQR	Regina, Canada	CYVV	YVV	Wiarton, Canada
CYQS	YQS	St Thomas Mun, Canada	CYVZ	YVZ	Deer Lake, Canada
CYQT	YQT	Thunder Bay, Canada	CYWA	YWA	Petawawa, Canada
CYQU	YQU	Grande Prairie, Canada	CYWG	YWG	Winnipeg Intl, Canada
CYQV	YQV	Yorkton, Canada	CYWJ	YWJ	Deline, Canada
CYQW	YQW	North Battleford, Canada	CYWK	YWK	Wabush, Canada
CYQX	YQX	Gander Intl, Canada	CYWL	YWL	Williams Lake, Canada
CYQY	YQY	Sydney, Canada	CYWP	YWP	Webequie, Canada
CYQZ	YQZ	Quesnel, Canada	CYWY	YWY	Wrigley, Canada
CYRB	YRB	Resolute Bay, Canada	CYXC	YXC	Cranbrook, Canada
CYRI	YRI	Riviere, Canada	CYXD	YXD	Edmonton City Center, Canada
CYRJ	YRJ	Roberval, Canada	CYXE	YXE	Saskatoon/Diefenbaker, Canada
CYRL	YRL	Red Lake, Canada	CYXH	YXH	Medicine Hat, Canada
CYRM	YRM	Rocky Mountain House, Canada	CYXJ	YXJ	Ft St John, Canada
CYRO	YRO	Rockcliffe, Canada	CYXK	YXK	Rimouski, Canada
CYRQ	YRQ	Trois, Canada	CYXL	YXL	Sioux Lookout, Canada
CYRS	YRS	Red Sucker Lake, Canada	CYXN	YXN	Whale Cove, Canada
CYRT	YRT	Rankin Inlet, Canada	CYXP	YXP	Pangnirtung, Canada
CYRV	YRV	Revelstoke, Canada	CYXQ	YXQ	Beaver Creek, Canada

ICAO	IATA	Airport & Country	ICAO	IATA	Airport & Country
CYXR	YXR	Earlton, Canada	CZMN	PIW	Pikwitonei, Canada
CYXS	YXS	Prince George, Canada	CZMT	ZMT	Masset, Canada
CYXT	YXT	Terrace, Canada	CZNG	XPP	Poplar River, Canada
CYXU	YXU	London, Canada	CZPB	ZPB	Sachigo Lake, Canada
CYXX	YXX	Abbotsford, Canada	CZPC	WPC	Pincher Creek, Canada
CYXY	YXY	Whitehorse Intl, Canada	CZPO	ZPO	Pinehouse Lake, Canada
CYXZ	YXZ	Wawa, Canada	CZRJ	ZRJ	Weagamow Lake, Canada
CYYB	YYB	North Bay, Canada	CZSJ	ZSJ	Sandy Lake, Canada
CYYC	YYC	Calgary Intl, Canada	CZSN	XSI	South Indian Lake, Canada
CYYD	YYD	Smithers, Canada	CZST	ZST	Stewart, Canada
CYYE	YYE	Ft Nelson, Canada	CZTA	YDV	Bloodvein River, Canada
CYYF	YYF	Penticton, Canada	CZTM	ZTM	Shamattawa, Canada
CYYG	YYG	Charlottetown, Canada	CZUC	ZUC	Ignace Mun, Canada
CYYH	YYH	Taloyoak, Canada	CZUM	ZUM	Churchill Falls, Canada
CYYJ	YYJ	Victoria Intl, Canada	CZWH	XLB	Lac Brochet, Canada
CYYL	YYL	Lynn Lake, Canada	CZWL	ZWL	Wollaston Lake, Canada
CYYM	YYM	Cowley, Canada	DAAD	BUJ	Bou Saada Airport, Algeria
CYYN	YYN	Swift Current, Canada	DAAE	BJA	Bejaia, Algeria
CYYQ	YYQ	Churchill, Canada	DAAG	ALG	Algiers, Algeria
CYYR	YYR	Goose, Canada	DAAJ	DJG	Djanet, Algeria
CYYT	YYT	St John's Intl, Canada	DAAT	TMR	Tamanrasset, Algeria
CYYU	YYU	Kapuskasing, Canada	DAAV	GJL	Jijel, Algeria
CYYW	YYW	Armstrong, Canada	DABB	AAE	Annaba, Algeria
CYYY	YYY	Mont, Canada	DABC	CZL	Constantine, Algeria
CYYZ	YYZ	Toronto, Canada	DABS	TEE	Tebessa, Algeria
CYZE	YZE	Gore Bay, Canada	DAOB	TID	Tiaret, Algeria
CYZF	YZF	Yellowknife, Canada	DAON	TLM	Messali El Hadj Airport, Algeria
CYZG	YZG	Salluit, Canada	DAOO	ORN	Ouran, Algeria
CYZH	YZH	Slave Lake, Canada	DATG	INF	In Guezzam Airport, Algeria
CYZP	YZP	Sandspit, Canada	DATM	BMW	Bordj Mokhtar Airport, Algeria
CYZR	YZR	Sarnia, Canada	DAUA	AZR	Adrar, Algeria
CYZS	YZS	Coral Harbour, Canada	DAUE	ELG	El Golea Airport, Algeria
CYZT	YZT	Port Hardy, Canada	DAUG	GHA	Ghardaia, Algeria
CYZU	YZU	Whitecourt, Canada	DAUH	HME	Hassi Messaoud, Algeria
CYZV	YZV	Sept, Canada	DAUI	INZ	In Salah Airport, Algeria
CYZW	YZW	Teslin, Canada	DAUK	TGR	Touggourt, Algeria
CYZX	YZX	Greenwood, Canada	DAUO	ELU	El Oued, Algeria
CZAC	ZAC	York Landing, Canada	DAUT	TMX	Timimoun Airport, Algeria
CZAM	YSN	Salmon Arm, Canada	DAUU	OGX	Ouargla Airport, Algeria
CZBB	YDT	Boundary Bay, Canada	DAUZ	IAM	Zarzaitine, Algeria
CZBD	ILF	Ilford, Canada	DBBB	COO	Cadjehoun, Benin
CZBF	ZBF	Bathurst, Canada	DBBK	KDC	Kandi, Benin
CZBM	ZBM	Bromont, Canada	DBBN	NAE	Natitingou, Benin
CZEE	KES	Kelsey, Canada	DBBP	PKO	Parakou, Benin
CZEM	ZEM	Eastmain River, Canada	DFCC	OUG	Ouahigouya, Burkina Faso
CZFA	ZFA	Faro, Canada	DFEE	DOR	Dori Ville, Burkina Faso
CZFD	ZFD	Fond, Canada	DFEF	FNG	Fada N'Gourma, Burkina Faso
CZFG	XPK	Pukatawagan, Canada	DFEG	XGG	Gorom, Burkina Faso
CZFM	ZFM	Ft Mc Pherson, Canada	DFEL	XKA	Kantchari, Burkina Faso
CZFN	ZFN	Tulita, Canada	DFET	TEG	Tenkodogo, Burkina Faso
CZGF	ZGF	Grand Forks, Canada	DFFD	OUA	Ouagadougou, Burkina Faso
CZGI	ZGI	Gods River, Canada	DFOB	BNR	Banfora, Burkina Faso
CZGR	ZGR	Little Grand Rapids, Canada	DFOD	DGU	Dedougou, Burkina Faso
CZHP	ZHP	High Prairie, Canada	DFOG	XGA	Amilcar Cabral, Burkina Faso
CZJG	ZJG	Jenpeg, Canada	DFON	XNU	Nouna, Burkina Faso
CZJN	ZJN	Swan River, Canada	DFOO	BOY	Bobo, Burkina Faso
CZKE	ZKE	Kashechewan, Canada	DGAA	ACC	Kotoka Intl, Ghana
CZLQ	YTD	Thicket Portage, Canada	DGLE	TML	Tamale, Ghana
CZMD	MSA	Muskrat Dam, Canada	DGSI	KMS	Kumasi, Ghana
CZML	ZMH	108 Mile, Canada	DGSN	NYI	Sunyani, Ghana

Appendix E

ICAO	IATA	Airport & Country	ICAO	IATA	Airport & Country
DGTK	TKD	Takoradi, Ghana	EDAH	HDF	Heringsdorf, Germany
DIAP	ABJ	Felix Houphouet, Ivory Coast	EDAH	HDF	Heringsdorf, Germany
DIAU	OGO	Abengourou, Ivory Coast	EDBH	BBH	Ostseeflughafen Stralsund, Germany
DIBI	BXI	Boundiali, Ivory Coast	EDBM	ZMG	Magdeburg, Germany
DIBK	BYK	Bouake, Ivory Coast	EDDB	SXF	Berlin Schonefeld, Germany
DIBN	BQO	Tehini, Ivory Coast	EDDC	DRS	Flughafen Dresden, Germany
DIBU	BDK	Soko, Ivory Coast	EDDE	ERF	Erfurt, Germany
DIDK	DIM	Dimbokro Ville, Ivory Coast	EDDF	FRA	Frankfurt am Main, Germany
DIDL	DJO	Daloa, Ivory Coast	EDDG	FMO	Munster, Germany
DIFK	FEK	Ferkessedougou, Ivory Coast	EDDH	HAM	Hamburg, Germany
DIGL	GGO	Guiglo, Ivory Coast	EDDI	THF	Berlin Tempelhof, Germany
DIGN	BBV	Nero, Ivory Coast	EDDK	CGN	Koln – Bonn, Germany
DIKO	HGO	Korhogo, Ivory Coast	EDDL	DUS	Dusseldorf International, Germany
DIMN	MJC	Man, Ivory Coast	EDDM	MUC	Munchen, Germany
DIOD	KEO	Odienne, Ivory Coast	EDDN	NUE	Nurnberg, Germany
DIOF	OFI	Ouango Fitini, Ivory Coast	EDDP	LEJ	Leipzig, Germany
DISP	SPY	San Pedro, Ivory Coast	EDDR	SCN	Saarbrucken, Germany
DISS	ZSS	Sassandra, Ivory Coast	EDDS	STR	Stuttgart, Germany
DITB	TXU	Tabou, Ivory Coast	EDDT	TXL	Berlin Tegel, Germany
DITM	TOZ	Mahana, Ivory Coast	EDDV	HAJ	Hannover, Germany
DIYO	ASK	Yamoussoukro, Ivory Coast	EDDW	BRE	Airport Bremen, Germany
DNAA	ABV	Nnamdi Azikiwe, Nigeria	EDFH	HHN	Frankfurt, Germany
DNAK	AKR	Akure, Nigeria	EDFM	MHG	Mannheim, Germany
DNBE	BNI	Benin, Nigeria	EDGS	SGE	Siegerlandflughafen, Germany
DNCA	CBQ	Calabar, Nigeria	EDHI	XFW	Finkenwerder, Germany
DNEN	ENU	Enugu, Nigeria	EDHK	KEL	Kiel, Germany
DNIB	IBA	New Ibadan, Nigeria	EDHL	LBC	Lubeck, Germany
DNIL	ILR	Ilorin, Nigeria	EDJA	FMM	Allgau Airport / Memmingen, Germany
DNJO	JOS	Jos, Nigeria	EDLE	ESS	Essen – Mulheim, Germany
DNKA	KAD	New Kaduna, Nigeria	EDLI	BFE	Bielefeld, Germany
DNKN	KAN	Mallam Aminu Kano, Nigeria	EDLN	MGL	Monchengladbach, Germany
DNMA	MIU	Maiduguri, Nigeria	EDLP	PAD	Paderborn, Germany
DNMM	LOS	Murtala Muhammed, Nigeria	EDLV	NRN	Airport Weeze, Germany
DNMN	MXJ	Minna, Nigeria	EDLW	DTM	Dortmund Airport 21, Germany
DNPO	PHC	Port Harcourt, Nigeria	EDMA	AGB	Augsburg, Germany
DNSO	SKO	Siddiq Abubakar Iii Intl, Nigeria	EDMO	OBI	Oberpfaffenhofen, Germany
DNYO	YOL	Yola, Nigeria	EDNY	FDH	Friedrichshafen, Germany
DNZA	ZAR	Zaria, Nigeria	EDOP	SZW	Schwerin – Parchim, Germany
DRRM	MFQ	Maradi, Niger	EDQD	BYU	Bayreuth, Germany
DRRN	NIM	Diori Hamani, Niger	EDQM	HOQ	Hof, Germany
DRRT	THZ	Tahoua, Niger	EDSB	FKB	Karlsruhe/Baden, Germany
DRZA	AJY	Manu Dayak, Niger	EDTL	LHA	Lahr, Germany
DRZR	ZND	Zinder, Niger	EDVE	BWE	Flughafen Braunschweig, Germany
DTKA	TBJ	Tabarka 7 Novembre, Tunesia	EDVK	KSF	Kassel, Germany
DTMB	MIR	Habib Bourguiba, Tunesia	EDWB	BRV	Bremerhaven, Germany
DTTA	TUN	Carthage, Tunesia	EDWE	EME	Emden, Germany
DTTF	GAF	Ksar, Tunesia	EDWG	AGE	Wangerooge, Germany
DTTJ	DJE	Zarzis, Tunesia	EDWI	WVN	Wilhelmshaven, Germany
DTTR	EBM	El Borma, Tunesia	EDWJ	JUI	Juist, Germany
DTTX	SFA	Thyna, Tunesia	EDWR	BMK	Borkum, Germany
DTTZ	TOE	Nefta, Tunesia	EDWY	NRD	Norderney, Germany
DXNG	LRL	Niamtougou, Togo	EDXB	HEI	Heide, Germany
DXXX	LFW	Tokoin, Togo	EDXH	HGL	Helgoland, Germany
EBAW	ANR	Deurne Airport, Belgium	EDXW	GWT	Sylt, Germany
EBBR	BRU	Brussels National Airport, Belgium	EDXY	OHR	Wyk, Germany
EBCI	CRL	Brussels South Charleroi, Belgium	EEKA	KDL	Kardla Airport, Estonia
EBKT	KJK	Kortrijk, Belgium	EEKE	URE	Kuressaare Airport, Estonia
EBLG	LGG	Liege Airport, Belgium	EEPU	EPU	Parnu Airport, Estonia
EBOS	OST	Ostend, Belgium	EETN	TLL	Tallinn Airport, Estonia
EDAC	AOC	Altenburg, Germany	EETU	TAY	Tartu Airport, Estonia

PlanePlotter User Guide

ICAO	IATA	Airport & Country
EFET	ENF	Enontekio Airport, Finland
EFHF	HEM	Helsinki, Finland
EFHK	HEL	Helsinki, Finland
EFHV	HYV	Hyvinkaa Airport, Finland
EFIL	SJY	Seinajoki Airport, Finland
EFIT	KTQ	Kitee Airport, Finland
EFIV	IVL	Ivalo Airport, Finland
EFJO	JOE	Joensuu Airport, Finland
EFJY	JYV	Jyvaskyla Airport, Finland
EFKA	KAU	Kauhava Airport, Finland
EFKE	KEM	Kemi, Finland
EFKI	KAJ	Kajaani Airport, Finland
EFKJ	KHJ	Kauhajoki Airport, Finland
EFKK	KOK	Kruunupyy Airport, Finland
EFKS	KAO	Kuusamo Airport, Finland
EFKT	KTT	Kittila Airport, Finland
EFKU	KUO	Kuopio Airport, Finland
EFLP	LPP	Lappeenranta Airport, Finland
EFMA	MHQ	Mariehamn Airport, Finland
EFMI	MIK	Mikkeli Airport, Finland
EFOU	OUL	Oulu Airport, Finland
EFPO	POR	Pori Airport, Finland
EFRO	RVN	Rovaniemi Airport, Finland
EFSA	SVL	Savonlinna Airport, Finland
EFSO	SOT	Sodankyla Airport, Finland
EFTP	TMP	Tampere, Finland
EFTU	TKU	Turku Airport, Finland
EFVA	VAA	Vaasa Airport, Finland
EFVR	VRK	Varkaus Airport, Finland
EFYL	YLI	Ylivieska Airport, Finland
EGAA	BFS	Belfast Intl, United Kingdom
EGAB	ENK	Enniskillen, United Kingdom
EGAC	BHD	Belfast City Airport, United Kingdom
EGAE	LDY	City of Derry Airport, United Kingdom
EGBB	BHX	Birmingham Intl, United Kingdom
EGBE	CVT	Coventry Airport, United Kingdom
EGBJ	GLO	Gloucester Airport, United Kingdom
EGBK	ORM	Sywell Aerodrome, United Kingdom
EGBN	NQT	Nottingham Airport, United Kingdom
EGCC	MAN	Manchester Intl, United Kingdom
EGCN	DSA	Doncaster Sheffield, United Kingdom
EGDJ	UPV	Upavon Airfield, United Kingdom
EGDQ	NQY	Newquay Cornwall, United Kingdom
EGEC	CAL	Campbeltown Airport, United Kingdom
EGEH	WHS	Whalsay Airport, United Kingdom
EGEN	NRL	North Ronaldsay, Isle of Man
EGEP	PPW	Papa Westray Airport, Orkney isles
EGET	LWK	Tingwall Airport, United Kingdom
EGEW	WRY	Westray Airport, United Kingdom
EGFE	HAW	Haverfordwest, United Kingdom
EGFF	CWL	Cardiff Intl, United Kingdom
EGFH	SWS	Swansea Airport, United Kingdom
EGGD	BRS	Bristol Intl, United Kingdom
EGGP	LPL	Liverpool, United Kingdom
EGGW	LTN	London Luton Airport, United Kingdom
EGHC	LEQ	Land's End Airport, United Kingdom
EGHD	PLH	Plymouth City Airport, United Kingdom
EGHE	ISC	St. Mary's Airport, United Kingdom
EGHH	BOH	Bournemouth Airport, United Kingdom
EGHI	SOU	Southampton Airport, United Kingdom

ICAO	IATA	Airport & Country
EGHJ	BBP	Bembridge Airport, United Kingdom
EGHL	QLA	Lasham Airfield, United Kingdom
EGJA	ACI	Alderney Airport, United Kingdom
EGJB	GCI	Guernsey Airport, United Kingdom
EGJJ	JER	Jersey Airport, United Kingdom
EGKA	ESH	Shoreham Airport, United Kingdom
EGKB	BQH	Biggin Hill Airport, United Kingdom
EGKK	LGW	London Gatwick, United Kingdom
EGKR	KRH	Redhill Airport, United Kingdom
EGLC	LCY	London City Airport, United Kingdom
EGLF	FAB	Farnborough Airfield, United Kingdom
EGLK	BBS	Blackbushe Airport, United Kingdom
EGLL	LHR	London Heathrow, United Kingdom
EGMC	SEN	Southend Airport, United Kingdom
EGMD	LYX	Ashford Airport, United Kingdom
EGMH	MSE	Kent Intl, United Kingdom
EGNC	CAX	Carlisle Airport, United Kingdom
EGNH	BLK	Blackpool Intl, United Kingdom
EGNJ	HUY	Humberside Airport, United Kingdom
EGNL	BWF	Barrow/, United Kingdom
EGNM	LBA	Leeds Bradford Intl, United Kingdom
EGNR	CEG	Hawarden Airport, United Kingdom
EGNS	IOM	Isle of Man Airport, United Kingdom
EGNT	NCL	Newcastle Airport, United Kingdom
EGNV	MME	Durham Tees Valley, United Kingdom
EGNX	EMA	East Midlands Airport, United Kingdom
EGPA	KOI	Kirkwall Airport, United Kingdom
EGPB	LSI	Sumburgh Airport, United Kingdom
EGPC	WIC	Wick Airport, United Kingdom
EGPD	ABZ	Aberdeen Airport, United Kingdom
EGPE	INV	Inverness Airport, United Kingdom
EGPF	GLA	Glasgow Intl, United Kingdom
EGPH	EDI	Edinburgh Airport, United Kingdom
EGPI	ILY	Islay Airport, United Kingdom
EGPK	PIK	Glasgow Prestwick, United Kingdom
EGPL	BEB	Benbecula Airport, United Kingdom
EGPM	SCS	Scatsta Airport, United Kingdom
EGPN	DND	Dundee Airport, United Kingdom
EGPO	SYY	Stornoway Airport, United Kingdom
EGPR	BRR	Barra Airport, United Kingdom
EGPT	PSL	Perth Airport , United Kingdom
EGPU	TRE	Tiree Airport, United Kingdom
EGSC	CBG	Cambridge City, United Kingdom
EGSH	NWI	Norwich Intl, United Kingdom
EGSS	STN	London Stansted, United Kingdom
EGSY	SZE	Sheffield City Airport, United Kingdom
EGTE	EXT	Exeter Intl, United Kingdom
EGTG	FZO	Filton Aerodrome, United Kingdom
EGTK	OXF	Oxford Airport, United Kingdom
EGTO	RCS	Rochester Airport, United Kingdom
EGYP	MPN	Mount Pleasant, Falkland Islands
EHAM	AMS	Amsterdam Schiphol, Netherlands
EHBK	MST	Maastricht Aachen, Netherlands
EHGG	GRQ	Groningen Airport Eelde, Netherlands
EHLE	LEY	Lelystad Airport, Netherlands
EHRD	RTM	Rotterdam Airport, Netherlands
EHTW	ENS	Enschede Airport Twente, Netherlands
EICA	NNR	Connemara Regional Airport, Ireland
EICK	ORK	Cork Intl, Ireland

Appendix E

ICAO	IATA	Airport & Country	ICAO	IATA	Airport & Country
EICM	GWY	Galway Airport, Ireland	ENRO	RRS	Roros Airport, Norway
EIDL	CFN	Donegal Airport, Ireland	ENRS	RET	Rost Airport, Norway
EIDW	DUB	Dublin Intl, Ireland	ENRY	RYG	Moss Airport, Rygge, Norway
EIKN	NOC	Knock Intl, Ireland	ENSB	LYR	Svalbard Airport, Norway
EIKY	KIR	Kerry Airport, Ireland	ENSD	SDN	Sandane Airport, Norway
EINN	SNN	Shannon Airport, Ireland	ENSG	SOG	Sogndal Airport, Norway
EISG	SXL	Sligo Airport, Ireland	ENSH	SVJ	Svolvae Airport, Norway
EIWF	WAT	Waterford Airport, Ireland	ENSK	SKN	Stokmarknes Airport, Norway
EKAH	AAR	Aarhus Airport, Denmark	ENSN	SKE	Skien Airport, Norway
EKBI	BLL	Billund Airport, Denmark	ENSO	SRP	Stord Airport, Sorstokken, Norway
EKCH	CPH	Copenhagen Kastrup Intl, Denmark	ENSR	SOJ	Sorkjosen Airport, Norway
EKEB	EBJ	Esbjerg Airport, Denmark	ENSS	VAW	Vardo Airport, Norway
EKKA	KRP	Karup Airport, Denmark	ENST	SSJ	Sandnessjoen Airport, Norway
EKOD	ODE	Odense Airport, Denmark	ENTC	TOS	Tromso Airport, Norway
EKRK	RKE	Copenhagen Roskilde, Denmark	ENTO	TRF	Sandefjord Airport, Norway
EKRN	RNN	Ronne Bornholm Airport, Denmark	ENVA	TRD	Trondheim Airport, Norway
EKSB	SGD	Soenderborg Airport, Denmark	ENVD	VDS	Vadso Airport, Norway
EKTS	TED	Thisted Airport, Denmark	ENZV	SVG	Stavanger Airport, Sola, Norway
EKVG	FAE	Vagar Airprot, Faroe Islands	EPBY	BZG	Bydgoszcz Ignacy, Poland
EKVJ	STA	Stauning Airport, Denmark	EPGD	GDN	Gdansk Lech Walesa Airport, Poland
EKYT	AAL	Aalborg Airport, Denmark	EPKK	KRK	John Paul II Intl Krakow, Poland
ELLX	LUX	Luxembourg Intl, Luxembourg	EPKT	KTW	Katowice Intl, Poland
ENAL	AES	Alesund Airport, Norway	EPLL	LCJ	Lodz Wladyslaw Reymont, Poland
ENAN	ANX	Andoya Airport, Norway	EPPO	POZ	Poznan, Poland
ENAT	ALF	Alta Airport, Norway	EPRZ	RZE	Rzeszow, Poland
ENBL	FDE	F, Norway	EPSC	SZZ	Szczecin, Poland
ENBN	BNN	Bronnoysund Airport, Norway	EPSY	SZY	Szczytno, Poland
ENBO	BOO	Bodo Airport, Norway	EPWA	WAW	Warsaw Frederic Chopin, Poland
ENBR	BGO	Bergen Airport, Flesland, Norway	EPWR	WRO	Copernicus Airport Wroclaw, Poland
ENBS	BJF	Batsfjord Airport, Norway	EPZG	IEG	Zielona Gora Intl, Poland
ENBV	BVG	Berlevag Airport, Norway	ESDF	RNB	Ronneby, Sweden
ENCN	KRS	Kristiansand Airport, Norway	ESGG	GOT	Landvetter, Sweden
ENDI	DLD	Geilo Airport, Norway	ESGJ	JKG	Jonkoping, Sweden
ENDU	BDU	Bardufoss Airport, Norway	ESGL	LDK	Lidkoping, Sweden
ENEV	EVE	Harstad, Norway	ESGP	GSE	Goteborg City Airport, Sweden
ENFG	VDB	Fagernes Airport, Norway	ESGR	KVB	Skovde, Sweden
ENFL	FRO	Floro Airport, Norway	ESGT	THN	Trollhattan, Sweden
ENGM	OSL	Oslo Gardermoen Airport, Norway	ESKK	KSK	Karlskoga, Sweden
ENHA	HMR	Hamar Airport, Norway	ESKM	MXX	Siljan, Sweden
ENHD	HAU	Haugesund Airport, Norway	ESKN	NYO	Skavsta, Sweden
ENHF	HFT	Hammerfest Airport, Norway	ESMK	KID	Kristianstad, Sweden
ENHK	HAA	Hasvik Airport, Norway	ESMO	OSK	Oskarshamn, Sweden
ENHV	HVG	Honningsvag Airport, Norway	ESMQ	KLR	Kalmar, Sweden
ENKA	QKX	Kautokeino Airport, Norway	ESMS	MMX	Sturup, Sweden
ENKB	KSU	Kristiansund Airport, Norway	ESMT	HAD	Halmstad, Sweden
ENKL	GLL	Gol Airport, Norway	ESMX	VXO	Kronoberg, Sweden
ENKR	KKN	Kirkenes Airport, Norway	ESND	EVG	Sveg, Sweden
ENLI	FAN	Farsund Airport, Norway	ESNG	GEV	Gallivare, Sweden
ENLK	LKN	Leknes Airport, Norway	ESNH	HUV	Hudiksvall, Sweden
ENMH	MEH	Mehamn Airport, Norway	ESNK	KRF	Kramfors, Sweden
ENML	MOL	Molde Airport, Norway	ESNL	LYC	Lycksele, Sweden
ENMS	MJF	Mosjoen Airport, Norway	ESNN	SDL	Sundsvall, Sweden
ENNA	LKL	Lakselv Airport, Norway	ESNO	OER	Ornskoldsvik, Sweden
ENNK	NVK	Narvik Airport, Norway	ESNQ	KRN	Kiruna, Sweden
ENNM	OSY	Namsos Airport, Norway	ESNS	SFT	Skelleftea, Sweden
ENNO	NTB	Notodden Airport, Norway	ESNU	UME	Umea, Sweden
ENOL	OLA	Orland Main Air Station, Norway	ESNV	VHM	Vilhelmina, Sweden
ENOV	HOV	Orsta, Norway	ESNX	AJR	Arvidsjaur, Sweden
ENRA	MQN	Mo i Rana Airport, Norway	ESNY	SOO	Soderhamn, Sweden
ENRM	RVK	Rorvik Airport, Norway	ESOE	ORB	Orebro, Sweden

ICAO	IATA	Airport & Country	ICAO	IATA	Airport & Country
ESOH	HFS	Hagfors, Sweden	FANS	NLP	Nelspruit, South Africa
ESOK	KSD	Karlstad, Sweden	FAOH	OUH	Oudtshoorn, South Africa
ESOW	VST	Hasslo, Sweden	FAPA	AFD	Port Alfred, South Africa
ESPA	LLA	Kallax, Sweden	FAPE	PLZ	Port Elizabeth, South Africa
ESPC	OSD	Froson, Sweden	FAPG	PBZ	Plettenberg Bay, South Africa
ESSA	ARN	Arlanda, Sweden	FAPH	PHW	Hendrik Van Eck, South Africa
ESSB	BMA	Bromma, Sweden	FAPI	PTG	Pietersburg, South Africa
ESSD	BLE	Borlange, Sweden	FAPK	PRK	Prieska, South Africa
ESSF	HLF	Hultsfred, Sweden	FAPM	PZB	Pietermaritzburg, South Africa
ESSK	GVX	Gavle, Sweden	FAPN	NTY	Pilanesberg, South Africa
ESSL	LPI	Saab, Sweden	FAQT	UTW	Queenstown, South Africa
ESSP	NRK	Kungsangen, Sweden	FARB	RCB	Richards Bay, South Africa
ESST	TYF	Fryklanda, Sweden	FARI	RVO	Reivilo, South Africa
ESSU	EKT	Eskilstuna, Sweden	FARS	ROD	Robertson, South Africa
ESSV	VBY	Visby, Sweden	FASB	SBU	Springbok, South Africa
ESSW	VVK	Vastervik, Sweden	FASC	ZEC	Secunda, South Africa
ESTA	AGH	Angelholm, Sweden	FASS	SIS	Sishen, South Africa
ESUD	SQO	Storuman, Sweden	FASZ	SZK	Skukuza, South Africa
ESUE	IDB	Idre, Sweden	FATZ	LTA	Tzaneen, South Africa
ESUP	PJA	Pajala, Sweden	FAUL	ULD	Prince M.Buthelezi, South Africa
ESUT	HMV	Hemavan, Sweden	FAUP	UTN	Upington, South Africa
ETMN	FCN	Cuxhaven – Nordholz, Germany	FAUT	UTT	K.D. Matanzima, South Africa
ETNL	RLG	Rostock, Germany	FAVB	VRU	Vryburg, South Africa
ETNU	FNB	Neubrandenburg, Germany	FAVG	VIR	Virginia, South Africa
ETSI	IGS	Ingolstadt, Germany	FAVR	VRE	Vredendal, South Africa
EVLA	LPX	Liepaja Intl, Latvia	FAVY	VYD	Vryheid, South Africa
EVRA	RIX	Riga Intl, Latvia	FAWB	PRY	Wonderboom, South Africa
EVVA	VTS	Ventspils Intl, Latvia	FAWM	WEL	Welkom, South Africa
FAAB	ALJ	Alexander Bay, South Africa	FBFT	FRW	Francistown, Botsuana
FABE	BIY	Bisho, South Africa	FBGZ	GNZ	Ghanzi, Botsuana
FABL	BFN	Bloemfontein, South Africa	FBJW	JWA	Jwaneng, Botsuana
FACD	CDO	Cradock, South Africa	FBKE	BBK	Kasane, Botsuana
FACT	CPT	Cape Town Intl, South Africa	FBKR	KHW	Khwai River, Botsuana
FADN	DUR	Durban Intl, South Africa	FBMN	MUB	Maun, Botsuana
FAEL	ELS	East London, South Africa	FBOR	ORP	Orapa, Botsuana
FAEM	EMG	Empangeni, South Africa	FBSK	GBE	Sir Seretse Khama Intl, Botsuana
FAFB	FCB	Sentra Oes, South Africa	FBSP	PKW	Selebi, Botsuana
FAGC	GCJ	Grand Central, South Africa	FBSW	SWX	Shakawe, Botsuana
FAGG	GRJ	George, South Africa	FBTL	TLD	Tuli Block, Botsuana
FAGI	GIY	Giyani, South Africa	FBTS	TBY	Tshabong, Botsuana
FAGM	QRA	Rand, South Africa	FCBB	BZV	Maya, Congo
FAHL	HLW	Hluhluwe, South Africa	FCBD	DJM	Djambala, Congo
FAHR	HRS	Harrismith, South Africa	FCBK	KNJ	Kindamba, Congo
FAHT	HDS	Hoedspruit, South Africa	FCBL	LCO	Lague, Congo
FAJS	JNB	Johannesburg Intl, South Africa	FCBM	MUY	Mouyondzi, Congo
FAKD	KXE	Klerksdorp, South Africa	FCBS	SIB	Sibiti, Congo
FAKM	KIM	Kimberley, South Africa	FCBY	NKY	Yokangassi, Congo
FAKP	KOF	Komatipoort, South Africa	FCBZ	ANJ	Zanaga, Congo
FAKU	KMH	Johan Pienaar, South Africa	FCMM	MSX	Mossendjo, Congo
FAKZ	KLZ	Kleinsee, South Africa	FCOB	BOE	Boundji, Congo
FALA	HLA	Lanseria, South Africa	FCOE	EWO	Ewo, Congo
FALC	LMR	Finsch Mine, South Africa	FCOI	ION	Impfondo, Congo
FALO	LCD	Louis Trichardt, South Africa	FCOK	KEE	Kelle, Congo
FALW	SDB	Langebaanweg Ab, South Africa	FCOM	MKJ	Makoua, Congo
FALY	LAY	Ladysmith, South Africa	FCOO	FTX	Owando, Congo
FAMD	AAM	Malamala, South Africa	FCOS	SOE	Souanke, Congo
FAMG	MGH	Margate, South Africa	FCOT	BTB	Betou, Congo
FAMM	MBD	Mafikeng Intl, South Africa	FCOU	OUE	Ouesso, Congo
FAMN	LLE	Malelane, South Africa	FCPA	KMK	Makabana, Congo
FANC	NCS	Newcastle, South Africa	FCPL	DIS	Dolisie, Congo

Appendix E

ICAO	IATA	Airport & Country	ICAO	IATA	Airport & Country
FCPP	PNR	A. Neto, Congo	FMMG	WAQ	Antsalova, Madagascar
FDMS	MTS	Matsapha Airport, Swaziland	FMMH	VVB	Mahanoro, Madagascar
FEFB	MKI	M'boki, Central Africa	FMMI	TNR	Ivato, Madagascar
FEFF	BGF	M'poko, Central Africa	FMMK	JVA	Ankavandra, Madagascar
FEFG	BGU	Bangassou, Central Africa	FMML	BMD	Tsiribihina, Madagascar
FEFI	IRO	Birao, Central Africa	FMMN	ZVA	Miandrivazo, Madagascar
FEFM	BBY	Bambari, Central Africa	FMMO	MXT	Maintirano, Madagascar
FEFN	NDL	N'dele, Central Africa	FMMQ	ILK	Atsinanana, Madagascar
FEFO	BOP	Bouar, Central Africa	FMMR	TVA	Morafenobe, Madagascar
FEFR	BIV	Bria, Central Africa	FMMS	SMS	Sainte, Madagascar
FEFS	BSN	Bossangoa, Central Africa	FMMT	TMM	Toamasina, Madagascar
FEFT	BBT	Berberati, Central Africa	FMMU	WTA	Tambohorano, Madagascar
FEFY	AIG	Yalinga, Central Africa	FMMV	MOQ	Morondava, Madagascar
FEFZ	IMO	Zemio, Central Africa	FMMX	WTS	Tsiroanomandidy, Madagascar
FEGZ	BOZ	Bozoum, Central Africa	FMMY	VAT	Vatomandry, Madagascar
FGBT	BSG	Bata Airport, Equatorial Guinea	FMMZ	WAM	Ambatondrazaka, Madagascar
FGSL	SSG	Malabo Airport, Equatorial Guinea	FMNA	DIE	Arrachart, Madagascar
FIMP	MRU	Mauritius Intl, Mauritius	FMNC	WMR	Avaratra, Madagascar
FIMR	RRG	Plaine Corail, Mauritius	FMND	ZWA	Andapa, Madagascar
FKKB	KBI	Kribi, Cameroon	FMNE	AMB	Ambilobe, Madagascar
FKKC	TKC	Tiko, Cameroon	FMNF	WBD	Avaratra, Madagascar
FKKD	DLA	Douala, Cameroon	FMNG	WPB	Port Berge, Madagascar
FKKF	MMF	Mamfe, Cameroon	FMNH	ANM	Antsirabato, Madagascar
FKKI	OUR	Batouri, Cameroon	FMNJ	IVA	Ampapamena, Madagascar
FKKL	MVR	Salak, Cameroon	FMNL	HVA	Analalava, Madagascar
FKKM	FOM	Nkounja, Cameroon	FMNM	MJN	Amborovy, Madagascar
FKKN	NGE	N'gaoundere, Cameroon	FMNN	NOS	Fascene, Madagascar
FKKO	BTA	Bertoua, Cameroon	FMNO	DWB	Soalala, Madagascar
FKKR	GOU	Garoua, Cameroon	FMNQ	BPY	Besalampy, Madagascar
FKKS	DSC	Dschang, Cameroon	FMNR	WMN	Maroantsetra, Madagascar
FKKU	BFX	Bafoussam, Cameroon	FMNS	SVB	Sambava South, Madagascar
FKKV	BPC	Bamenda, Cameroon	FMNT	TTS	Tsaratanana, Madagascar
FKKW	EBW	Ebolowa, Cameroon	FMNV	VOH	Vohimarina, Madagascar
FKKY	YAO	Yaounde/Ville, Cameroon	FMNW	WAI	Ambalabe, Madagascar
FKYS	NSI	Nsimalen, Cameroon	FMNX	WMA	Mandritsara, Madagascar
FLCP	CIP	Chipata, Zambia	FMSB	WBO	Antsoa, Madagascar
FLKE	ZKP	Kasompe, Zambia	FMSC	WMD	Mandabe, Madagascar
FLKL	KLB	Kalabo, Zambia	FMSD	FTU	Tolagnaro, Madagascar
FLKO	KMZ	Kaoma, Zambia	FMSF	WFI	Fianarantsoa, Madagascar
FLKS	KAA	Kasama, Zambia	FMSG	RVA	Farafangana, Madagascar
FLKY	ZKB	Kasaba Bay, Zambia	FMSI	IHO	Ihosy, Madagascar
FLLI	LVI	Livingstone, Zambia	FMSJ	MJA	Manja, Madagascar
FLLK	LXU	Lukulu, Zambia	FMSK	WVK	Manakara, Madagascar
FLLS	LUN	Lusaka Intl, Zambia	FMSL	OVA	Bekily, Madagascar
FLMA	MNS	Mansa, Zambia	FMSM	MNJ	Mananjary, Madagascar
FLMF	MFU	Mfuwe, Zambia	FMSN	TDV	Samangoky, Madagascar
FLMG	MNR	Mongu, Zambia	FMSR	MXM	Morombe, Madagascar
FLNA	ZGM	Ngoma, Zambia	FMST	TLE	Toliara, Madagascar
FLND	NLA	Ndola, Zambia	FMSY	AMP	Ampanihy, Madagascar
FLSN	SXG	Senanga, Zambia	FMSZ	WAK	Ankazoabo, Madagascar
FLSO	KIW	Southdowns, Zambia	FNAM	AZZ	Ambriz, Angola
FLSW	SLI	Solwezi, Zambia	FNBC	SSY	M'banza Congo, Angola
FLZB	BBZ	Zambezi, Zambia	FNBG	BUG	Benguela, Angola
FMCH	HAH	Prince Said Ibrahim, Comoros	FNCA	CAB	Cabinda, Angola
FMCI	NWA	Moheli, Comoros	FNGI	NGV	N'giva, Angola
FMCN	YVA	Iconi, Comoros	FNHU	NOV	Albano Machado, Angola
FMCV	AJN	Ouani, Comoros	FNKU	SVP	Bie, Angola
FMEE	RUN	St Denis Gillot, Reunion	FNLU	LAD	Luanda, Angola
FMMC	WML	Malaimbandy, Madagascar	FNMA	MEG	Malange, Angola
FMME	ATJ	Antsirabe, Madagascar	FNME	SPP	Menongue, Angola

ICAO	IATA	Airport & Country	ICAO	IATA	Airport & Country
FNNG	GXG	Negage, Angola	FTTI	ATV	Ati, Chad
FNPA	PBN	Porto Amboim, Angola	FTTJ	NDJ	N'Djamena, Chad
FNSA	VHC	Saurimo, Angola	FTTK	BKR	Bokoro, Chad
FNSO	SZA	Soyo, Angola	FTTL	OTC	Berim, Chad
FNUB	SDD	Lubango, Angola	FTTM	MVO	Mongo, Chad
FNUE	LUO	Luena, Angola	FTTN	AMC	Am, Chad
FNUG	UGO	Uige, Angola	FTTP	PLF	Pala, Chad
FNWK	CEO	Wako Kungo, Angola	FTTS	OUT	Bousso, Chad
FNXA	XGN	Xangongo, Angola	FTTU	AMO	Mao, Chad
FNZE	ARZ	N'zeto, Angola	FTTY	FYT	Faya Largeau, Chad
FOGB	BGB	Booue, Gabon	FVBU	BUQ	Bulawayo, Zimbabwe
FOGE	KDN	Ndende, Gabon	FVCZ	BFO	Buffalo Range, Zimbabwe
FOGF	FOU	Fougamou, Gabon	FVFA	VFA	Victoria Falls, Zimbabwe
FOGG	MBC	Mbigou, Gabon	FVHA	HRE	Harare Intl, Zimbabwe
FOGI	MGX	Moabi, Gabon	FVKB	KAB	Kariba, Zimbabwe
FOGM	MJL	Mouila Ville, Gabon	FVMU	UTA	Mutare, Zimbabwe
FOGO	OYE	Oyem, Gabon	FVMV	MVZ	Masvingo, Zimbabwe
FOGQ	OKN	Okondja, Gabon	FVTL	GWE	Thornhill, Zimbabwe
FOGR	LBQ	Lambarene, Gabon	FVWN	HWN	Hwange National Park, Zimbabwe
FOGV	MVX	Minvoul, Gabon	FVWT	WKI	Hwange Town, Zimbabwe
FOOB	BMM	Bitam, Gabon	FWCD	CEH	Chelinda, Malawi
FOOD	MFF	Moanda, Gabon	FWCL	BLZ	Chileka, Malawi
FOOE	MKB	Mekambo, Gabon	FWCM	CMK	Club Makokola, Malawi
FOOG	POG	Port Gentil, Gabon	FWDW	DWA	Dwangwa, Malawi
FOOH	OMB	Omboue Hospital, Gabon	FWKA	KGJ	Karonga, Malawi
FOOK	MKU	Makokou, Gabon	FWKG	KBQ	Kasungu, Malawi
FOOL	LBV	Leon M'ba, Gabon	FWLI	LLW	Lilongwe Intl, Malawi
FOOM	MZC	Mitzic, Gabon	FWLK	LIX	Likoma, Malawi
FOON	MVB	M'vengue, Gabon	FWMG	MAI	Mangochi, Malawi
FOOR	LTL	Lastourville, Gabon	FWMY	MYZ	Monkey Bay, Malawi
FOOS	ZKM	Sette, Gabon	FWSM	LMB	Salima, Malawi
FOOT	TCH	Tchibanga, Gabon	FWUU	ZZU	Mzuzu, Malawi
FOOY	MYB	Mayumba, Gabon	FXLR	LRB	Leribe, Lesotho
FPPR	PCP	Principe, Sao Tome and Principe	FXMF	MFC	Mafeteng, Lesotho
FPST	TMS	Sao Tom, Sao Tome and Principe	FXMK	MKH	Mokhotlong, Lesotho
FQAG	ANO	Angoche, Mozambique	FXMM	MSU	Moshoeshoe I Intl, Lesotho
FQBR	BEW	Beira, Mozambique	FXQN	UNE	Qacha's Nek, Lesotho
FQCB	FXO	Cuamba, Mozambique	FXSH	SHK	Sehonghong, Lesotho
FQCH	VPY	Chimoio, Mozambique	FXSM	SOK	Semonkong, Lesotho
FQIN	INH	Inhambane, Mozambique	FYAR	ADI	Arandis, Namibia
FQLC	VXC	Lichinga, Mozambique	FYGF	GFY	Grootfontein, Namibia
FQMA	MPM	Maputo, Mozambique	FYKM	MPA	Katima Mulilo, Namibia
FQMP	MZB	Mocimboa Da Praia, Mozambique	FYKT	KMP	Keetmanshoop, Namibia
FQNC	MNC	Nacala, Mozambique	FYLZ	LUD	Luderitz, Namibia
FQNP	APL	Nampula, Mozambique	FYNA	NNI	Namutoni, Namibia
FQPB	POL	Pemba, Mozambique	FYOA	OND	Ondangwa, Namibia
FQQL	UEL	Quelimane, Mozambique	FYOG	OMD	Oranjemund, Namibia
FQTT	TET	Chingozi, Mozambique	FYOO	OKF	Okaukuejo, Namibia
FQVL	VNX	Vilankulu, Mozambique	FYRU	NDU	Rundu, Namibia
FSDR	DES	Desroches, Seychelles	FYSM	SWP	Swakopmund, Namibia
FSIA	SEZ	Seychelles Intl, Seychelles	FYTM	TSB	Tsumeb, Namibia
FSPP	PRI	Praslin, Seychelles	FYWB	WVB	Walvis Bay, Namibia
FSSB	BDI	Bird Island, Seychelles	FYWE	ERS	Eros, Namibia
FSSD	DEI	Denis Island, Seychelles	FYWH	WDH	Hosea Kutako Intl, Namibia
FSSF	FRK	Fregate, Seychelles	FZAA	FIH	N'djili Intl, Congo
FTTA	SRH	Sarh, Chad	FZAB	NLO	N'dolo, Congo
FTTB	OGR	Bongor, Chad	FZAJ	BOA	Boma, Congo
FTTC	AEH	Abeche, Chad	FZAL	LZI	Luozi, Congo
FTTD	MQQ	Moundou, Chad	FZAM	MAT	Matadi, Congo
FTTH	LTC	Lai, Chad	FZBA	INO	Inongo, Congo

Appendix E

ICAO	IATA	Airport & Country	ICAO	IATA	Airport & Country
FZBI	NIO	Nioki, Congo	GATB	TOM	Tombouctou, Mali
FZBO	FDU	Bandundu, Congo	GAYE	EYL	Yelimane, Mali
FZCA	KKW	Kikwit, Congo	GBYD	BJL	Banjul Intl, Gambia
FZCB	IDF	Idiofa, Congo	GCFV	FUE	Fuerteventura Airport, Spain
FZCE	LUS	Lusanga, Congo	GCGM	GMZ	La Gomera Airport, Spain
FZCV	MSM	Masi, Congo	GCHI	VDE	El Hierro Airport, Spain
FZEA	MDK	Mbandaka, Congo	GCLA	SPC	La Palma Airport, Spain
FZEN	BSU	Basankusu, Congo	GCLP	LPA	Gran Canaria Intl, Spain
FZFA	LIE	Libenge, Congo	GCRR	ACE	Lanzarote Airport, Spain
FZFD	BDT	Gbadolite, Congo	GCTS	TFS	Tenerife Reina Sofia Airport, Spain
FZFK	GMA	Gemena, Congo	GCXO	TFN	Tenerife Los Rodeos Airport, Spain
FZFP	KLI	Kotakoli, Congo	GECT	JCU	Ceuta Heliport, Spain
FZFU	BMB	Bumba, Congo	GEML	MLN	Melilla Airport, Spain
FZGA	LIQ	Lisala, Congo	GFBN	BTE	Bonthe, Sierra Leone
FZGN	BNB	Boende, Congo	GFBO	KBS	Bo, Sierra Leone
FZGV	IKL	Ikela, Congo	GFGK	GBK	Gbangbatoke, Sierra Leone
FZIC	FKI	Bangoka Intl, Congo	GFHA	HGS	Hastings, Sierra Leone
FZJH	IRP	Isiro, Congo	GFKB	KBA	Kabala, Sierra Leone
FZKA	BUX	Bunia, Congo	GFKE	KEN	Kenema, Sierra Leone
FZKJ	BZU	Buta, Congo	GFLL	FNA	Lungi, Sierra Leone
FZMA	BKY	Kavumu, Congo	GFYE	WYE	Yengema, Sierra Leone
FZNA	GOM	Goma Intl, Congo	GGOV	OXB	Osvaldo Vieira, Guinea Bissau
FZNP	BNC	Beni, Congo	GLBU	UCN	Buckanan, Liberia
FZOA	KND	Kindu, Congo	GLCP	CPA	Cape Palmas, Liberia
FZOD	KLY	Kalima, Congo	GLGE	SNI	Sinoe, Liberia
FZOK	KGN	Kasongo, Congo	GLMR	MLW	Spriggs Payne, Liberia
FZOP	PUN	Punia, Congo	GLNA	NIA	Nimba, Liberia
FZQA	FBM	Lubumbashi Intl, Congo	GLRB	ROB	Roberts Intl, Liberia
FZQC	PWO	Pweto, Congo	GLTN	THC	Tchien, Liberia
FZQG	KEC	Kasenga, Congo	GMAT	TTA	Plage Blanche, Morocco
FZQM	KWZ	Kolwezi, Congo	GMFF	FEZ	Saiss, Morocco
FZRA	MNO	Manono, Congo	GMFK	ERH	Moulay Ali Cherif, Morocco
FZRB	BDV	Moba, Congo	GMFO	OUD	Angads, Morocco
FZRF	FMI	Kalemie, Congo	GMMC	CAS	Anfa, Morocco
FZRM	KBO	Kabalo, Congo	GMME	RBA	Sale, Morocco
FZRQ	KOO	Kongolo, Congo	GMMF	SII	Sidi Ifni, Morocco
FZSA	KMN	Kamina Base, Congo	GMMN	CMN	Mohamed V, Morocco
FZSK	KAP	Kapanga, Congo	GMMS	SFI	Safi, Morocco
FZTK	KNM	Kaniama, Congo	GMMX	RAK	Menara, Morocco
FZUA	KGA	Kananga, Congo	GMMZ	OZZ	Ouarzazate, Morocco
FZUG	LZA	Luiza, Congo	GMTA	AHU	Cherif Al Idrissi, Morocco
FZUK	TSH	Tshikapa, Congo	GMTN	TTU	Sania Ramel, Morocco
FZVA	LJA	Lodja, Congo	GMTT	TNG	Ibn Batouta, Morocco
FZVI	LBO	Lusambo, Congo	GOGG	ZIG	Ziguinchor, Senegal
FZVM	MEW	Mweka, Congo	GOGK	KDA	Kolda, Senegal
FZVR	BAN	Basongo, Congo	GOGS	CSK	Cap Skiring, Senegal
FZVS	PFR	Ilebo, Congo	GOOK	KLC	Kaolack, Senegal
FZWA	MJM	Mbuji, Congo	GOOY	DKR	Leopold Sedar Senghor, Senegal
FZWC	GDJ	Gandajika, Congo	GOSM	MAX	Ouro Sogui, Senegal
FZWT	KBN	Tunta, Congo	GOSP	POD	Podor, Senegal
GABS	BKO	Senou, Mali	GOSR	RDT	Richard Toll, Senegal
GAGM	GUD	Goundam, Mali	GOSS	XLS	St Louis, Senegal
GAGO	GAQ	Korogoussou, Mali	GOTB	BXE	Bakel, Senegal
GAKA	KNZ	Kenieba, Mali	GOTK	KGG	Kedougou, Senegal
GAKO	KTX	Koutiala, Mali	GOTS	SMY	Simenti, Senegal
GAKY	KYS	Kayes, Mali	GOTT	TUD	Tambacounda, Senegal
GAMB	MZI	Barbe, Mali	GQNA	AEO	Aioun, Mauritania
GANK	NRM	Keibane, Mali	GQNB	OTL	Boutilimit, Mauritania
GANR	NIX	Nioro, Mali	GQNC	THI	Tichit, Mauritania
GASK	KSS	Sikasso, Mali	GQND	TIY	Tidjikja, Mauritania

ICAO	IATA	Airport & Country	ICAO	IATA	Airport & Country
GQNE	BGH	Bogue, Mauritania	HANK	NEK	Nekemte, Ethiopia
GQNF	KFA	Kiffa, Mauritania	HASD	SXU	Soddo, Ethiopia
GQNH	TMD	Timbedra, Mauritania	HASK	SKR	Shakiso, Ethiopia
GQNI	EMN	Nema, Mauritania	HASO	ASO	Asosa, Ethiopia
GQNJ	AJJ	Akjoujt, Mauritania	HATP	TIE	Tippi, Ethiopia
GQNK	KED	Kaedi, Mauritania	HAWC	WAC	Wacca, Ethiopia
GQNL	MOM	Letfotar, Mauritania	HBBA	BJM	Bujumbura Intl, Burundi
GQNN	NKC	Nouakchott, Mauritania	HBBE	GID	Gitega, Burundi
GQNS	SEY	Selibabi, Mauritania	HBBO	KRE	Kirundo, Burundi
GQNT	THT	Tamchakett, Mauritania	HCMA	ALU	Alula, Somalia
GQPA	ATR	Atar, Mauritania	HCMC	CXN	Candala, Somalia
GQPF	FGD	F'derik, Mauritania	HCME	HCM	Eil, Somalia
GQPP	NDB	Nouadhibou, Mauritania	HCMF	BSA	Bosaso, Somalia
GQPZ	OUZ	Tazadit, Mauritania	HCMG	GSR	Gardo, Somalia
GUCY	CKY	Gbessia, Guinea	HCMI	BBO	Berbera, Somalia
GUFA	FIG	Katourou, Guinea	HCMK	KMU	Kisimayu, Somalia
GUFH	FAA	Badala, Guinea	HCMM	MGQ	Mogadishu, Somalia
GUKU	KSI	Kissi, Guinea	HCMO	CMO	Obbia, Somalia
GULB	LEK	Tata, Guinea	HCMR	GLK	Galcaio, Somalia
GUMA	MCA	Macenta, Guinea	HCMS	CMS	Scusciuban, Somalia
GUNZ	NZE	Konia, Guinea	HCMU	ERA	Erigavo, Somalia
GUOK	BKJ	Baralande, Guinea	HCMV	BUO	Burao, Somalia
GUSB	SBI	Sambailo, Guinea	HEAR	AAC	El Arish Intl, Egypt
GUSI	GII	Siguiri, Guinea	HEAT	ATZ	Asyut Intl, Egypt
GUXD	KNN	Kankan, Guinea	HEAX	ALY	Alexandria Intl, Egypt
GVAC	SID	Amilcar Cabral, Cape Verde	HEBA	HBE	Borg El Arab Intl, Egypt
GVBA	BVC	Rabil, Cape Verde	HEBL	ABS	Abu Simbel Airport, Egypt
GVFM	RAI	Francisco Mendes, Cape Verde	HECA	CAI	Cairo Intl, Egypt
GVMA	MMO	Maio, Cape Verde	HEDK	DAK	Dakhla Airport, Egypt
GVSN	SNE	Preguica, Cape Verde	HEGN	HRG	Hurghada Intl, Egypt
GVSV	VXE	S. Pedro, Cape Verde	HEKG	UVL	El Kharga Airport, Egypt
HAAB	ADD	Addis Abeba Bole Intl, Ethiopia	HELX	LXR	Luxor Intl, Egypt
HAAM	AMH	Arba Minch, Ethiopia	HEMA	RMF	Marsa Alam Intl, Egypt
HAAX	AXU	Axum, Ethiopia	HEMM	MUH	Mersa Matruh Airport, Egypt
HABC	BCO	Baco, Ethiopia	HEOW	GSQ	Shark El Oweinat Intl, Egypt
HABD	BJR	Bahir Dar, Ethiopia	HEPS	PSD	Port Said Airport, Egypt
HABE	BEI	Beica, Ethiopia	HESC	SKV	St Catherine Intl, Egypt
HADC	DSE	Combolcha, Ethiopia	HESH	SSH	Sharm El Sheikh Intl, Egypt
HADD	DEM	Dembi Dollo, Ethiopia	HESN	ASW	Aswan Intl, Egypt
HADM	DBM	Debre Marcos, Ethiopia	HETB	TCP	Taba Intl, Egypt
HADR	DIR	A.T.D. Yilma Intl, Ethiopia	HETR	ELT	El Tor Airport, Egypt
HADT	DBT	Debre Tabor, Ethiopia	HFFF	JIB	Djibouti Ambouli, Djibouti
HAFN	FNH	Fincha, Ethiopia	HKAM	ASV	Amboseli, Kenya
HAGB	GOB	Robe, Ethiopia	HKED	EDL	Eldoret, Kenya
HAGH	GNN	Ghinnir, Ethiopia	HKES	EYS	Eliye Springs, Kenya
HAGM	GMB	Gambella, Ethiopia	HKFG	FER	Ferguson's Gulf, Kenya
HAGN	GDQ	Azezo, Ethiopia	HKGA	GAS	Garissa, Kenya
HAGR	GOR	Gore, Ethiopia	HKHO	HOA	Hola, Kenya
HAHU	HUE	Humera, Ethiopia	HKJK	NBO	Jomo Kenyatta, Kenya
HAJM	JIM	Aba Segud, Ethiopia	HKKI	KIS	Kisumu, Kenya
HAKD	ABK	Kebri Dehar, Ethiopia	HKKL	ILU	Kilaguni, Kenya
HALA	AWA	Awasa, Ethiopia	HKKR	KEY	Kericho, Kenya
HALL	LLI	Lalibela, Ethiopia	HKKT	KTL	Kitale, Kenya
HAMA	MKS	Mekane Salam, Ethiopia	HKLO	LOK	Lodwar, Kenya
HAMK	MQX	Alula Aba Nega, Ethiopia	HKLU	LAU	Lamu/Manda, Kenya
HAML	MZX	Masslo, Ethiopia	HKLY	LOY	Loyangalani, Kenya
HAMM	ETE	Metema, Ethiopia	HKMA	NDE	Mandera, Kenya
HAMN	NDM	Mendi, Ethiopia	HKMB	RBT	Marsabit, Kenya
HAMT	MTF	Mizan Teferi, Ethiopia	HKML	MYD	Malindi, Kenya
HANJ	NEJ	Nejjo, Ethiopia	HKMO	MBA	Moi, Kenya

Appendix E

ICAO	IATA	Airport & Country	ICAO	IATA	Airport & Country
HKMY	OYL	Oda, Kenya	HTTG	TGT	Tanga, Tanzania
HKNI	NYE	Nyeri, Kenya	HTZA	ZNZ	Kisauni, Tanzania
HKNK	NUU	Nakuru, Kenya	HUAR	RUA	Arua, Uganda
HKNW	WIL	Wilson, Kenya	HUEN	EBB	Entebbe Intl, Uganda
HKNY	NYK	Nanyuki, Kenya	HUGU	ULU	Gulu, Uganda
HKSB	UAS	Buffalo Spring, Kenya	HUJI	JIN	Jinja, Uganda
HKWJ	WJR	Waghala, Kenya	HUKS	KSE	Kasese, Uganda
HLKF	AKF	Kufra, Libya	HUMA	MBQ	Mbarara, Uganda
HLLS	SEB	Sebha, Libya	HUMI	KCU	Masindi, Uganda
HLMB	LMQ	Marsa Brega, Libya	HUPA	PAF	Pakuba, Uganda
HLON	HUQ	Hon, Libya	HUSO	SRT	Soroti, Uganda
HRYG	GYI	Gisenyi, Rwanda	HUTO	TRY	Tororo, Uganda
HRYI	BTQ	Butare, Rwanda	KABE	ABE	Lehigh Valley Intl, USA
HRYR	KGL	Gregoire Kayibanda, Rwanda	KABI	ABI	Abilene Regional Airport, USA
HRYU	RHG	Ruhengeri, Rwanda	KABQ	ABQ	Albuquerque Intl Sunport, USA
HRZA	KME	Kamembe, Rwanda	KABR	ABR	Aberdeen Regional Airport, USA
HSAT	ATB	Atbara, Sudan	KABY	ABY	Southwest Georgia Regional, USA
HSDN	DOG	Dongola, Sudan	KACK	ACK	Nantucket Memorial Airport, USA
HSFS	ELF	El Fasher, Sudan	KACT	ACT	Waco Regional Airport, USA
HSGF	GSU	Gedaref, Sudan	KACV	ACV	Arcata Airport, USA
HSGG	DNX	Galegu, Sudan	KACY	ACY	Atlantic City Intl, USA
HSGN	EGN	Geneina, Sudan	KAEX	AEX	Alexandria Intl, USA
HSKA	KSL	Kassala, Sudan	KAGS	AGS	Augusta Regional, USA
HSKI	KST	Kosti, Sudan	KAHN	AHN	Athens, USA
HSMR	MWE	Merowe, Sudan	KALB	ALB	Albany Intl, USA
HSNH	NUD	El Nahud, Sudan	KALO	ALO	Waterloo Regional Airport, USA
HSNL	UYL	Nyala, Sudan	KALW	ALW	Walla Walla Regional Airport, USA
HSNW	NHF	New Halfa, Sudan	KAMA	AMA	Rick Husband Amarillo Intl, USA
HSOB	EBD	El Obeid, Sudan	KAOO	AOO	Altoona, USA
HSPN	PZU	Port Sudan, Sudan	KAPF	APF	Naples Municipal Airport, USA
HSSJ	JUB	Juba, Sudan	KAPN	APN	Alpena County Regional Airport, USA
HSSM	MAK	Malakal, Sudan	KASE	ASE	Aspen, USA
HSSS	KRT	Khartoum, Sudan	KATL	ATL	Hartsfield, USA
HSSW	WHF	Wadi Halfa, Sudan	KATW	ATW	Outagamie County Regional, USA
HSWW	WUU	Wau, Sudan	KAUS	AUS	Austin, USA
HTAR	ARK	Arusha, Tanzania	KAVL	AVL	Asheville Regional Airport, USA
HTBU	BKZ	Bukoba, Tanzania	KAVP	AVP	Wilkes, USA
HTDA	DAR	Dar, Tanzania	KAZO	AZO	Kalamazoo/Battle Creek Intl, USA
HTDO	DOD	Dodoma, Tanzania	KBDL	BDL	Bradley Intl, USA
HTIR	IRI	Iringa, Tanzania	KBED	BED	Laurence G. Hanscom Field, USA
HTKA	TKQ	Kigoma, Tanzania	KBFI	BFI	King County Intl, USA
HTKI	KIY	Kilwa Masoko, Tanzania	KBFL	BFL	Meadows Field Airport, USA
HTKJ	JRO	Kilimanjaro Intl, Tanzania	KBGM	BGM	Greater Binghamton Airport, USA
HTLI	LDI	Lindi, Tanzania	KBGR	BGR	Bangor Intl, USA
HTLM	LKY	Lake Manyara, Tanzania	KBHB	BHB	Hancock County, USA
HTMA	MFA	Mafia, Tanzania	KBHM	BHM	Birmingham Intl, USA
HTMB	MBI	Mbeya, Tanzania	KBIL	BIL	Billings Logan Intl, USA
HTMD	MWN	Mwadui, Tanzania	KBIS	BIS	Bismarck Municipal Airport, USA
HTMI	XMI	Masasi, Tanzania	KBJI	BJI	Bemidji Regional Airport, USA
HTMT	MYW	Mtwara, Tanzania	KBLI	BLI	Bellingham Intl, USA
HTMU	MUZ	Musoma, Tanzania	KBMI	BMI	Central Illinois Regional Airport, USA
HTMW	MWZ	Mwanza, Tanzania	KBNA	BNA	Nashville Intl, USA
HTNA	NCH	Nachingwea, Tanzania	KBOI	BOI	Boise Air Terminal, USA
HTNJ	JOM	Njombe, Tanzania	KBOS	BOS	Gen. Edward Lawrence Logan, USA
HTPE	PMA	Pemba, Tanzania	KBPT	BPT	Southeast Texas Regional, USA
HTSN	SEU	Seronera, Tanzania	KBQK	BQK	Brunswick Golden Isles Airport, USA
HTSO	SGX	Songea, Tanzania	KBRD	BRD	Brainerd Lakes Regional Airport, USA
HTSU	SUT	Sumbawanga, Tanzania	KBRO	BRO	Brownsville/South Padre Island Intl, USA
HTSY	SHY	Shinyanga, Tanzania			
HTTB	TBO	Tabora, Tanzania	KBTM	BTM	Bert Mooney Airport, USA

ICAO	IATA	Airport & Country	ICAO	IATA	Airport & Country
KBTR	BTR	Baton Rouge Metropolitan, USA	KEWB	EWB	New Bedford Regional Airport, USA
KBTV	BTV	Burlington Intl, USA	KEWN	EWN	Craven County Regional Airport, USA
KBUF	BUF	Buffalo Niagara Intl, USA	KEWR	EWR	Newark Liberty Intl, USA
KBUR	BUR	Bob Hope Airport, USA	KEYW	EYW	Key West Intl, USA
KBWI	BWI	Baltimore, USA	KFAR	FAR	Hector Intl, USA
KBZN	BZN	Gallatin Field Airport, USA	KFAT	FAT	Fresno Yosemite Intl, USA
KCAE	CAE	Columbia Metropolitan Airport, USA	KFAY	FAY	Fayetteville Regional Airport, USA
KCAK	CAK	Akron, USA	KFHR	FHR	Friday Harbor Airport, USA
KCDC	CDC	Cedar City Regional Airport, USA	KFLG	FLG	Flagstaff Pulliam Airport, USA
KCEC	CEC	Jack McNamara Field Airport, USA	KFLL	FLL	Fort Lauderdale, USA
KCHA	CHA	Chattanooga Metropolitan, USA	KFLO	FLO	Florence Regional Airport, USA
KCHO	CHO	Charlottesville, USA	KFMN	FMN	Four Corners Regional Airport, USA
KCHS	CHS	Charleston Intl, USA	KFNT	FNT	Bishop Intl, USA
KCIC	CIC	Chico Municipal Airport, USA	KFOE	FOE	Forbes Field, USA
KCID	CID	The Eastern Iowa Airport, USA	KFSD	FSD	Sioux Falls Regional Airport, USA
KCIU	CIU	Chippewa County Intl, USA	KFSM	FSM	Fort Smith Regional Airport, USA
KCKB	CKB	Marion Regional Airport, USA	KFWA	FWA	Fort Wayne Intl, USA
KCLE	CLE	Cleveland, USA	KGCC	GCC	Gillette, USA
KCLL	CLL	Easterwood Airport, USA	KGCN	GCN	Grand Canyon National Park, USA
KCLM	CLM	William R. Fairchild Intl, USA	KGEG	GEG	Spokane Intl, USA
KCLT	CLT	Charlotte/Douglas Intl, USA	KGFK	GFK	Grand Forks Intl, USA
KCMH	CMH	Port Columbus Intl, USA	KGGG	GGG	East Texas Regional Airport, USA
KCMI	CMI	University of Illinois, USA	KGJT	GJT	Walker Field Airport, USA
KCMX	CMX	Houghton County Memorial, USA	KGNV	GNV	Gainesville Regional Airport, USA
KCOD	COD	Yellowstone Regional Airport, USA	KGPI	GPI	Glacier Park Intl, USA
KCOS	COS	Colorado Springs Municipal, USA	KGPT	GPT	Gulfport, USA
KCOU	COU	Columbia Regional Airport, USA	KGRB	GRB	Austin Straubel Intl, USA
KCPR	CPR	Natrona County Intl, USA	KGRK	GRK	Killeen, USA
KCRP	CRP	Corpus Christi Intl, USA	KGRR	GRR	Gerald R. Ford Intl, USA
KCRQ	CRQ	McClellan, USA	KGSO	GSO	Piedmont Triad Intl, USA
KCRW	CRW	Yeager Airport, USA	KGSP	GSP	Greenville, USA
KCSG	CSG	Columbus Metropolitan Airport, USA	KGTF	GTF	Great Falls Intl, USA
KCVG	CVG	Northern Kentucky Intl, USA	KGTR	GTR	Golden Triangle Regional Airport, USA
KCVX	CVX	Charlevoix Municipal Airport, USA	KGUC	GUC	Gunnison, USA
KCWA	CWA	Central Wisconsin Airport, USA	KHDN	HDN	Yampa Valley Airport, USA
KCYS	CYS	Cheyenne Regional Airport, USA	KHGR	HGR	Hagerstown Regional Airport, USA
KDAB	DAB	Daytona Beach Intl, USA	KHLN	HLN	Helena Regional Airport, USA
KDAL	DAL	Dallas Love Field, USA	KHOU	HOU	William P. Hobby Airport, USA
KDAY	DAY	James M. Cox Dayton Intl, USA	KHPN	HPN	Westchester County Airport, USA
KDBQ	DBQ	Dubuque Regional Airport, USA	KHRL	HRL	Valley Intl, USA
KDCA	DCA	Washington National Airport, USA	KHSV	HSV	Huntsville Intl, USA
KDEC	DEC	Decatur Airport, USA	KHTS	HTS	Tri, USA
KDEN	DEN	Denver Intl, USA	KHVN	HVN	Tweed, USA
KDFW	DFW	Dallas, USA	KHXD	HXD	Hilton Head Airport, USA
KDHN	DHN	Dothan Regional Airport, USA	KHYA	HYA	Barnstable Municipal Airport, USA
KDLH	DLH	Duluth Intl, USA	KIAD	IAD	Washington Dulles Intl, USA
KDRO	DRO	Durango, USA	KIAH	IAH	George Bush Intercontinental, USA
KDSM	DSM	Des Moines Intl, USA	KICT	ICT	Wichita Mid, USA
KDTW	DTW	Detroit Metropolitan, USA	KIDA	IDA	Idaho Falls Regional Airport, USA
KDUJ	DUJ	DuBois, USA	KIFP	IFP	Laughlin/Bullhead Intl, USA
KEAT	EAT	Pangborn Memorial Airport, USA	KILM	ILM	Wilmington Intl, USA
KEAU	EAU	Chippewa Valley Regional, USA	KIND	IND	Indianapolis Intl, USA
KEFD	EFD	Ellington Field, USA	KINL	INL	Falls Intl, USA
KEGE	EGE	Eagle County Regional Airport, USA	KIPL	IPL	Imperial County Airport, USA
KEKO	EKO	Elko Regional Airport, USA	KIPT	IPT	Williamsport Regional Airport, USA
KELM	ELM	Elmira/Corning Regional Airport, USA	KISP	ISP	Long Island MacArthur Airport, USA
KELP	ELP	El Paso Intl, USA	KITH	ITH	Ithaca Tompkins Regional, USA
KERI	ERI	Erie Intl, USA	KIYK	IYK	Inyokern Airport, USA
KEUG	EUG	Eugene Airport, USA	KJAC	JAC	Jackson Hole Airport, USA
KEVV	EVV	Evansville Regional Airport, USA	KJAN	JAN	Jackson, USA

Appendix E

ICAO	IATA	Airport & Country	ICAO	IATA	Airport & Country
KJAX	JAX	Jacksonville Intl, USA	KOAJ	OAJ	Albert J. Ellis Airport, USA
KJFK	JFK	John F. Kennedy Intl, USA	KOAK	OAK	Metropolitan Oakland Intl, USA
KJHW	JHW	Chautauqua County Airport, USA	KOKC	OKC	Will Rogers World Airport, USA
KJLN	JLN	Joplin Regional Airport, USA	KOMA	OMA	Eppley Airfield, USA
KJST	JST	John Murtha Johnstown, USA	KONT	ONT	Ontario Intl, USA
KLAF	LAF	Purdue University Airport, USA	KORD	ORD	Chicago O'Hare Intl, USA
KLAN	LAN	Capital City Airport, USA	KORF	ORF	Norfolk Intl, USA
KLAS	LAS	McCarran Intl, USA	KORH	ORH	Worcester Regional Airport, USA
KLAW	LAW	Lawton, USA	KOTH	OTH	Southwest Oregon Regional, USA
KLAX	LAX	Los Angeles Intl, USA	KOXR	OXR	Oxnard Airport, USA
KLBB	LBB	Lubbock Preston Smith Intl, USA	KPAH	PAH	Barkley Regional Airport, USA
KLBE	LBE	Arnold Palmer Regional Airport, USA	KPBI	PBI	Palm Beach Intl, USA
KLCH	LCH	Lake Charles Regional Airport, USA	KPCW	PCW	Erie, USA
KLEX	LEX	Blue Grass Airport, USA	KPDT	PDT	Eastern Oregon Regional Airport, USA
KLFT	LFT	Lafayette Regional Airport, USA	KPDX	PDX	Portland Intl, USA
KLGA	LGA	LaGuardia Airport, USA	KPFN	PFN	Panama City, USA
KLGB	LGB	Long Beach Municipal Airport, USA	KPGA	PGA	Page Municipal Airport, USA
KLIT	LIT	Little Rock National Airport, USA	KPGV	PGV	Pitt, USA
KLMT	LMT	Klamath Falls Airport, USA	KPHF	PHF	Williamsburg Intl, USA
KLNK	LNK	Lincoln Airport, USA	KPHL	PHL	Philadelphia Intl, USA
KLNS	LNS	Lancaster Airport, USA	KPHX	PHX	Phoenix Sky Harbor Intl, USA
KLRD	LRD	Laredo Intl, USA	KPIA	PIA	Greater Peoria Regional Airport, USA
KLSE	LSE	La Crosse Municipal Airport, USA	KPIB	PIB	Hattiesburg, USA
KLWS	LWS	Lewiston, USA	KPIE	PIE	St. Petersburg, USA
KLYH	LYH	Lynchburg Regional Airport, USA	KPIH	PIH	Pocatello Regional Airport, USA
KMAF	MAF	Midland Intl, USA	KPIR	PIR	Pierre Regional Airport, USA
KMBS	MBS	MBS Intl, USA	KPIT	PIT	Pittsburgh Intl, USA
KMCI	MCI	Kansas City Intl, USA	KPKB	PKB	Mid, USA
KMCN	MCN	Middle Georgia Regional Airport, USA	KPLN	PLN	Pellston Regional Airport, USA
KMCO	MCO	Orlando Intl, USA	KPNS	PNS	Pensacola Regional Airport, USA
KMCW	MCW	Mason City Municipal Airport, USA	KPQI	PQI	Northern Maine Regional Airport, USA
KMDT	MDT	Harrisburg Intl, USA	KPSC	PSC	Tri, USA
KMDW	MDW	Chicago Midway Intl, USA	KPSM	PSM	Pease Intl Tradeport Airport, USA
KMEI	MEI	Meridian Regional Airport, USA	KPSP	PSP	Palm Springs Intl, USA
KMEM	MEM	Memphis Intl, USA	KPUW	PUW	Pullman, USA
KMFE	MFE	McAllen, USA	KPVC	PVC	Provincetown Municipal Airport, USA
KMFR	MFR	Rogue Valley Intl, USA	KPVD	PVD	Theodore Francis Green Statet, USA
KMGM	MGM	Montgomery Regional Airport, USA	KPWM	PWM	Portland Intl Jetport, USA
KMGW	MGW	Morgantown Municipal Airport, USA	KRAP	RAP	Rapid City Regional Airport, USA
KMHK	MHK	Manhattan Regional Airport, USA	KRDD	RDD	Redding Municipal Airport, USA
KMHT	MHT	Manchester, USA	KRDG	RDG	Reading Regional Airport, USA
KMIA	MIA	Miami Intl, USA	KRDM	RDM	Redmond Municipal Airport, USA
KMKE	MKE	General Mitchell Intl, USA	KRDU	RDU	Raleigh, USA
KMKG	MKG	Muskegon County Airport, USA	KRHI	RHI	Rhinelander, USA
KMKT	MKT	Mankato Regional Airport, USA	KRIC	RIC	Richmond Intl, USA
KMLB	MLB	Melbourne Intl, USA	KRIW	RIW	Riverton Regional Airport, USA
KMLI	MLI	Quad City Intl, USA	KRKD	RKD	Knox County Regional Airport, USA
KMLU	MLU	Monroe Regional Airport, USA	KRNO	RNO	Tahoe Intl, USA
KMOB	MOB	Mobile Regional Airport, USA	KROA	ROA	Roanoke Regional Airport, USA
KMOD	MOD	Modesto City, USA	KROC	ROC	Greater Rochester Intl, USA
KMOT	MOT	Minot Intl, USA	KRST	RST	Rochester Intl, USA
KMRY	MRY	Monterey Peninsula Airport, USA	KRSW	RSW	Southwest Florida Intl, USA
KMSN	MSN	Dane County Regional Airport, USA	KSAF	SAF	Santa Fe Municipal Airport, USA
KMSO	MSO	Missoula Intl, USA	KSAN	SAN	San Diego Intl, USA
KMSP	MSP	Minneapolis, USA	KSAT	SAT	San Antonio Intl, USA
KMSY	MSY	Louis Armstrong New Orleans, USA	KSAV	SAV	Savannah Intl, USA
KMTJ	MTJ	Montrose Regional Airport, USA	KSAW	SAW	Sawyer Intl, USA
KMVY	MVY	Martha's Vineyard Airport, USA	KSBA	SBA	Santa Barbara Municipal Airport, USA
KMWA	MWA	Williamson County Regional, USA	KSBN	SBN	South Bend Regional Airport, USA
KMYR	MYR	Myrtle Beach Intl, USA	KSBP	SBP	San Luis Obispo County, USA

PlanePlotter User Guide

ICAO	IATA	Airport & Country	ICAO	IATA	Airport & Country
KSBY	SBY	Salisbury, USA	LDOS	OSI	Osijek Airport, Croatia
KSCK	SCK	Stockton Metropolitan Airport, USA	LDPL	PUY	Pula Airport, Croatia
KSDF	SDF	Louisville Intl, USA	LDRI	RJK	Rijeka Airport, Croatia
KSDY	SDY	Sidney, USA	LDSB	BWK	Airport Brac, Croatia
KSEA	SEA	Seattle, USA	LDSP	SPU	Airport Split, Croatia
KSFB	SFB	Orlando Sanford Intl, USA	LDZA	ZAG	Zagreb Airport, Croatia
KSFO	SFO	San Francisco Intl, USA	LDZD	ZAD	Zadar Airport, Croatia
KSGF	SGF	Springfield, USA	LEAB	ABC	Albacete Airport, Spain
KSGU	SGU	St. George Municipal Airport, USA	LEAL	ALC	Alicante Airport, Spain
KSHR	SHR	Sheridan County Airport, USA	LEAM	LEI	Almeria Intl, Spain
KSHV	SHV	Shreveport Regional Airport, USA	LEAS	OVD	Asturias Airport, Spain
KSJC	SJC	Norman Y. Mineta San Jos, USA	LEBA	ODB	Cordoba Airport, Spain
KSJT	SJT	San Angelo Regional Airport, USA	LEBB	BIO	Bilbao Airport, Spain
KSLC	SLC	Salt Lake City Intl, USA	LEBL	BCN	Barcelona Intl, Spain
KSMF	SMF	Sacramento Intl, USA	LEBZ	BJZ	Badajoz Airport, Spain
KSMX	SMX	Santa Maria Public Airport, USA	LECO	LCG	A Coruna Airport, Spain
KSNA	SNA	John Wayne Airport, USA	LEGE	GRO	Girona, Spain
KSPI	SPI	Abraham Lincoln Capital Airport, USA	LEGR	GRX	Granada, Spain
KSPS	SPS	Sheppard Air Force Base, USA	LEIB	IBZ	Ibiza Airport, Spain
KSRQ	SRQ	Sarasota, USA	LEJR	XRY	Jerez Airport, Spain
KSTC	STC	St. Cloud Regional Airport, USA	LELC	MJV	Murcia, Spain
KSTL	STL	Lambert, USA	LELL	QSA	Sabadell Airport, Spain
KSUN	SUN	Friedman Memorial Airport, USA	LELN	LEN	Leon Airport, Spain
KSUX	SUX	Sioux Gateway Airport, USA	LELO	RJL	Logrono, Spain
KSWF	SWF	Stewart Intl, USA	LEMD	MAD	Madrid Barajas Intl, Spain
KSYR	SYR	Syracuse Hancock Intl, USA	LEMG	AGP	Malaga Airport, Spain
KTEX	TEX	Telluride Regional Airport, USA	LEMH	MAH	Menorca Airport, Spain
KTLH	TLH	Tallahassee Regional Airport, USA	LEPA	PMI	Palma de Mallorca Airport, Spain
KTOL	TOL	Toledo Express Airport, USA	LEPP	PNA	Pamplona Airport, Spain
KTPA	TPA	Tampa Intl, USA	LERS	REU	Reus Airport, Spain
KTRI	TRI	Tri, USA	LESA	SLM	Salamanca Airport, Spain
KTTN	TTN	Trenton, USA	LESO	EAS	San Sebastian Airport, Spain
KTUL	TUL	Tulsa Intl, USA	LEST	SCQ	Santiago Airport, Spain
KTUP	TUP	Tupelo Regional Airport, USA	LETO	TOJ	Madrid Torrejon Airport, Spain
KTUS	TUS	Tucson Intl, USA	LEVC	VLC	Valencia Airport, Spain
KTVC	TVC	Cherry Capital Airport, USA	LEVD	VLL	Valladolid Airport, Spain
KTWF	TWF	Magic Valley Regional Airport, USA	LEVT	VIT	Vitoria Airport, Spain
KTXK	TXK	Texarkana Regional Airport, USA	LEVX	VGO	Vigo Airport, Spain
KTYR	TYR	Tyler Pounds Regional Airport, USA	LEXJ	SDR	Santander Airport, Spain
KTYS	TYS	McGhee Tyson Airport, USA	LEZG	ZAZ	Zaragoza Airport, Spain
KUIN	UIN	Quincy Regional Airport, USA	LEZL	SVQ	Seville Airport, Spain
KUNV	UNV	University Park Airport, USA	LFAB	DPE	Dieppe, France
KVCT	VCT	Victoria Regional Airport, USA	LFAC	CQF	Calais, France
KVCV	VCV	Southern California Logistics, USA	LFAT	LTQ	Le Touquet, France
KVGT	VGT	North Las Vegas Airport, USA	LFBA	AGF	Agen, France
KVLD	VLD	Valdosta Regional Airport, USA	LFBD	BOD	Bordeaux Airport Airport, France
KVPS	VPS	Okaloosa Regional Airport, USA	LFBE	EGC	Bergerac, France
KXNA	XNA	Northwest Arkansas Regional, USA	LFBG	CNG	Cognac, France
KYKM	YKM	Yakima Air Terminal, USA	LFBH	LRH	La Rochelle, France
KYUM	YUM	Yuma Intl / MCAS Yuma, USA	LFBI	PIS	Poitiers, France
LATI	TIA	Rinas Mother Teresa Airport, Albania	LFBK	MCU	Montlucon, France
LBBG	BOJ	Burgas Intl, Bulgaria	LFBL	LIG	Limoges Airport, France
LBGO	GOZ	Gorna Oryahovitsa Airport, Bulgaria	LFBN	NIT	Souche Airport, France
LBPD	PDV	Plovdiv Intl, Bulgaria	LFBO	TLS	Toulouse, France
LBSF	SOF	Sofia Airport, Bulgaria	LFBP	PUF	Pau, France
LBWN	VAR	Varna Intl, Bulgaria	LFBT	LDE	Tarbes, France
LCEN	ECN	Ercan Intl, Cyprus	LFBU	ANG	Angouleme, France
LCLK	LCA	Larnaca Intl, Cyprus	LFBV	BVE	Brive, France
LCPH	PFO	Paphos Intl, Cyprus	LFBX	PGX	Perigueux, France
LDDU	DBV	Dubrovnik Airport, Croatia	LFBZ	BIQ	Biarritz, France

Appendix E

ICAO	IATA	Airport & Country	ICAO	IATA	Airport & Country
LFCC	ZAO	Cahors, France	LFPG	CDG	Paris, France
LFCI	LBI	Albi, France	LFPN	TNF	Toussus – le Noble Airport, France
LFCK	DCM	Castres, France	LFPO	ORY	Paris, France
LFCR	RDZ	Rodez, France	LFPT	POX	Pontoise, France
LFCY	RYN	Royan, France	LFQD	QRV	Arras, France
LFDH	AUC	Auch, France	LFQF	XXG	Autun, France
LFDN	RCO	Rochefort, France	LFQG	NVS	Nevers, France
LFEY	IDY	d'Yeu, France	LFQM	QBQ	Besancon, France
LFGA	CMR	Colmar, France	LFQQ	LIL	Lille, France
LFGJ	DLE	Dole, France	LFQT	HZB	Merville, France
LFHM	MVV	Megeve Airport, France	LFRB	BES	Brest Bretagne Airport, France
LFHO	OBS	Aubenas Ardeche Airport, France	LFRC	CER	Cherbourg, France
LFHP	LPY	Le Puy, France	LFRD	DNR	Dinard, France
LFJL	ETZ	Metz, France	LFRE	LBY	Labaule, France
LFJR	ANE	Angers Loire Aeroport Airport, France	LFRF	GFR	Granville, France
LFKB	BIA	Bastia, France	LFRG	DOL	Deauville, France
LFKC	CLY	Calvi, France	LFRH	LRT	Lorient, France
LFKF	FSC	Figari, France	LFRI	EDM	La Roche sur Yon, France
LFKJ	AJA	Campo Dell Oro Airport, France	LFRK	CFR	Caen, France
LFKO	PRP	Tavaria Airport, France	LFRM	LME	Le Mans, France
LFKS	SOZ	Solenzara, France	LFRN	RNS	Rennes, France
LFKX	MFX	Meribel Airport, France	LFRO	LAI	Lannion, France
LFLA	AUF	Auxerre, France	LFRQ	UIP	Quimper, France
LFLB	CMF	Chambery, France	LFRS	NTE	Nantes Atlantique Airport, France
LFLC	CFE	Clermont, France	LFRT	SBK	Saint, France
LFLD	BOU	Bourges Airport, France	LFRU	MXN	Morlaix, France
LFLJ	CVF	Courchevel Airport, France	LFRV	VNE	Vannes, France
LFLL	LYS	Lyon Saint Exupery Airport, France	LFRZ	SNR	Saint, France
LFLO	RNE	Roanne, France	LFSB	MLH	EuroAirport Basel, France
LFLP	NCY	Annecy, France	LFSD	DIJ	Dijon, France
LFLS	GNB	Grenoble, France	LFSF	MZM	Metz, France
LFLU	VAF	Valence, France	LFSG	EPL	Epinal, France
LFLV	VHY	Vichy, France	LFSN	ENC	Nancy, France
LFLW	AUR	Aurillac Airport, France	LFSR	RHE	Reims, France
LFLX	CHR	Chateauroux, France	LFST	SXB	Strasbourg Airport, France
LFLY	LYN	Lyon, France	LFSZ	VTL	Vittel, France
LFMD	CEQ	Cannes, France	LFVP	FSP	St. Pierre Airport, Saint
LFMH	EBU	Saint, France	LFXC	VTL	Vittel, France
LFMK	CCF	Carcassonne en Pays, France	LGAL	AXD	Alexandroupolis Intl, Greece
LFML	MRS	Marseille, France	LGAV	ATH	Athens Intl, Greece
LFMN	NCE	Nice, France	LGBL	VOL	Aghialos National Airport, Greece
LFMP	PGF	Perpignan, France	LGHI	JKH	Chios Island National Airport, Greece
LFMQ	CTT	Le Castellet Airport, France	LGIK	JIK	Ikaria Island National Airport, Greece
LFMT	MPL	Montpellier, France	LGIO	IOA	Ioannina National Airport, Greece
LFMU	BZR	Beziers, France	LGIR	HER	Heraklion Intl, Greece
LFMV	AVN	Avignon Airport, France	LGKA	KSO	Kastoria National Airport, Greece
LFNA	GAT	Tallard Airport, France	LGKC	KIT	Kithira Island National Airport, Greece
LFNB	MEN	Mende, France	LGKF	EFL	Kefalonia Island Intl, Greece
LFOB	BVA	Beauvais, France	LGKJ	KZS	Kastelorizo Island Public, Greece
LFOE	EVX	Evreux, France	LGKL	KLX	Kalamata Intl, Greece
LFOH	LEH	Le Havre, France	LGKO	KGS	Kos Island Intl, Greece
LFOI	XAB	Abbeville, France	LGKP	AOK	Karpathos Island National, Greece
LFOJ	ORE	Orleans, France	LGKR	CFU	Corfu Intl, Greece
LFOP	URO	Rouen, France	LGKS	KSJ	Kassos Island Public Airport, Greece
LFOQ	XBQ	Blois, France	LGKV	KVA	Kavala Intl, Greece
LFOT	TUF	Tours, France	LGKZ	KZI	Kozani National Airport, Greece
LFOU	CET	Cholet Lepontreau Airport, France	LGLM	LXS	Lemnos Intl, Greece
LFOV	LVA	Laval, France	LGMK	JMK	Mykonos Island National, Greece
LFPB	LBG	Paris, France	LGML	MLO	Milos Island National Airport, Greece
LFPC	CSF	Creil Airport, France	LGMT	MJT	Mytilene Intl, Greece

PlanePlotter User Guide

ICAO	IATA	Airport & Country	ICAO	IATA	Airport & Country
LGNX	JNX	Naxos Island National Airport, Greece	LIPZ	VCE	Marco Polo Venice Airport, Italy
LGPA	PAS	Paros National Airport, Greece	LIQL	LCV	Tassignano Airport, Italy
LGPL	JTY	Astypalaia Island National, Greece	LIQS	SAY	Siena Airport, Italy
LGPZ	PVK	Aktion National Airport, Greece	LIRA	CIA	Ciampino Airport, Italy
LGRP	RHO	Rhodes Intl, Greece	LIRF	FCO	Fiumicino, Italy
LGRX	GPA	Araxos National Airport, Greece	LIRJ	EBA	Aeroporto Marina di Campo, Italy
LGSA	CHQ	Chania Intl, Greece	LIRL	QLT	Latina Airport, Italy
LGSK	JSI	Skiathos Island National, Greece	LIRN	NAP	Naples Intl, Italy
LGSM	SMI	Samos Intl, Greece	LIRP	PSA	Galileo Galilei Airport, Italy
LGSO	JSY	Syros Island National Airport, Greece	LIRQ	FLR	A. Vespucci Florence Airport, Italy
LGSR	JTR	Santorini National Airport, Greece	LIRS	GRS	Grosseto Airport, Italy
LGST	JSH	Sitia Public Airport, Greece	LIRZ	PEG	San Egidio Airport, Italy
LGSY	SKU	Skyros Island National Airport, Greece	LJLJ	LJU	Aerodrom Ljubljana, Slovenia
LGTS	SKG	Thessaloniki Intl, Greece	LJMB	MBX	Maribor Airport, Slovenia
LGZA	ZTH	Zakynthos Intl, Greece	LJPZ	POW	Portoroz Airport, Slovenia
LHBP	BUD	Budapest Ferihegy Intl, Hungary	LKKU	UHE	Kunovice Airport, Czech Republic
LHDC	DEB	Debrecen Intl, Hungary	LKKV	KLV	Karlovy Vary Intl, Czech Republic
LHMC	MCQ	Miskolc Airport, Hungary	LKMT	OSR	Ostrava, Czech Republic
LHPR	QGY	Gyor, Hungary	LKPD	PED	Pardubice Airport, Czech Republic
LHSM	SOB	S, Hungary	LKPR	PRG	Prague Ruzyne Intl, Czech Republic
LIBC	CRV	Crotone Airport, Italy	LKTB	BRQ	Brno Turany Airport, Czech Republic
LIBD	BRI	Palese Macchie Airport, Italy	LLBG	TLV	Ben Gurion Intl, Israel
LIBF	FOG	Gino Lisa Airport, Italy	LLBS	BEV	Teyman Airport, Israel
LIBG	TAR	Grottaglie Airport, Italy	LLET	ETH	Eilat Airport, Israel
LIBN	LCC	Lecce Airport, Italy	LLEY	EIY	Ein Yahav Airfield, Israel
LIBP	PSR	Abruzzo Intl, Italy	LLHA	HFA	Haifa Airport, Israel
LIBR	BDS	Casale Airport, Italy	LLIB	RPN	Rosh Pina Airport, Israel
LICA	SUF	Lamezia Terme Intl, Italy	LLKS	KSW	Qiryat Shemona Airport, Israel
LICC	CTA	Catania, Italy	LLMR	MIP	Mitzpe Ramon Airfield, Israel
LICD	LMP	Lampedusa Airport, Italy	LLMZ	MTZ	Masada Airfield, Israel
LICG	PNL	Pantelleria Airport, Italy	LLOV	VDA	Ovda Intl, Israel
LICJ	PMO	Palermo Intl, Italy	LLSD	SDV	Sde Dov Airport, Israel
LICR	REG	Reggio Calabria Airport, Italy	LMML	MLA	Luqa, Malta
LICT	TPS	Vincenzo Florio Airport, Italy	LNMC	MCM	Monaco Heliport, Monaco
LIDB	BLX	Belluno Airport, Italy	LOWG	GRZ	Graz, Austria
LIDR	RAN	Ravenna Airport, Italy	LOWI	INN	Innsbruck, Austria
LIEA	AHO	Alghero Airport, Italy	LOWK	KLU	Klagenfurt, Austria
LIEE	CAG	Cagliari, Italy	LOWL	LNZ	Linz, Austria
LIEO	OLB	Olbia, Italy	LOWS	SZG	Salzburg, Austria
LIET	TTB	Tortoli, Italy	LOWW	VIE	Wien, Austria
LIMC	MXP	Malpensa Intl, Italy	LPAZ	SMA	Santa Maria Airport, Portugal
LIME	BGY	Orio al Serio Intl, Italy	LPBG	BGC	Braganca Airport, Portugal
LIMF	TRN	Torino Caselle Airport, Italy	LPBR	BGZ	Braga Airport, Portugal
LIMG	ALL	Villanova d'Albenga Intl, Italy	LPCH	CHV	Chaves Airport, Portugal
LIMJ	GOA	Genoa Cristoforo Colombo, Italy	LPCO	CBP	Coimbra Airport, Portugal
LIML	LIN	Linate Airport, Italy	LPCR	CVU	Corvo Airport, Portugal
LIMP	PMF	Parma Airport, Italy	LPCV	COV	Covilha Airport, Portugal
LIMW	AOT	Aosta Airport, Italy	LPFL	FLW	Flores Airport, Portugal
LIMZ	CUF	Cuneo Levaldigi Airport, Italy	LPFR	FAO	Faro Airport, Portugal
LIPB	BZO	Bolzano Dolomiti Airport, Italy	LPGR	GRW	Graciosa Airport, Portugal
LIPD	UDN	Campoformido Airport, Italy	LPHR	HOR	Horta Airport, Portugal
LIPE	BLQ	Aeroporto G. Marconi di Bologna, Italy	LPMA	FNC	Madeira Airport, Portugal
LIPH	TSF	Treviso Airport, Italy	LPPD	PDL	Joao Paulo II Airport, Portugal
LIPK	FRL	Forli Airport, Italy	LPPI	PIX	Pico Airport, Portugal
LIPO	VBS	Montichiari Airport, Italy	LPPM	PRM	Portimao Airport, Portugal
LIPQ	TRS	Friuli Venezia Giulia Airport, Italy	LPPR	OPO	Francisco Sa Carneiro, Portugal
LIPR	RMI	Federico Fellini Intl, Italy	LPPS	PXO	Porto Santo Airport, Portugal
LIPT	VIC	Vicenza Trissino Airport, Italy	LPPT	LIS	Portela Airport, Portugal
LIPX	VRN	Verona Airport, Italy	LPSI	SIE	Sines Airport, Portugal
LIPY	AOI	Ancona Falconara Airport, Italy	LPSJ	SJZ	Sao Jorge Airport, Portugal

Appendix E

ICAO	IATA	Airport & Country
LPVR	VRL	Vila Real Airport, Portugal
LPVZ	VSE	Viseu Airport, Portugal
LQBK	BNX	Banja Luka Intl, Bosnia
LQMO	OMO	Mostar Intl, Bosnia
LQSA	SJJ	Sarajevo Intl, Bosnia
LQTZ	TZL	Tuzla Intl, Bosnia
LRAR	ARW	Arad Intl, Romania
LRBC	BCM	Bacau Intl, Romania
LRBM	BAY	Baia Mare Airport, Romania
LRBS	BBU	Bucharest Intl, Romania
LRCK	CND	Constanta Intl, Romania
LRCL	CLJ	Cluj, Romania
LRCS	CSB	Caransebes Airport, Romania
LRCV	CRA	Craiova Airport, Romania
LRIA	IAS	Iasi Intl, Romania
LROD	OMR	Oradea Intl, Romania
LROP	OTP	Bucharest Intl, Romania
LRSB	SBZ	Sibiu Intl, Romania
LRSM	SUJ	Satu Mare Intl, Romania
LRSV	SCV	Suceava Airport, Romania
LRTC	TCE	Tulcea Airport, Romania
LRTM	TGM	Targu Mures Intl, Romania
LRTR	TSR	Timisoara Intl, Romania
LSGG	GVA	Geneva, Switzerland
LSGS	SIR	Sion Airport, Switzerland
LSZA	LUG	Lugano Airport, Switzerland
LSZB	BRN	Bern, Switzerland
LSZH	ZRH	Zurich Airport, Switzerland
LSZR	ACH	St Gallen, Switzerland
LSZS	SMV	Samedan Airport, Switzerland
LTAC	ESB	Ankara Esenboga Intl, Turkey
LTAF	ADA	Adana Airport, Turkey
LTAI	AYT	Antalya Airport, Turkey
LTAJ	GZT	Oguzeli Airport, Turkey
LTAN	KYA	Konya Airport, Turkey
LTAR	VAS	Sivas Airport, Turkey
LTAT	MLX	Malatya Erhac Airport, Turkey
LTAU	ASR	Kayseri Erkilet Intl, Turkey
LTAY	DNZ	Denizli Cardak Airport, Turkey
LTAZ	NAV	Nevsehir Kapadokya Airport, Turkey
LTBA	IST	Istanbul Ataturk Intl, Turkey
LTBH	CKZ	Canakkale Airport, Turkey
LTBJ	ADB	Izmir Adnan Menderes Airport, Turkey
LTBR	YEI	Bursa Yenisehir Airport, Turkey
LTBS	DLM	Dalaman Airport, Turkey
LTBU	TEQ	Corlu Airport, Turkey
LTCA	EZS	Elazig Airport, Turkey
LTCC	DIY	Diyarbakir Airport, Turkey
LTCD	ERC	Erzincan Airport, Turkey
LTCE	ERZ	Erzurum Airport, Turkey
LTCF	KSY	Kars Airport, Turkey
LTCG	TZX	Trabzon Airport, Turkey
LTCH	SFQ	Sanliurfa Airport, Turkey
LTCI	VAN	Van Ferit Melen Airport, Turkey
LTCJ	BAL	Batman Airport, Turkey
LTCK	MSR	Mus Airport, Turkey
LTCL	SXZ	Siirt Airport, Turkey
LTCN	KCM	Kahramanmaras Airport, Turkey
LTCO	AJI	Agri Airport, Turkey
LTCP	ADF	Adiyaman Airport, Turkey

ICAO	IATA	Airport & Country
LTCR	MQM	Mardin Airport, Turkey
LTFC	ISE	Suleyman Demirel Airport, Turkey
LTFD	EDO	Edremit Korfez Airport, Turkey
LTFE	BJV	Milas, Turkey
LTFH	SZF	Samsun Carsamba Airport, Turkey
LTFJ	SAW	Istanbul Sabiha Gokcen Intl, Turkey
LUKK	KIV	Chisinau, Moldova
LWOH	OHD	Ohrid, Macedonia
LWOH	OHD	Ohrid Airport, Macedonia
LWSK	SKP	Skopje, Macedonia
LWSK	SKP	Skopje Airport, Macedonia
LXGB	GIB	Gibraltar, Gibraltar
LYBE	BEG	Belgrade Nikola Tesla Airport, Serbia
LYNI	INI	Nis Constantine the Great Intl, Serbia
LYNS	QND	Novi Sad Airport, Serbia
LYPA	QBG	Pancevo Airport, Serbia
LYPG	TGD	Podgorica, Montenegro
LYPR	PRN	Pristina Intl, Serbia
LYTV	TIV	Tivat Airport, Montenegro
LYUZ	ZZE	Uzice Airport, Serbia
LYVA	QWV	Valjevo Airport, Serbia
LZIB	BTS	Bratislava, Slovakia
LZKZ	KSC	Kosice Airport, Slovakia
LZPP	PZY	Piestany Airport, Slovakia
LZSL	SLD	Sliac Airport, Slovakia
LZTT	TAT	Tatry Airport, Slovakia
LZZI	ILZ	Zilina Airport, Slovakia
MBGT	GDT	Grand Turk Intl, Turks & Caicos
MDBH	BRX	Maria Montez Intl, Dominican Repub
MDCR	CBJ	Cabo Rojo Airport, Dominican Repub
MDCZ	COZ	Constanza Airport, Dominican Repub
MDHE	HEX	Herrera Intl, Dominican Republic
MDLI	SLI	La Isabela Intl, Dominican Republic
MDLR	LRM	La Romana Intl, Dominican Republic
MDPC	PUJ	Punta Cana Intl, Dominican Republic
MDPO	EPS	El Portillo Airport, Dominican Republic
MDPP	POP	Gregorio Luperon, Dominican Repub
MDSB	SNX	Sabana de la Mar, Dominican Repub
MDSD	SDQ	Las Americas, Dominican Republic
MDSJ	SJM	San Juan de la Maguana, Dom Repub
MDST	STI	Cibao Intl, Dominican Republic
MGCB	CBV	Coban Airport, Guatemala
MGCR	CMM	Carmelita Airport, Guatemala
MGCT	CTF	Coatepeque Airport, Guatemala
MGGT	GUA	Guatemala La Aurora Intl, Guatemala
MGHT	HUG	Huehuetenango Airport, Guatemala
MGMM	MCR	Melchor de Mencos, Guatemala
MGPB	PBR	Puerto Barrios Airport, Guatemala
MGPP	PON	Poptun Airport, Guatemala
MGQC	AQB	Quiche Airport, Guatemala
MGQZ	AAZ	Quezaltenango Airport, Guatemala
MGRB	RUV	Rubelsanto Airport, Guatemala
MGRT	RER	Retalhuleu Airport, Guatemala
MGTK	FRS	Mundo Maya Intl, Guatemala
MHCA	CAA	Catacamas, Honduras
MHLC	LCE	Goloson Intl, Honduras
MHLE	LEZ	La Esperanza, Honduras
MHLM	SAP	La Mesa Intl, Honduras
MHPL	PEU	Puerto Lempira, Honduras
MHRO	RTB	Roatan Intl, Honduras

ICAO	IATA	Airport & Country	ICAO	IATA	Airport & Country
MHTE	TEA	Tela, Honduras	MMZC	ZCL	La Calera Airport, Mexico
MHTG	TGU	Toncontin Intl, Honduras	MMZH	ZIH	Ixtapa, Mexico
MKJP	KIN	Norman Manley Intl, Jamaica	MMZO	ZLO	Playa de Oro Intl, Mexico
MKJS	MBJ	Sangster Intl, Jamaica	MNBL	BEF	Bluefields, Nicaragua
MMAA	ACA	General Juan N. Alvarez Intl, Mexico	MNMG	MGA	Augusto Cesar Sandino, Nicaragua
MMAS	AGU	Lic. Jesus Teran Peredo Intl, Mexico	MNPC	PUZ	Puerto Cabezas, Nicaragua
MMBT	HUX	Bah, Mexico	MPBO	BOC	Bocas Del Toro Intl, Panama
MMCB	CVJ	General Mariano Matamoros, Mexico	MPCH	CHX	Cap. Manuel Nino Intl, Panama
MMCE	CME	Ciudad del Carmen Intl, Mexico	MPDA	DAV	Enrique Malek Intl, Panama
MMCL	CUL	Federal de Bachigualato Intl, Mexico	MPEJ	ONX	Enrique Adolfo Jimenez, Panama
MMCM	CTM	Chetumal Intl, Mexico	MPJE	JQE	Jaque, Panama
MMCN	CEN	Ciudad Obregon Intl, Mexico	MPTO	PTY	Tocumen Intl, Panama
MMCP	CPE	Ing. Alberto Acuna Ongay Intl, Mexico	MRAN	FON	La Fortuna Airport, Costa Rica
MMCS	CJS	Abraham Gonzalez Intl, Mexico	MRAO	TTQ	Tortuguero Airport, Costa Rica
MMCU	CUU	Gen Roberto Fierro Villalobos, Mexico	MRBA	BAI	Buenos Aires Airport, Costa Rica
MMCV	CVM	General Pedro J. Mendez, Mexico	MRBC	BCL	Barra del Colorado Airport, Costa Rica
MMCZ	CZM	Cozumel Intl, Mexico	MRCC	OTR	Coto 47 Airport, Costa Rica
MMDO	DGO	General Guadalupe Victoria, Mexico	MRCR	RIK	Carrillo Airport, Costa Rica
MMEP	TPQ	Amado Nervo National Airport, Mexico	MRDK	DRK	Drake Bay Airport, Costa Rica
MMGL	GDL	Don Miguel Hidalgo y Costilla, Mexico	MRGF	GLF	Golfito Airport, Costa Rica
MMGM	GYM	Gen Jose Maria Yanez Intl, Mexico	MRGP	GPL	Guapiles Airport, Costa Rica
MMGR	GUB	Guerrero Negro Airport, Mexico	MRIA	PBP	Punta Islita Airport, Costa Rica
MMHO	HMO	Ignacio Pesqueira Garcia, Mexico	MRLB	LIR	Daniel Oduber Intl, Costa Rica
MMIA	CLQ	Lic. Miguel de Madrid Airport, Mexico	MRLC	LSL	Los Chiles Airport, Costa Rica
MMIO	SLW	Plan de Guadalupe Intl, Mexico	MRLM	LIO	Limon Intl, Costa Rica
MMJA	JAL	El Lencero Airport, Mexico	MRNS	NOM	Nosara Airport, Costa Rica
MMLC	LZC	Lazaro Cardenas Airport, Mexico	MROC	SJO	Juan Santamaria Intl, Costa Rica
MMLM	LMM	Federal del Valle del Fuerte, Mexico	MRPJ	PJM	Puerto Jimenez Airport, Costa Rica
MMLO	BJX	Del Bajio Intl, Mexico	MRPM	PMZ	Palmar Sur Airport, Costa Rica
MMLP	LAP	Manuel Marquez de Leon Intl, Mexico	MRPV	SYQ	Tobias Bolanos Intl, Costa Rica
MMLT	LTO	Loreto Intl, Mexico	MRQP	XQP	La Managua Airport, Costa Rica
MMMA	MAM	General Servando Canales, Mexico	MRTM	TNO	Tamarindo Airport, Costa Rica
MMMD	MID	Manuel Crescencio Rej, Mexico	MRTR	TMU	Tambor Airport, Costa Rica
MMML	MXL	General Rodolfo S, Mexico	MRUP	UPL	Upala Airport, Costa Rica
MMMM	MLM	General Francisco J. Mujica, Mexico	MTCH	CAP	Cap Haitien Intl, Haiti
MMMT	MTT	Coatzacoalcos National, Mexico	MTJA	JAK	Jacmel, Haiti
MMMV	LOV	Venustiano Carranza Intl, Mexico	MTJE	JEE	Jeremie, Haiti
MMMX	MEX	Mexico City Intl, Mexico	MTPP	PAP	Port, Haiti
MMMY	MTY	General Mariano Escobedo, Mexico	MTPX	PAX	Port De Paix, Haiti
MMMZ	MZT	General Rafael Buelna Intl, Mexico	MUBA	BCA	Gustavo Rizo Airport, Cuba
MMNL	NLD	Quetzalcoatl Intl, Mexico	MUBY	BYM	Carlos Manuel de Cespedes, Cuba
MMOX	OAX	Xoxocotlan Intl, Mexico	MUCA	AVI	Maximo Gomez Intl, Cuba
MMPA	PAZ	El Tajin National Airport, Mexico	MUCF	CFG	Jaime Gonzalez Airport, Cuba
MMPB	PBC	Hermanos Serdan Intl, Mexico	MUCL	CYO	Vilo Acuna Airport, Cuba
MMPG	PDS	Piedras Negras Intl, Mexico	MUCM	CMW	Ignacio Agramonte Intl, Cuba
MMPN	UPN	Gen. Ignacio Lopez Rayon, Mexico	MUCU	SCU	Antonio Maceo Intl, Cuba
MMPR	PVR	Lic. Gustavo Diaz Ordaz Intl, Mexico	MUGT	GAO	Mariana Grajales Airport, Cuba
MMPS	PXM	Puerto Escondido Intl, Mexico	MUHA	HAV	Jose Marti Intl, Cuba
MMRX	REX	General Lucio Blanco Intl, Mexico	MUHG	HOG	Frank Pais Airport, Cuba
MMSD	SJD	Los Cabos Intl, Mexico	MUKW	VRO	Kawama Airport, Cuba
MMSP	SLP	Ponciano Arriaga Intl, Mexico	MULM	LCL	La Coloma Airport, Cuba
MMTC	TRC	Torreon Intl, Mexico	MUMO	MOA	Orestes Acosta Airport, Cuba
MMTG	TGZ	Llano San Juan Airport, Mexico	MUMZ	MZO	Sierra Maestra Airport, Cuba
MMTJ	TIJ	General Abelardo, Mexico	MUNB	QSN	San Nicolas de Bari Airport, Cuba
MMTM	TAM	Gen Francisco Javier Mina, Mexico	MUNC	ICR	Nicaro Airport, Cuba
MMTO	TLC	Lic. Adolfo Lopez Mateos, Mexico	MUNG	GER	Rafael Cabrera Airport, Cuba
MMTP	TAP	Tapachula Intl, Mexico	MUOC	CCC	Cayo Coco Airport, Cuba
MMUN	CUN	Cancun Intl, Mexico	MUPR	QPD	Pinar del R, Cuba
MMVA	VSA	Carlos Rovirosa Perez, Mexico	MUSC	SNU	Abel Santa Maria Airport, Cuba
MMVR	VER	General Heriberto Jara Intl, Mexico	MUSN	SZJ	Siguanea Airport, Cuba

Appendix E

ICAO	IATA	Airport & Country	ICAO	IATA	Airport & Country
MUTD	TND	Alberto Delgado Airport, Cuba	NWWR	MEE	La Roche Aero, New Caledonia
MUVR	VRA	Juan Gualberto Gomez, Cuba	NWWU	TOU	Touho, New Caledonia
MUVT	VTU	Hermanos Ameijeiras Airport, Cuba	NWWV	UVE	Ouloup, New Caledonia
MWCB	CYB	Gerrard, Cayman Islands	NWWW	NOU	Tontouta, New Caledonia
MWCR	GCM	Roberts Intl, Cayman Islands	NZAA	AKL	Auckland Intl, New Zealand
MYAB	MAY	Clarence Bain Airport, Bahamas	NZAP	TUO	Taupo, New Zealand
MYAF	ASD	Andros Town Intl, Bahamas	NZAS	ASG	Ashburton, New Zealand
MYAK	COX	Congo Town Airport, Bahamas	NZCH	CHC	Christchurch Intl, New Zealand
MYAM	MHH	Marsh Harbour Airport, Bahamas	NZCI	CHT	Chatham Is/Tuuta, New Zealand
MYAN	SAQ	San Andros Airport, Bahamas	NZDN	DUD	Dunedin, New Zealand
MYAP	AXP	Spring Point Airport, Bahamas	NZGB	GBZ	Great Barrier, New Zealand
MYAT	TCB	Treasure Cay Airport, Bahamas	NZGM	GMN	Greymouth, New Zealand
MYBC	CCZ	Chub Cay Intl, Bahamas	NZGS	GIS	Gisborne, New Zealand
MYBG	GHC	Great Harbour Cay Airport, Bahamas	NZGT	GTN	Glentanner, New Zealand
MYBS	BIM	South Bimini Airport, Bahamas	NZHK	HKK	Hokitika, New Zealand
MYCA	ATC	Arthurs Town Airport, Bahamas	NZHN	HLZ	Hamilton Intl, New Zealand
MYCB	CAT	New Bight Airport, Bahamas	NZKI	KBZ	Kaikoura, New Zealand
MYCI	CRI	Colonial Hill Airport, Bahamas	NZKK	KKE	Kerikeri, New Zealand
MYEF	GGT	Exuma Intl, Bahamas	NZKO	KKO	Kaikohe, New Zealand
MYEH	ELH	North Eleuthera Airport, Bahamas	NZKT	KAT	Kaitaia, New Zealand
MYEM	GHB	Governor's Harbour Airport, Bahamas	NZLX	ALR	Alexandra, New Zealand
MYEN	NMC	Norman's Cay Airport, Bahamas	NZMA	MTA	Matamata, New Zealand
MYER	RSD	Rock Sound Intl, Bahamas	NZMC	MON	Mount Cook, New Zealand
MYES	TYM	Staniel Cay Airport, Bahamas	NZMF	MFN	Milford Sound, New Zealand
MYGF	FPO	Grand Bahama Intl, Bahamas	NZMS	MRO	Masterton, New Zealand
MYIG	IGA	Inagua Airport, Bahamas	NZNP	NPL	New Plymouth, New Zealand
MYLD	LGI	Deadman's Cay Airport, Bahamas	NZNR	NPE	Napier, New Zealand
MYLS	SML	Stella Maris Airport, Bahamas	NZNS	NSN	Nelson, New Zealand
MYMM	MYG	Mayaguana Airport, Bahamas	NZNV	IVC	Invercargill, New Zealand
MYNN	NAS	Nassau Intl, Bahamas	NZOH	OHA	Ohakea, New Zealand
MYPI	PID	New Providence Airport, Bahamas	NZOU	OAM	Oamaru, New Zealand
MYSM	ZSA	San Salvador Airport, Bahamas	NZPM	PMR	Palmerston North, New Zealand
MZBZ	BZE	Philip S W Goldson Intl, Belize	NZPP	PPQ	Paraparaumu, New Zealand
N55	UIT	Jaluit, Marshall Islands	NZQN	ZQN	Queenstown, New Zealand
NCAI	AIT	Avarua, Cook Islands	NZRA	RAG	Raglan, New Zealand
NCMK	MUK	Akatoka Manava, Cook Islands	NZRO	ROT	Rotorua, New Zealand
NCPY	PYE	Tongareva, Cook Islands	NZTG	TRG	Tauranga, New Zealand
NCRG	RAR	Rarotonga Intl, Cook Islands	NZTH	TMZ	Thames, New Zealand
NFFN	NAN	Nadi Intl, Fiji	NZTU	TIU	Timaru, New Zealand
NFNA	SUV	Nausori Intl, Fiji	NZWB	BHE	Woodbourne, New Zealand
NFNL	LBS	Labasa, Fiji	NZWF	WKA	Wanaka, New Zealand
NFNR	RTA	Rotuma, Fiji	NZWK	WHK	Whakatane, New Zealand
NFTF	TBU	Fua'amotu Intl, Tonga	NZWN	WLG	Wellington Intl, New Zealand
NFTL	HPA	Ha'apai, Tonga	NZWO	WIR	Wairoa, New Zealand
NFTP	NTT	Niuatoputapu, Tonga	NZWR	WRE	Whangarei, New Zealand
NFTV	VAV	Vava'u, Tonga	NZWS	WSZ	Westport, New Zealand
NGFU	FUN	Funafuti Intl, Tuvalu	NZWU	WAG	Wanganui, New Zealand
NGTA	TRW	Bonriki Intl, Kiribati	OABN	BIM	Bamyan, Afghanistan
NGTU	BBG	Butaritari, Kiribati	OABT	BST	Bost, Afghanistan
NTTB	BOB	Bora Bora Airport, French Polynesia	OACC	CCN	Charkhcharan, Afghanistan
NTTG	RGI	Rangiroa Airport, French Polynesia	OAFR	FAH	Farah, Afghanistan
NTTR	RFP	Raiatea Airport, French Polynesia	OAFZ	FBD	Faizabad, Afghanistan
NVSS	SON	Santo, Vanuatu	OAGZ	GRG	Gardez, Afghanistan
NVVV	VLI	Bauerfield, Vanuatu	OAHN	KWH	Khwahan, Afghanistan
NWWD	KNQ	Kone, New Caledonia	OAHR	HEA	Herat, Afghanistan
NWWE	ILP	Moue, New Caledonia	OAJL	JAA	Jalalabad, Afghanistan
NWWH	HLU	Nesson, New Caledonia	OAKB	KBL	Kabul Intl, Afghanistan
NWWK	KOC	Koumac, New Caledonia	OAKN	KDH	Kandahar, Afghanistan
NWWL	LIF	Ouanaham, New Caledonia	OAKS	KHT	Khost, Afghanistan
NWWM	GEA	Magenta, New Caledonia	OAMN	MMZ	Maimama, Afghanistan

ICAO	IATA	Airport & Country	ICAO	IATA	Airport & Country
OAMS	MZR	Mazar, Afghanistan	OITT	TBZ	Tabriz Intl, Iran
OASN	SGA	Sheghnan, Afghanistan	OIYY	AZD	Shahid Sadooghi Airport, Iran
OAUZ	UND	Kunduz, Afghanistan	OIZB	ACZ	Zabol Airport, Iran
OAZJ	ZAJ	Zaranj, Afghanistan	OIZC	ZBR	Konarak Airport, Iran
OBBI	BAH	Bahrain Intl, Bahrain	OIZH	ZAH	Zahedan Intl, Iran
OEAB	AHB	Abha Regional Airport, Saudi Arabia	OJAI	AMM	Queen Alia Intl, Jordan
OEBH	BHH	Bisha Domestic Airport, Saudi Arabia	OJAM	ADJ	Marka Intl, Jordan
OEDF	DMM	King Fahd Intl, Saudi Arabia	OJAQ	AQJ	Aqaba Intl, Jordan
OEDW	DWD	Dawadmi Domestic, Saudi Arabia	OKBK	KWI	Kuwait Intl, Kuwait
OEGN	GIZ	Gizan Regional Airport, Saudi Arabia	OLBA	BEY	Beirut Intl, Lebanon
OEGS	ELQ	Gassim Regional Airport, Saudi Arabia	OLKA	KYE	Rene Mouawad AB, Lebanon
OEGT	URY	Gurayat Domestic, Saudi Arabia	OMAA	AUH	Abu Dhabi Intl, United Arab Emirates
OEHL	HAS	Hail Regional Airport, Saudi Arabia	OMAD	AZI	Bateen, United Arab Emirates
OEJN	JED	King Abdulaziz Intl, Saudi Arabia	OMAL	AAN	Al Ain Intl, United Arab Emirates
OEKK	HBT	Hafr al, Saudi Arabia	OMDB	DXB	Dubai Intl, United Arab Emirates
OEMA	MED	Moh Bin Abdulaziz, Saudi Arabia	OMFJ	FJR	Fujairah Intl, United Arab Emirates
OENG	EAM	Najran Domestic Airport, Saudi Arabia	OMRK	RKT	Khaimah Intl, United Arab Emirates
OEPA	EPA	Qaisumah Domestic, Saudi Arabia	OMSJ	SHJ	Sharjah Intl, United Arab Emirates
OERF	RAH	Rafha Domestic Airport, Saudi Arabia	OOKB	KHS	Khasab AB, Oman
OERK	RUH	King Khalid Intl, Saudi Arabia	OOMA	MSH	Masirah AB, Oman
OERR	RAE	Arar Domestic Airport, Saudi Arabia	OOMS	MCT	Seeb Intl, Oman
OESH	ESH	Sharorah Domestic, Saudi Arabia	OONR	OMM	Marmul, Oman
OETB	TUU	Tabuk Regional Airport, Saudi Arabia	OOSA	SLL	Salalah, Oman
OETF	TIF	Taif Regional Airport, Saudi Arabia	OOTH	TTH	Thumrait AB, Oman
OETR	TUI	Turaif Domestic Airport, Saudi Arabia	OPBN	BNP	Bannu Airport, Pakistan
OEWD	EWD	Wadi al, Saudi Arabia	OPBW	BHV	Bahawalpur Airport, Pakistan
OEWJ	EJH	Wedjh Domestic Airport, Saudi Arabia	OPCH	CJL	Chitral Airport, Pakistan
OEYN	YNB	Yanbu Domestic Airport, Saudi Arabia	OPCL	CHB	Chilas Airport, Pakistan
OIAA	ABD	Abadan Airport, Iran	OPDB	DBA	Dalbandin Airport, Pakistan
OIAM	MRX	Mahshahr Airport, Iran	OPDG	DEA	Dera Ghazi Khan Airport, Pakistan
OIAW	AWZ	Ahwaz Airport, Iran	OPDI	DSK	Dera Ismail Khan Airport, Pakistan
OIBB	BUZ	Bushehr Airport, Iran	OPFA	LYP	Faisalabad Intl, Pakistan
OIBI	YEH	Asalouyeh Airport, Iran	OPGD	GWD	Gwadar Intl, Pakistan
OIBK	KIH	Kish Airport, Iran	OPGT	GIL	Gilgit Airport, Pakistan
OIBL	BDH	Bandar Lengeh Airport, Iran	OPJI	JIW	Jiwani Airport, Pakistan
OIBQ	KHK	Khark Airport, Iran	OPKC	KHI	Jinnah Intl, Pakistan
OIBS	SXI	Sirri Island Airport, Iran	OPKD	HDD	Hyderabad Airport, Pakistan
OIBV	LVP	Lavan Airport, Iran	OPKH	KDD	Khuzdar Airport, Pakistan
OICC	KSH	Shahid Ashrafi Esfahani Airport, Iran	OPLA	LHE	Allama Iqbal Intl, Pakistan
OICK	KHD	Khoram Abad Airport, Iran	OPMA	XJM	Mangla Airport, Pakistan
OICS	SDG	Sanandaj Airport, Iran	OPMF	MFG	Muzaffarabad Airport, Pakistan
OIFM	IFN	Isfahan Intl, Iran	OPMJ	MJD	Moenjodaro Airport, Pakistan
OIGG	RAS	Rasht Airport, Iran	OPMP	MPD	Sindhri Airport, Pakistan
OIHH	HDM	Hamadan Airport, Iran	OPMT	MUX	Multan Intl, Pakistan
OIIE	IKA	Imam Khomeini Intl, Iran	OPNH	WNS	Nawabshah Airport, Pakistan
OIII	THR	Mehrabad Intl, Iran	OPOR	ORW	Ormara Airport, Pakistan
OIKB	BND	Bandar Abbass Intl, Iran	OPPC	PAJ	Parachinar Airport, Pakistan
OIKK	KER	Kerman Airport, Iran	OPPG	PJG	Panjgur Airport, Pakistan
OIKM	BXR	Bam Airport, Iran	OPPI	PSI	Pasni Airport, Pakistan
OIKR	RJN	Rafsanjan Airport, Iran	OPPS	PEW	Peshawar Intl, Pakistan
OIKY	SYJ	Sirjan Airport, Iran	OPQT	UET	Quetta Intl, Pakistan
OIMB	XBJ	Birjand Airport, Iran	OPRK	RYK	Shaikh Zayed Intl, Pakistan
OIMC	CKT	Sarakhs Airport, Iran	OPRN	ISB	Islamabad Intl, Pakistan
OIMM	MHD	Mashhad Intl, Iran	OPRT	RAZ	Rawalakot Airport, Pakistan
OIMN	BJB	Bojnord Airport, Iran	OPSB	SBQ	Sibi Airport, Pakistan
OIMT	TCX	Tabas Airport, Iran	OPSD	KDU	Skardu Airport, Pakistan
OINR	RZR	Ramsar Airport, Iran	OPSK	SKZ	Sukkur Airport, Pakistan
OINZ	SRY	Dasht, Iran	OPSN	SYW	Sehwan Sharif Airport, Pakistan
OISS	SYZ	Shiraz Intl, Iran	OPSS	SDT	Saidu Sharif Airport, Pakistan
OITL	ADU	Ardabil Airport, Iran	OPSU	SUL	Sui Airport, Pakistan

Appendix E

ICAO	IATA	Airport & Country	ICAO	IATA	Airport & Country
OPTA	TLB	Tarbela Dam Airport, Pakistan	PHOG	OGG	Kahului Airport, USA
OPTU	TUK	Turbat Airport, Pakistan	PHTO	ITO	Hilo Intl, USA
OPZB	PZH	Zhob Airport, Pakistan	PKMA	ENT	Enewetak Aux Af, Marshall Islands
OSAP	ALP	Aleppo Intl, Syria	PKMJ	MAJ	Marshall Is Intl, Marshall Islands
OSDI	DAM	Damascus Intl, Syria	PKWA	KWA	Bucholz Aa, Marshall Islands
OSDZ	DEZ	Deir Zzor, Syria	PLCH	CXI	Cassidy Intl, Kiribati
OSKL	KAC	Kamishly, Syria	PTKK	TKK	Chuuk Intl, Micronesia
OSLK	LTK	Bassel Al, Syria	PTPN	PNI	Pohnpei Int, Micronesia
OSPR	PMS	Palmyra, Syria	PTRO	ROR	Babelthuap/Koror, Micronesia
OTBD	DOH	Doha Intl, Qatar	PTRO	ROR	Koror, Palau
OYAA	ADE	Aden Intl, Yemen	PTSA	KSA	Kosrae, Micronesia
OYAB	EAB	Abbs, Yemen	PTYA	YAP	Yap Intl, Micronesia
OYAT	AXK	Ataq, Yemen	RCKH	KHH	Kaohsiung Intl, Taiwan
OYBN	BHN	Beihan, Yemen	RCSS	TSA	Taipei Songshan Airport, Taiwan
OYBQ	BUK	Al, Yemen	RCTP	TPE	Chiang Kai, Taiwan
OYGD	AAY	Al, Yemen	RJAA	NRT	Narita Intl, Japan
OYHD	HOD	Hodeidah Intl, Yemen	RJAF	MMJ	Matsumoto Airport, Japan
OYMB	MYN	Marib, Yemen	RJBB	KIX	Kansai Intl, Japan
OYMK	UKR	Mukeiras, Yemen	RJBD	SHM	Nanki, Japan
OYQN	IHN	Qishn, Yemen	RJBE	UKB	Kobe Airport, Japan
OYRN	RIY	Riyan, Yemen	RJCB	OBO	Obihiro Airport, Japan
OYSH	SYE	Saadah, Yemen	RJCC	CTS	New Chitose Airport, Japan
OYSN	SAH	Sanaa Intl, Yemen	RJCH	HKD	Hakodate Airport, Japan
OYSQ	SCT	Moori, Yemen	RJCK	KUH	Kushiro Airport, Japan
OYSY	GXF	Sayun, Yemen	RJCM	MMB	Memanbetsu Airport, Japan
OYTZ	TAI	Ganed, Yemen	RJCN	SHB	Nakashibetsu Airport, Japan
PABE	BET	Bethel Airport, USA	RJCR	RBJ	Rebun Airport, Japan
PABR	BRW	Wiley Post, USA	RJCW	WKJ	Wakkanai Airport, Japan
PACV	CDV	Merle K. Mudhole Smith Airport, USA	RJDB	IKI	Iki Airport, Japan
PADL	DLG	Dillingham Airport, USA	RJDC	UBJ	Yamaguchi Ube Airport, Japan
PADQ	ADQ	Kodiak Airport, USA	RJDT	TSJ	Tsushima Airport, Japan
PADU	DUT	Unalaska Airport, USA	RJEB	MBE	Monbetsu Airport, Japan
PAEN	ENA	Kenai Municipal Airport, USA	RJEC	AKJ	Asahikawa Airport, Japan
PAFA	FAI	Fairbanks Intl, USA	RJEO	OIR	Okushiri Airport, Japan
PAGA	GAL	Edward G. Pitka Sr. Airport, USA	RJER	RIS	Rishiri Airport, Japan
PAHO	HOM	Homer Airport, USA	RJFC	KUM	Yakushima Airport, Japan
PAJN	JNU	Juneau Intl, USA	RJFE	FUJ	Goto, Japan
PAKN	AKN	King Salmon Airport, USA	RJFF	FUK	Fukuoka Airport, Japan
PAKT	KTN	Ketchikan Intl, USA	RJFG	TNE	Tanegashima Airport, Japan
PALH	LHD	Lake Hood Seaplane Base, USA	RJFK	KOJ	Kagoshima Airport, Japan
PAMR	MRI	Merrill Field, USA	RJFM	KMI	Miyazaki Airport, Japan
PANC	ANC	Ted Stevens Anchorage Intl, USA	RJFO	OIT	Oita Airport, Japan
PANI	ANI	Aniak Airport, USA	RJFR	KKJ	New Kitakyushu Airport, Japan
PAOM	OME	Nome Airport, USA	RJFS	HSG	Saga Airport, Japan
PAOT	OTZ	Ralph Wien Memorial Airport, USA	RJFT	KMJ	Kumamoto Airport, Japan
PAPG	PSG	Petersburg James A. Johnson, USA	RJFU	NGS	Nagasaki Airport, Japan
PASC	SCC	Deadhorse Airport, USA	RJGG	NGO	Chubu Intl, Japan
PASI	SIT	Sitka Rocky Gutierrez Airport, USA	RJKA	ASJ	Amami Airport, Japan
PAVD	VDZ	Valdez Airport, USA	RJKB	OKE	Okinoerabu Airport, Japan
PAYA	YAK	Yakutat Airport, USA	RJKI	KKX	Kikai Airport, Japan
PCIS	CIS	Canton Afld, Kiribati	RJKN	TKN	Tokunoshima Airport, Japan
PGRO	ROP	Rota Island, Northern Mariana Islands	RJNA	NKM	Nagoya Airport, Japan
PGSN	SPN	Obyan Intl, Northern Mariana Islands	RJNF	FKJ	Fukui Airport, Japan
PGUM	GUM	Guam Intl, Guam	RJNO	OKI	Oki Airport, Japan
PHHN	HNM	Hana Airport, USA	RJNT	TOY	Toyama Airport, Japan
PHKO	KOA	Kona Intl at Keahole, USA	RJNW	NTQ	Noto Airport, Japan
PHLI	LIH	Lihue Airport, USA	RJOA	HIJ	Hiroshima Airport, Japan
PHMK	MKK	Molokai Airport, USA	RJOB	OKJ	Okayama Airport, Japan
PHNL	HNL	Honolulu Intl, USA	RJOC	IZO	Izumo Airport, Japan
PHNY	LNY	Lanai Airport, USA	RJOK	KCZ	Kochi Airport, Japan

ICAO	IATA	Airport & Country	ICAO	IATA	Airport & Country
RJOM	MYJ	Matsuyama Airport, Japan	RPMC	CBO	Awang Airport, Philippines
RJOO	ITM	Osaka Intl , Japan	RPMD	DVO	Francisco Bangoy Intl, Philippines
RJOR	TTJ	Tottori Airport, Japan	RPME	BXU	Bancasi Airport, Philippines
RJOT	TAK	Takamatsu Airport, Japan	RPMF	BPH	Bislig Airport, Philippines
RJOW	IWJ	Iwami Airport, Japan	RPMG	DPL	Dipolog Airport, Philippines
RJSA	AOJ	Aomori Airport, Japan	RPMH	CGM	Camiguin Airport, Philippines
RJSC	GAJ	Yamagata Airport, Japan	RPMI	IGN	Maria Cristina Airport, Philippines
RJSD	SDS	Sado Airport, Japan	RPMJ	JOL	Jolo Airport, Philippines
RJSF	FKS	Fukushima Airport, Japan	RPML	CGY	Lumbia Airport, Philippines
RJSI	HNA	Hanamaki Airport, Japan	RPMM	MLP	Malabang Airport, Philippines
RJSK	AXT	Akita Airport, Japan	RPMN	SGS	Sanga, Philippines
RJSN	KIJ	Niigata Airport, Japan	RPMO	OZC	Labo Airport, Philippines
RJSR	ONJ	Odate, Japan	RPMP	PAG	Pagadian Airport, Philippines
RJSS	SDJ	Sendai Airport, Japan	RPMQ	MXI	Imelda R. Marcos Airport, Philippines
RJSY	SYO	Shonai Airport, Japan	RPMR	GES	General Santos Intl, Philippines
RJTH	HAC	Hachijojima Airport, Japan	RPMS	SUG	Surigao Airport, Philippines
RJTO	OIM	Oshima Airport, Japan	RPMU	CDY	Cagayan de Sulu Airport, Philippines
RJTQ	MYE	Miyakejima Airport, Japan	RPMV	IPE	Ipil Airport, Philippines
RJTT	HND	Tokyo Intl , Japan	RPMW	TDG	Tandag Airport, Philippines
RKJJ	KWJ	Kwangju Aero, South Korea	RPMX	LLY	Liloy Airport, Philippines
RKJK	KUV	Kunsan Ab, South Korea	RPMZ	ZAM	Zamboanga Intl, Philippines
RKJM	MPK	Mokpo, South Korea	RPNS	SIA	Siargao Airport, Philippines
RKJY	RSU	Yeosu, South Korea	RPUB	BAG	Loakan Airport, Philippines
RKND	SHO	Sokcho Aero, South Korea	RPUD	DTE	Daet Airport, Philippines
RKNN	KAG	Kangnung Aero, South Korea	RPUH	SJI	McGuire Field, Philippines
RKNW	WJU	Wonju, South Korea	RPUK	CPP	Calapan Airport, Philippines
RKPC	CJU	Cheju Intl, South Korea	RPUM	MBO	Mamburao Airport, Philippines
RKPE	CHF	R, South Korea	RPUN	WNP	Naga Airport, Philippines
RKPK	PUS	Kimhae Intl, South Korea	RPUO	BSO	Basco Airport, Philippines
RKPU	USN	Ulsan, South Korea	RPUQ	VGN	Mindoro Airport, Philippines
RKSM	SSN	Seoul Ab, South Korea	RPUR	BQA	Baler Airport, Philippines
RKSO	OSN	Osan Ab, South Korea	RPUS	SFE	San Fernando Airport, Philippines
RKSS	SEL	Kimpo Intl, South Korea	RPUT	TUG	Tuguegarao Airport, Philippines
RKSW	SWU	Suwon, South Korea	RPUV	VRC	Virac Airport, Philippines
RKTH	KPO	Pohang Aero, South Korea	RPUW	MRQ	Marinduque Airport, Philippines
RKTN	TAE	Taegu Aero, South Korea	RPUY	CYZ	Cauayan Airport, Philippines
RKTU	CJJ	Chongju Intl, South Korea	RPUZ	BGN	Bagabag Airport, Philippines
RKTY	YEC	Yechon Aero, South Korea	RPVA	TAC	Daniel Z. Romualdez, Philippines
ROAH	OKA	Naha Airport, Japan	RPVB	BCD	Bacolod City Domestic, Philippines
ROIG	ISG	Ishigaki Airport, Japan	RPVC	CYP	Calbayog Airport, Philippines
ROKJ	UEO	Kumejima Airport, Japan	RPVD	DGT	Sibulan Airport, Philippines
ROKR	KJP	Kerama Airport, Japan	RPVE	MPH	Godofredo P. Ramos, Philippines
ROMD	MMD	Minami, Japan	RPVF	CRM	Catarman National Airport, Philippines
ROMY	MMY	Miyako Airport, Japan	RPVG	SAA	Guiuan Airport, Philippines
RORA	AGJ	Aguni Airport, Japan	RPVH	HIL	Hilongos Airport, Philippines
RORE	IEJ	Iejima Airport, Japan	RPVI	ILO	Mandurriao Airport, Philippines
RORH	HTR	Hateruma Airport, Japan	RPVJ	MBT	Masbate Airport, Philippines
RORK	KTD	Kita, Japan	RPVK	KLO	Kalibo Airport, Philippines
RORS	SHI	Shimojishima Airport, Japan	RPVM	CEB	Mactan, Philippines
RORT	TRA	Tarama Airport, Japan	RPVO	OMC	Ormoc Airport, Philippines
RORY	RNJ	Yoron Airport, Japan	RPVP	PPS	Puerto Princesa Airport, Philippines
ROYN	OGN	Yonaguni Airport, Japan	RPVR	RXS	Roxas Airport, Philippines
RPLB	SFS	Subic Bay Intl, Philippines	RPVS	EUQ	Evelio Javier Airport, Philippines
RPLC	CRK	Diosdado Macapagal Intl, Philippines	RPVT	TAG	Tagbilaran Airport, Philippines
RPLI	LAO	Laoag Intl, Philippines	RPVU	TBH	Tugdan Airport, Philippines
RPLL	MNL	Ninoy Aquino Intl, Philippines	RPVV	USU	Busuanga Airport, Philippines
RPLO	CYU	Cuyo Airport, Philippines	SAAP	PRA	General Urquiza Airport, Argentina
RPLP	LGP	Legazpi Airport, Philippines	SAAR	ROS	Rosario Intl, Argentina
RPLU	LBX	Lubang Airport, Philippines	SABE	AEP	Aeroparque Jorge Newbery, Argentina
RPMA	AAV	Allah Valley Airport, Philippines	SACO	COR	Ingeniero Ambrosio Intl, Argentina

Appendix E

ICAO	IATA	Airport & Country	ICAO	IATA	Airport & Country
SADF	FDO	San Fernando Airport, Argentina	SBMQ	MCP	Macapa Intl, Brazil
SAEZ	EZE	Ezeiza Intl, Argentina	SBNF	NVT	Ministro Victor Konder Intl, Brazil
SAME	MDZ	El Plumerillo Intl, Argentina	SBNT	NAT	Augusto Severo Intl, Brazil
SAMM	LGS	Comodoro Ricardo, Argentina	SBPA	POA	Salgado Filho Intl, Brazil
SAMR	AFA	Suboficial Aytes Germano, Argentina	SBPJ	PMW	Palmas Airport, Brazil
SANC	CTC	Coronel Felipe Varela, Argentina	SBPP	PMG	Ponta Pora Intl, Brazil
SANE	SDE	Vicecomodoro Aragonez, Argentina	SBPS	BPS	Porto Seguro Airport, Brazil
SANL	IRJ	Capit, Argentina	SBRB	RBR	Presidente Medici Airport, Brazil
SANT	TUC	Benjamin Matienzo Intl, Argentina	SBRF	REC	Guararapes Intl, Brazil
SANU	UAQ	San Juan Airport, Argentina	SBRJ	SDU	Santos Dumont Regional, Brazil
SAOC	RCU	Las Higueras Airport, Argentina	SBRP	RAO	Leite Lopes Airport, Brazil
SAOR	VME	Villa Reynolds Airport, Argentina	SBSL	SLZ	Marechal Cunha Machado Intl, Brazil
SAOU	LUQ	San Luis Airport, Argentina	SBSN	STM	Santarem Airport, Brazil
SARE	RES	Resistencia Intl, Argentina	SBSP	CGH	Congonhas Intl, Brazil
SARF	FMA	Formosa Intl, Argentina	SBSR	SJP	Sao Jose do Rio Preto Airport, Brazil
SARI	IGR	Cataratas del Iguazu Intl, Argentina	SBSV	SSA	Dep Luis Eduardo Magalhaes, Brazil
SARP	PSS	Gen Jose de San Martin, Argentina	SBTE	THE	Senador Petronio Portella, Brazil
SASA	SLA	El Aybal Airport, Argentina	SBUL	UDI	Uberlandia Airport, Brazil
SATR	RCQ	Daniel Jurkic Airport, Argentina	SBVT	VIX	Goiabeiras Airport, Brazil
SAVC	CRD	General Enrique Mosconi, Argentina	SCAC	ZUD	Pupelde, Chile
SAVE	EQS	Esquel Airport, Argentina	SCAN	LOB	San Rafael, Chile
SAVR	ARR	Alto Rio Senguer Airport, Argentina	SCAR	ARI	Chacalluta Intl, Chile
SAVV	VDM	Gobernador Castello, Argentina	SCAS	WPA	Cabo Juan Roman, Chile
SAVY	PMY	El Tehuelche Airport, Argentina	SCBA	BBA	Balmaceda, Chile
SAWE	RGA	Hermes Quijada Airport, Argentina	SCBE	TOQ	Barriles, Chile
SAWG	RGL	Piloto Civil N. Fernandez, Argentina	SCCC	CCH	Chile Chico, Chile
SAWH	USH	Ushuaia Intl, Argentina	SCCF	CJC	El Loa, Chile
SAZB	BHI	Comandante Espora, Argentina	SCCH	YAI	Gen. Bernardo O'higgins, Chile
SAZG	GPO	General Pico Airport, Argentina	SCCI	PUQ	Carlos Ibanez Del Campo Intl, Chile
SAZM	MDQ	Mar Del Plata Intl, Argentina	SCCY	GXQ	Teniente Vidal, Chile
SAZR	RSA	Santa Rosa Airport, Argentina	SCDA	IQQ	Diego Aracena Intl, Chile
SAZS	BRC	San Carlos de Bariloche Intl, Argentina	SCEL	SCL	Arturo Merino Benitez Intl, Chile
SBAR	AJU	Santa Maria Airport, Brazil	SCES	ESR	El Salvador Bajo, Chile
SBBE	BEL	Val de C, Brazil	SCFA	ANF	Cerro Moreno Intl, Chile
SBBH	PLU	Pampulha Domestic Airport, Brazil	SCFM	WPR	Capitan Fuentes Martinez, Chile
SBBR	BSB	Pres Juscelino Kubitschek, Brazil	SCFT	FFU	Futaleufu, Chile
SBBV	BVB	Boa Vista Intl, Brazil	SCGE	LSQ	Maria Dolores, Chile
SBCF	CNF	Tancredo Neves Intl, Brazil	SCGZ	WPU	Guardiamarina Zanartu, Chile
SBCG	CGR	Campo Grande Intl, Brazil	SCHA	CPO	Chamonate, Chile
SBCR	CMG	Corumb, Brazil	SCHR	LGR	Cochrane, Chile
SBCT	CWB	Afonso Pena Intl, Brazil	SCIE	CCP	Carriel Sur Intl, Chile
SBCY	CGB	Marechal Rondon Airport, Brazil	SCJO	ZOS	Canal Bajo/Carlos H Siebert, Chile
SBCZ	CZS	Cruzeiro do Sul Intl, Brazil	SCLL	VLR	Vallenar, Chile
SBEG	MAO	Eduardo Gomes Intl, Brazil	SCLN	ZLR	Linares, Chile
SBFI	IGU	Foz do Iguacu Intl, Brazil	SCNT	PNT	Teniente Julio Gallardo, Chile
SBFL	FLN	Herc, Brazil	SCPC	ZPC	Pucon, Chile
SBFZ	FOR	Pinto Martins Intl, Brazil	SCRA	CNR	Chanaral, Chile
SBGL	GIG	Galeao, Brazil	SCRG	QRC	De La Independencia, Chile
SBGO	GYN	Santa Genoveva Airport, Brazil	SCSB	SMB	Franco Bianco, Chile
SBGR	GRU	Guarulhos Intl, Brazil	SCSE	LSC	La Florida, Chile
SBIL	IOS	Jorge Amado Airport, Brazil	SCSF	SSD	Victor Lafon, Chile
SBIZ	IMP	Prefeito Renato Moreira Airport, Brazil	SCTC	ZCO	Maquehue, Chile
SBJP	JPA	Presidente Castro Pinto Intl, Brazil	SCTE	PMC	El Tepual Intl, Chile
SBJV	JOI	Joinville, Brazil	SCTI	ULC	Los Cerrillos, Chile
SBKP	VCP	Viracopos Intl, Brazil	SCTL	TLX	Panguilemo, Chile
SBLO	LDB	Londrina Airport, Brazil	SCTN	WCH	Chaiten, Chile
SBMA	MAB	Maraba Airport, Brazil	SCTO	ZIC	Victoria, Chile
SBMG	MGF	Maringa Domestic Airport, Brazil	SCTT	TTC	Las Breas, Chile
SBMO	MCZ	Zumbi dos Palmares Airport, Brazil	SCVD	ZAL	Pichoy, Chile
			SCVM	KNA	Vina Del Mar, Chile

ICAO	IATA	Airport & Country	ICAO	IATA	Airport & Country
SEAM	ATF	Chachoan, Ecuador	SKMD	EOH	Olaya Herrera, Colombia
SECO	OCC	Francisco De Orellana, Ecuador	SKMF	MFS	Miraflores, Colombia
SECU	CUE	Mariscal Lamar, Ecuador	SKMG	MGN	Baracoa, Colombia
SEGS	GPS	Seymour, Ecuador	SKMP	MMP	San Bernardo, Colombia
SEGS	GPS	Seymour Airport, Ecuador	SKMR	MTR	Los Garzones, Colombia
SEGU	GYE	Simon Bolivar Intl, Ecuador	SKMU	MVP	Fabio A Leon Bentley, Colombia
SELA	LGQ	Lago Agrio, Ecuador	SKNV	NVA	Benito Salas, Colombia
SEMA	MRR	J M Velasco Ibarra, Ecuador	SKOC	OCV	Aguas Claras, Colombia
SEMC	XMS	Macas, Ecuador	SKOT	OTU	Otu, Colombia
SEMH	MCH	General Serrano, Ecuador	SKPC	PCR	Puerto Carreno, Colombia
SEMT	MEC	Eloy Alfaro Intl, Ecuador	SKPD	PDA	Obando, Colombia
SEPV	PVO	Reales Tamarindos, Ecuador	SKPE	PEI	Matecana, Colombia
SEQU	UIO	Mariscal Sucre Intl, Ecuador	SKPI	PTX	Pitalito, Colombia
SESA	SNC	Gen Ulpiano Paez, Ecuador	SKPP	PPN	Guillermo Leon Valencia, Colombia
SESC	SUQ	Sucua, Ecuador	SKPQ	PAL	German Olano Ab, Colombia
SESM	PTZ	Rio Amazonas, Ecuador	SKPR	PBE	Puerto Berrio, Colombia
SEST	SCY	San Cristobal, Ecuador	SKPS	PSO	Antonio Narino, Colombia
SESV	BHA	Los Perales, Ecuador	SKPV	PVA	Providencia, Colombia
SETH	TSC	Taisha, Ecuador	SKPZ	PZA	Paz De Ariporo, Colombia
SETI	TPN	Tiputini, Ecuador	SKQU	MQU	Mariquita, Colombia
SETN	ESM	General Rivadeneira, Ecuador	SKRG	MDE	Jose Maria Cordova, Colombia
SETR	TPC	Tarapoa, Ecuador	SKRH	RCH	Almirante Padilla, Colombia
SETU	TUA	El Rosal, Ecuador	SKSA	RVE	Los Colonizadores, Colombia
SGAS	ASU	Asuncion Silvio Pettirossi, Paraguay	SKSJ	SJE	Jorge E Gonzalez, Colombia
SGEN	ENO	Quiteria, Paraguay	SKSM	SMR	Simon Bolivar, Colombia
SKAC	ACR	Araracuara, Colombia	SKSO	SOX	Alberto Lleras Camargo, Colombia
SKAO	MCJ	San Jose De Maicao, Colombia	SKSV	SVI	San Vicente Del Caguan, Colombia
SKAR	AXM	El Eden, Colombia	SKTB	TIB	Tibu, Colombia
SKAS	PUU	Tres De Mayo, Colombia	SKTD	TDA	Trinidad, Colombia
SKBC	ELB	El Banco Apt., Colombia	SKTM	TME	Tame, Colombia
SKBG	BGA	Palonegro, Colombia	SKTQ	TQS	Tres Esquinas, Colombia
SKBO	BOG	Eldorado Intl, Colombia	SKTU	TRB	Gonzalo Mejia, Colombia
SKBQ	BAQ	Ernesto Cortissoz, Colombia	SKUC	AUC	Santiago Perez, Colombia
SKBS	BSC	Jose Celestino Mutis, Colombia	SKUI	UIB	El Carano, Colombia
SKBU	BUN	Buenaventura, Colombia	SKUL	ULQ	Farfan, Colombia
SKCC	CUC	Camilo Daza Intl, Colombia	SKVG	VGZ	Villagarzon, Colombia
SKCD	COG	Mandinga, Colombia	SKVP	VUP	Alfonso Lopez, Colombia
SKCG	CTG	Rafael Nunez, Colombia	SKVV	VVC	Vanguardia, Colombia
SKCL	CLO	Alfonso Bonilla Aragon Intl, Colombia	SKYP	EYP	El Yopal, Colombia
SKCM	CIM	Cimitarra, Colombia	SLAG	MHW	Monteagudo, Bolivia
SKCN	RAV	Cravo Norte, Colombia	SLAP	APB	Apolo, Bolivia
SKCO	TCO	La Florida, Colombia	SLBJ	BJO	Bermejo, Bolivia
SKCR	CUO	Caruru, Colombia	SLCA	CAM	Camiri, Bolivia
SKCU	CAQ	Caucasia, Colombia	SLCB	CBB	Jorge Wilsterman, Bolivia
SKCV	CVE	Covenas, Colombia	SLCO	CIJ	Cobija, Bolivia
SKCZ	CZU	Las Brujas, Colombia	SLCP	CEP	Concepcion, Bolivia
SKEB	EBG	El Bagre, Colombia	SLET	SRZ	El Trompillo, Bolivia
SKEJ	EJA	Yariguies, Colombia	SLGY	GYA	Cap Av Emilio Beltran, Bolivia
SKFL	FLA	Gustavo Artunduaga P, Colombia	SLJO	SJB	San Joaquin, Bolivia
SKGI	GIR	Santiago Vila, Colombia	SLJV	SJV	San Javier, Bolivia
SKGO	CRC	Santa Ana, Colombia	SLLP	LPB	El Alto Intl, Bolivia
SKGP	GPI	Guapi, Colombia	SLMG	MGD	Magdalena, Bolivia
SKHA	CPL	Gen Navas Pardo, Colombia	SLOR	ORU	Juan Mendoza, Bolivia
SKHC	HTZ	Hato Corozal, Colombia	SLPS	PSZ	Salvador Ogaya, Bolivia
SKIB	IBE	Perales, Colombia	SLRA	SRD	San Ramon, Bolivia
SKIG	IGO	Chigorodo, Colombia	SLRB	RBO	Robore, Bolivia
SKIP	IPI	San Luis, Colombia	SLRI	RIB	Cap De Av Selin Zeitun Lopez, Bolivia
SKLG	LQM	Caucaya, Colombia	SLRQ	RBQ	Rurrenabaque, Bolivia
SKLP	LPD	La Pedrera, Colombia	SLRY	REY	Reyes, Bolivia
SKLT	LET	Alfredo Vasquez Cobo, Colombia	SLSB	SRJ	Capitan German Quiroga, Bolivia

Appendix E

ICAO	IATA	Airport & Country	ICAO	IATA	Airport & Country
SLSM	SNM	San Ignacio De Moxos, Bolivia	SVAN	AAO	Anaco, Venezuela
SLSU	SRE	Juana Azurduy De Padilla, Bolivia	SVBC	BLA	Jose Antonio Anzoategui, Venezuela
SLTJ	TJA	Capt Oriel Lea Plaza, Bolivia	SVBI	BNS	Barinas, Venezuela
SLTR	TDD	Jorge Henrich Arauz, Bolivia	SVBM	BRM	Barquisimeto, Venezuela
SLVG	VAH	Vallegrande, Bolivia	SVCB	CBL	Ciudad Bolivar, Venezuela
SLVM	VLM	Rafael Pabon, Bolivia	SVCL	CLZ	Calabozo, Venezuela
SLVR	VVI	Viru Viru Intl, Bolivia	SVCN	CAJ	Canaima, Venezuela
SLYA	BYC	Yacuiba, Bolivia	SVCO	VCR	Carora, Venezuela
SMJP	PBM	J.A. Pengel Intl, Suriname	SVCP	CUP	Jose Francisco Bermudez, Venezuela
SMPA	OEM	Vincent Faiks, Suriname	SVCR	CZE	Jose L Chirinos Intl, Venezuela
SMZO	ORG	Zorg En Hoop, Suriname	SVCU	CUM	Antonio Jose De Sucre, Venezuela
SOCA	CAY	Cayenne, French Guiana	SVED	EOR	El Dorado, Venezuela
SPAR	ALD	Alerta, Peru	SVEZ	EOZ	Elorza, Venezuela
SPBB	MBP	Moyobamba, Peru	SVGD	GDO	Guasdualito, Venezuela
SPBL	BLP	Huallaga, Peru	SVGI	GUI	Guiria, Venezuela
SPBR	IBP	Iberia, Peru	SVGU	GUQ	Guanare, Venezuela
SPCL	PCL	David Abenzur Rengifo, Peru	SVIC	ICA	Icabaru, Venezuela
SPEO	CHM	Teniente Jaime A De Montreuil, Peru	SVKA	KAV	Bolivar, Venezuela
SPGM	TGI	Tingo Maria, Peru	SVLF	LFR	La Fria, Venezuela
SPHI	CIX	Jose Abelardo, Peru	SVMC	MAR	La Chinita Intl, Venezuela
SPHO	AYP	Col Alfredo Mendivil Duarte, Peru	SVMD	MRD	Alberto Carnevalli, Venezuela
SPHY	ANS	Andahuaylas, Peru	SVMG	PMV	Del Caribe Intl, Venezuela
SPIL	UMI	Quincemil, Peru	SVMI	CCS	Simon Bolivar Intl, Venezuela
SPIM	LIM	Jorge Chavez Intl, Peru	SVMT	MUN	Maturin Intl, Venezuela
SPJA	RIJ	Rioja, Peru	SVPA	PYH	Puerto Ayacucho, Venezuela
SPJI	JJI	Juanjui, Peru	SVPC	PBL	Gen. Bartolome Salom, Venezuela
SPJJ	JAU	Jauja, Peru	SVPM	SCI	Paramillo, Venezuela
SPJL	JUL	Inca Manco Capac, Peru	SVPR	CGU	Manuel Carlos Piar, Venezuela
SPJN	SJA	San Juan De Marcona, Peru	SVPT	PTM	Palmarito, Venezuela
SPJR	CJA	Armando Revoredo Iglesias, Peru	SVSE	SNV	Santa Elena De Uairen, Venezuela
SPLN	RIM	San Nicolas, Peru	SVSO	STD	Mayor Buenaventura Vivas, Venezuela
SPME	TBP	Pedro Canga, Peru	SVSP	SNF	Sub Teniente Nestor Arias, Venezuela
SPMS	YMS	Moises Benzaquen Rengifo, Peru	SVSR	SFD	San Fernando De Apure, Venezuela
SPNC	HUU	Alferez David Figueroa Fernan, Peru	SVST	SOM	San Tome, Venezuela
SPOV	SYC	Shiringayoc O Hda Mejia, Peru	SVTC	TUV	Tucupita, Venezuela
SPPY	CHH	Chachapoyas, Peru	SVTM	TMO	Tumeremo, Venezuela
SPQT	IQT	Col Francisco Secada Vignetta, Peru	SVUM	URM	Uriman, Venezuela
SPQU	AQP	Rodriguez Ballon, Peru	SVVA	VLN	Arturo Michelena Intl, Venezuela
SPRN	SMG	Santa Maria De Fatima, Peru	SVVG	VIG	Juan Pablo Perez Alfonzo, Venezuela
SPRU	TRU	Cap Carlos Martinez De Pinill, Peru	SVVL	VLV	Dr. Antonio Nicolas Briceno, Venezuela
SPSO	PIO	Pisco, Peru	SVVP	VDP	Valle De La Pascua, Venezuela
SPST	TPP	Cdte.Guillermo D Castillo Par, Peru	SYAH	AHL	Aishalton, Guyana
SPTN	TCQ	Col Carlos Ciriani Santa Rosa, Peru	SYAN	NAI	Annai, Guyana
SPTU	PEM	Padre Aldamiz, Peru	SYCJ	GEO	Cheddi Jagan Intl, Guyana
SPUR	PIU	Capt Guillermo Concha Iberico, Peru	SYIB	IMB	Imbaimadai, Guyana
SPYL	TYL	Capitan Montes, Peru	SYKM	KAR	Kamarang, Guyana
SPZO	CUZ	Velazco Astete, Peru	SYKR	KRM	Karanambo, Guyana
SUAG	ATI	Artigas Intl, Uruguay	SYKT	KTO	Kato, Guyana
SUCA	CYR	Colonia International, Uruguay	SYLP	LUB	Lumid Pau, Guyana
SUDU	DZO	Santa Bernardina Intl, Uruguay	SYLT	LTM	Lethem, Guyana
SULS	PDP	Capitan Curbelo Intl, Uruguay	SYMD	MHA	Mahdia, Guyana
SUMO	MLZ	Cerro Largo Intl, Uruguay	SYOR	ORJ	Orinduik, Guyana
SUMU	MVD	Carrasco Intl, Uruguay	TAPA	ANU	VC Bird Intl, Antigua and Barbuda
SUPU	PDU	Tydeo Larre Borges Intl, Uruguay	TAPH	BBQ	Codrington Airport, Antigua & Barbuda
SURV	RVY	Rivera Intl, Uruguay	TBPB	BGI	Grantley Adams Intl, Barbados
SUSO	STY	Salto Nueva Hesperides Intl, Uruguay	TDCF	DCF	Canefield Airport, Dominica
SUTB	TAW	Tacuarembo, Uruguay	TDPD	DOM	Melville Hall Airport, Dominica
SUTR	TYT	Treinta Y Tres, Uruguay	TFFA	DSD	Grande Anse Airport, Guadeloupe
SUVO	VCH	Vichadero, Uruguay	TFFB	BBR	Baillif Airport, Guadeloupe
SVAC	AGV	Oswaldo Guevara Mujica, Venezuela	TFFC	SFC	Saint, Guadeloupe

ICAO	IATA	Airport & Country	ICAO	IATA	Airport & Country
TFFF	FDF	Le Lamentin, Martinique	UHMA	DYR	Ugolny Airport, Russia
TFFG	SFG	Grand Case Airport, Guadeloupe	UHMD	PVS	Provideniya Bay Airport, Russia
TFFJ	SBH	Saint, Guadeloupe	UHMM	GDX	Sokol Airport, Russia
TFFM	GBJ	Marie Galante Airport, Guadeloupe	UHMP	PWE	Pevek Airport, Russia
TFFR	PTP	Pointe, Guadeloupe	UHPP	PKC	Yelizovo Airport, Russia
TFFS	LSS	Les Saintes Airport, Guadeloupe	UHSS	UUS	Yuzhno, Russia
TGPY	GND	Point Salines Intl, Grenada	UHWW	VVO	Vladivostok Intl, Russia
TIST	STT	King, US Virgin Islands	UIAA	HTA	Kadala Airport, Russia
TISX	STX	Henry E Rohlsen, US Virgin Islands	UIBB	BTK	Bratsk Airport, Russia
TJAB	ABO	Antonio Juarbe Pol, Puerto Rico	UIUU	UUD	Mukhino Airport, Russia
TJBQ	BQN	Rafael Hernandez Airport, Puerto Rico	UKBB	KBP	Boryspil Intl, Ukraine
TJCP	CPX	Benjamin Rivera Noriega, Puerto Rico	UKCC	DOK	Donetsk Airport, Ukraine
TJFA	FAJ	Diego Jimenez Torres, Puerto Rico	UKDD	DNK	Dnipropetrovsk Airport, Ukraine
TJIG	SIG	Isla Grande Airport, Puerto Rico	UKDE	OZH	Zaporizhzhia Intl, Ukraine
TJMZ	MAZ	Eugenio Mar, Puerto Rico	UKDR	KWG	Lozuvatka Intl, Ukraine
TJPS	PSE	Mercedita Airport, Puerto Rico	UKFF	SIP	Simferopol Airport, Ukraine
TJSJ	SJU	Luis Munoz Marin Intl, Puerto Rico	UKHH	HRK	Osnova Airport, Ukraine
TJVQ	VQS	Vieques Airport, Puerto Rico	UKKK	IEV	Zhuliany Intl, Ukraine
TLPC	SLU	George F.L. Charles, Saint Lucia	UKLI	IFO	Ivano, Ukraine
TLPL	UVF	Hewanorra Intl, Saint Lucia	UKLL	LWO	Lviv Airport, Ukraine
TNCA	AUA	Queen Beatrix Intl, Aruba	UKLR	RWN	Rivne Airport, Ukraine
TNCB	BON	Flamingo Intl, Netherlands Antilles	UKLU	UDJ	Uzhhorod Airport, Ukraine
TNCC	CUR	Hato Intl, Netherlands Antilles	UKON	NLV	Mykolaiv Airport, Ukraine
TNCE	EUX	F.D. Roosevelt, Netherlands Antilles	UKOO	ODS	Odessa Airport, Ukraine
TNCM	SXM	Princess Juliana, Netherlands Antilles	UKWW	VIN	Vinnytsia Airport, Ukraine
TNCS	SAB	Juancho, Netherlands Antilles	ULAA	ARH	Talagi Airport, Russia
TQPF	AXA	Anguilla Wallblake, Anguilla	ULAM	NNM	Naryan, Russia
TT02	ULI	Ulithi, Micronesia	ULDD	AMV	Amderma Airport, Russia
TTPP	POS	Piarco Intl, Trinidad & Tobago	ULKK	KSZ	Kotlas Airport, Russia
TUPJ	EIS	Beef I Intl, British Virgin Islands	ULLI	LED	Pulkovo Airport, Russia
TUPW	VIJ	Virgin Gorda, British Virgin Islands	ULMM	MMK	Murmansk Airport, Russia
TVSU	UNI	Union I. Intl, St. Vincent & Grenadines	ULOO	PKV	Pskov Airport, Russia
TVSV	SVD	E T Joshua, St. Vincent & Grenadines	ULPB	PES	Besovets Airport, Russia
TXKF	BDA	Hamilton Bermuda Intl, Bermuda	UMBB	BQT	Brest, Belarus
UAAA	ALA	Almaty, Kazakhstan	UMGG	GME	Gomel, Belarus
UAAH	BXH	Balkhash, Kazakhstan	UMII	VTB	Vitebsk, Belarus
UACC	TSE	Astana, Kazakhstan	UMKK	KGD	Khrabrovo Airport, Russia
UAFM	FRU	Manas, Kyrgyzstan	UMMG	GNA	Grodno, Belarus
UAFO	OSS	Osh, Kyrgyzstan	UMMM	MHP	Minsk, Belarus
UAKD	DZN	Zhezkazgan, Kazakhstan	UMMS	MSQ	Minsk, Belarus
UAKK	KGF	Karaganda, Kazakhstan	UNBB	BAX	Barnaul Airport, Russia
UARR	URA	Uralsk, Kazakhstan	UNEE	KEJ	Kemorovo Airport, Russia
UATG	GUW	Atyrau, Kazakhstan	UNKL	KJA	Krasnoyarsk Yemelyanovo, Russia
UATT	AKX	Aktyubinsk, Kazakhstan	UNKY	KYZ	Kyzyl Airport, Russia
UAUU	KSN	Narimanovka, Kazakhstan	UNNT	OVB	Novosibirsk Tolmachevo, Russia
UBBB	BAK	Bina, Azerbaijan	UNOO	OMS	Tsentralny Airport, Russia
UBBG	KVD	Gyandzha, Azerbaijan	UNWW	NOZ	Novokuznetsk Spichenkovo, Russia
UEAA	ADH	Aldan Airport, Russia	UOOO	NSK	Norilsk Airport, Russia
UEEE	YKS	Yakutsk Airport, Russia	URKA	AAQ	Vityazevo Airport, Russia
UELL	CNN	Chulman Airport, Russia	URKK	KRR	Pashkovsky Airport, Russia
UERP	PYJ	Polyarny Airport, Russia	URMM	MRV	Mineralnye Vody Airport, Russia
UERR	MJZ	Mirny Airport, Russia	URMN	NAL	Nalchik Airport, Russia
UESO	CKH	Chokurdakh Airport, Russia	URMO	OGZ	Beslan Airport, Russia
UGEE	EVN	Zvartnots, Armenia	URMT	STW	Shpakovskoye Airport, Russia
UGGG	TBS	Lochini, Georgia	URRR	ROV	Rostov, Russia
UGGG	TBS	Tbilisi Intl, Georgia	URWA	ASF	Astrakhan Airport, Russia
UGSS	SUI	Babushara Airport, Georgia	URWW	VOG	Gumrak Airport, Russia
UHBB	BQS	Ignatyevo Airport, Russia	USCC	CEK	Balandino Airport, Russia
UHBI	GDG	Magdagachi Airport, Russia	USCM	MQF	Magnitogorsk Airport, Russia
UHHH	KHV	Novy Airport, Russia	USKK	KVX	Kirov Airport, Russia

Appendix E

ICAO	IATA	Airport & Country	ICAO	IATA	Airport & Country
USNN	NJC	Nizhnevartovsk Airport, Russia	VCCA	ADP	Anuradhapura, Sri Lanka
USPP	PEE	Bolshoye Savino Airport, Russia	VCCB	BTC	Batticaloa AB, Sri Lanka
USRR	SGC	Surgut Airport, Russia	VCCC	RML	Ratmalana, Sri Lanka
USSS	SVX	Koltsovo Intl, Russia	VCCG	GOY	Amparai, Sri Lanka
UTAA	ASB	Ashgabat, Turkmenistan	VCCJ	JAF	Kankesanturai AB, Sri Lanka
UTDD	DYU	Dushanbe, Tajikistan	VCCT	TRR	China Bay AB, Sri Lanka
UTDL	LBD	Khudzhand, Tajikistan	VDBG	BBM	Battambang, Cambodia
UTNU	UGC	Urgench, Uzbekistan	VDPP	PNH	Pochentong Intl, Cambodia
UTSB	BHK	Bukhara, Uzbekistan	VDSR	REP	Siem Reap, Cambodia
UTSS	SKD	Samarkand, Uzbekistan	VE46	JGB	Jagdalpur Airport, India
UTST	TMJ	Termez, Uzbekistan	VEAN	IXV	Along Airport, India
UTTT	TAS	Tashkent Yuzhny, Uzbekistan	VEAT	IXA	Agartala Airport, India
UUDD	DME	Domodedovo Intl, Russia	VEBD	IXB	Bagdogra Airport, India
UUEE	SVO	Sheremetyevo Intl, Russia	VEBG	RGH	Balurghat Airport, India
UUII	IKT	Irkutsk Intl, Russia	VEBI	SHL	Shillong Airport, India
UUOB	EGO	Belgorod Airport, Russia	VEBS	BBI	Biju Patnaik Airport, India
UUOO	VOZ	Chertovitskoye Airport, Russia	VEC1	DIB	Chabua Airport, India
UUWW	VKO	Vnukovo Airport, Russia	VECC	CCU	Netaji Subhash Airport Intl, India
UUYH	UCT	Ukhta Airport, Russia	VECO	COH	Cooch Behar Airport, India
UUYY	SCW	Syktyvkar Airport, Russia	VEGT	GAU	Guwahati Intl, India
UWGG	GOJ	Strigino Airport, Russia	VEGY	GAY	Gaya Airport, India
UWKD	KZN	Kazan Airport, Russia	VEIM	IMF	Imphal Airport, India
UWKS	CSY	Cheboksary Airport, Russia	VEJS	IXW	Jamshedpur Airport, India
UWOO	REN	Tsentralny Airport, Russia	VEJT	JRH	Jorhat Airport, India
UWOR	OSW	Orsk Airport, Russia	VEKM	IXQ	Kamalpur Airport, India
UWPP	PEZ	Penza Airport, Russia	VEKR	IXH	Kailashahar Airport, India
UWSS	RTW	Tsentralny Airport, Russia	VEKU	IXS	Silchar Airport, India
UWUU	UFA	Ufa Airport, Russia	VEKW	IXN	Khowai Airport, India
UWWW	KUF	Kurumoch Airport, Russia	VELR	IXI	Lilabari Airport, India
VA1A	HBX	Hubli Airport, India	VEMH	LDA	Malda Airport, India
VA1P	DIU	Diu Airport, India	VEMR	DMU	Dimapur Airport, India
VAAH	AMD	Sardar Vallabhbhai Patel Intl, India	VEMZ	MZU	Muzzafarpur Airport, India
VAAK	AKD	Akola Airport, India	VEPG	IXT	Pasighat Airport, India
VAAU	IXU	Aurangabad Airport, India	VEPT	PAT	Patna Airport, India
VABB	BOM	Chhatrapati Shivaji Intl, India	VEPU	PUI	Purnea Airport, India
VABI	PAB	Bilaspur Airport, India	VERC	IXR	Birsa Munda Airport, India
VABJ	BHJ	Bhuj Rudra Mata Airport, India	VERK	RRK	Rourkela Airport, India
VABM	IXG	Belgaum Airport, India	VERU	RUP	Rupsi Airport, India
VABO	BDQ	Vadodara Airport, India	VETJ	TEI	Tezu Airport, India
VABP	BHO	Bhopal Airport, India	VETZ	TEZ	Tezpur Airport, India
VABV	BHU	Bhavnagar Airport, India	VEVZ	VTZ	Visakhapatnam Airport, India
VADN	NMB	Daman Airport, India	VEZO	ZER	Zero Airport, India
VAGO	GOI	Dabolim Airport, India	VGBR	BZL	Barisal Airport, Bangladesh
VAID	IDR	Devi Ahilyabai Holkar Intl, India	VGCB	CXB	Cox's Bazar Airport, Bangladesh
VAJB	JLR	Jabalpur Airport, India	VGCM	CLA	Comilla Airport, Bangladesh
VAJM	JGA	Jamnagar Airport, India	VGEG	CGP	Shah Amanat Intl, Bangladesh
VAKE	IXY	Kandla Airport, India	VGJR	JSR	Jessore Airport, Bangladesh
VAKJ	HJR	Khajuraho Airport, India	VGRJ	RJH	Rajshahi Airport, Bangladesh
VAKP	KLH	Kolhapur Airport, India	VGSD	SPD	Saidpur Airport, Bangladesh
VAKS	IXK	Junagadh Airport, India	VGSG	TKR	Thakuragaon Airport, Bangladesh
VANP	NAG	Dr. Babasaheb Ambedkar Intl, India	VGSY	ZYL	Osmani Intl, Bangladesh
VAPO	PNQ	Pune Airport, India	VGZR	DAC	Zia Intl, Bangladesh
VAPR	PBD	Porbandar Airport, India	VIAG	AGR	Agra Airport, India
VARG	RTC	Ratnagiri Airport, India	VIAR	ATQ	Raja Sansi Intl, India
VARK	RAJ	Rajkot Airport, India	VIBN	VNS	Varanasi Airport, India
VARP	RPR	Raipur Airport, India	VIBR	KUU	Bhuntar Airport, India
VASL	SSE	Sholapur Airport, India	VICG	IXC	Chandigarh Airport, India
VASU	STV	Surat Airport, India	VIDN	DED	Jolly Grant Airport, India
VAUD	UDR	Udaipur Airport, India	VIDP	DEL	Indira Gandhi Intl, India
VCBI	CMB	Bandaranaike Intl, Sri Lanka	VIGG	DHM	Gaggal Airport, India

ICAO	IATA	Airport & Country	ICAO	IATA	Airport & Country
VIGR	GWL	Gwalior Airport, India	VOBI	BEP	Bellary Airport, India
VIJO	JDH	Jodhpur Airport, India	VOBZ	VGA	Vijayawada Airport, India
VIJP	JAI	Jaipur Airport, India	VOCB	CJB	Coimbatore Airport, India
VIJR	JSA	Jaisalmer Airport, India	VOCL	CCJ	Calicut Intl, India
VIJU	IXJ	Jammu Airport, India	VOCP	CDP	Cuddapah Airport, India
VIKA	KNU	Kanpur Airport, India	VOHY	HYD	Hyderabad Airport, India
VIKO	KTU	Kota Airport, India	VOMD	IXM	Madurai Airport, India
VILD	LUH	Sahnewal Airport, India	VOML	IXE	Mangalore Airport, India
VILK	LKO	Amausi Airport, India	VOMM	MAA	Chennai Intl, India
VIPT	PGH	Pantnagar Airport, India	VOMY	MYQ	Mysore Airport, India
VISM	SLV	Shimla Airport, India	VOPB	IXZ	Port Blair Airport, India
VISR	SXR	Srinagar Airport, India	VOPC	PNY	Pondicherry Airport, India
VIST	TNI	Satna Airport, India	VOPN	BEK	Sri Sathya Sai Airport, India
VLAP	AOU	Attopeu, Laos	VORY	RJA	Rajahmundry Airport, India
VLHS	OUI	Ban Houeisay, Laos	VOSM	SXV	Salem Airport, India
VLLB	LPQ	Luang Prabang, Laos	VOTP	TIR	Tirupati Airport, India
VLLN	LXG	Luang Namtha, Laos	VOTR	TRZ	Tiruchirapalli Airport, India
VLOS	ODY	Oudomsay, Laos	VOTV	TRV	Thiruvananthapuram Intl, India
VLPS	PKZ	Pakse, Laos	VOWA	WGC	Warangal Airport, India
VLSB	ZBY	Sayaboury, Laos	VQPR	PBH	Paro Airport, Bhutan
VLSK	ZVK	Savannakhet, Laos	VRMG	GAN	Gan, Maldives
VLSN	NEU	Sam Neua, Laos	VRMH	HAQ	Hanimaadhoo, Maldives
VLSV	VNA	Saravane, Laos	VRMM	MLE	Male Intl, Maldives
VLTK	THK	Thakhek, Laos	VRMT	KDO	Kaadedhdhoo, Maldives
VLVT	VTE	Wattay Intl, Laos	VTBD	BKK	Bangkok Intl, Thailand
VLXK	XKH	Xieng Khouang, Laos	VTBS	QHI	Sattahip Airport, Thailand
VNBG	BJH	Bajhang, Nepal	VTBU	UTP	U, Thailand
VNBJ	BHP	Bhojpur, Nepal	VTCC	CNX	Chiang Mai Intl, Thailand
VNBL	BGL	Baglung, Nepal	VTCH	HGN	Mae Hong Son Airport, Thailand
VNBP	BHR	Bharatpur, Nepal	VTCL	LPT	Lampang Airport, Thailand
VNBR	BJU	Bajura, Nepal	VTCN	NNT	Nan Airport, Thailand
VNBT	BIT	Baitadi, Nepal	VTCP	PRH	Phrae Airport, Thailand
VNBW	BWA	Bhairahawa, Nepal	VTCR	N/A	Chiang Rai Airport, Thailand
VNDG	DNP	Dang, Nepal	VTCT	CEI	Chiang Rai Intl, Thailand
VNDH	DHI	Dhangadhi, Nepal	VTPH	HHQ	Hua Hin Airport, Thailand
VNDL	DAP	Darchula, Nepal	VTPM	MAQ	Mae Sot Airport, Thailand
VNDP	DOP	Dolpa, Nepal	VTPO	THS	Sukhothai Airport, Thailand
VNGK	GKH	Gorkha, Nepal	VTPP	PHS	Phitsanulok Airport, Thailand
VNJL	JUM	Jumla, Nepal	VTPT	TKT	Tak Airport, Thailand
VNJP	JKR	Janakpur, Nepal	VTPU	UTR	Uttaradit Airport, Thailand
VNJS	JMO	Jomsom, Nepal	VTSB	URT	Surat Thani Airport, Thailand
VNKT	KTM	Tribhuvan Intl, Nepal	VTSC	NAW	Narathiwat Airport, Thailand
VNLD	LDN	Lamidada, Nepal	VTSG	KBV	Krabi Airport, Thailand
VNLK	LUA	Lukla, Nepal	VTSH	SGZ	Songkhla Airport, Thailand
VNMA	NGX	Manang, Nepal	VTSK	PAN	Pattani Airport, Thailand
VNMG	MEY	Meghauli, Nepal	VTSM	USM	Samui Airport, Thailand
VNNG	KEP	Nepalgunj, Nepal	VTSN	NST	Cha Ian Airport, Thailand
VNPK	PKR	Pokhara, Nepal	VTSP	HKT	Phuket Intl, Thailand
VNRB	RJB	Rajbiraj, Nepal	VTSR	UNN	Ranong Airport, Thailand
VNRC	RHP	Ramechhap, Nepal	VTSS	HDY	Hat Yai Intl, Thailand
VNRK	RUK	Rukumkot, Nepal	VTST	TST	Trang Airport, Thailand
VNRT	RUM	Rumjatar, Nepal	VTUD	UTH	Udon Thani Airport, Thailand
VNSI	SIF	Simara, Nepal	VTUD	UTP	Udon Thani Intl, Thailand
VNSK	SKH	Surkhet, Nepal	VTUI	SNO	Sakon Nakhon Airport, Thailand
VNSR	FEB	Sanfebagar, Nepal	VTUK	KKC	Khon Kaen Airport, Thailand
VNST	IMK	Simikot, Nepal	VTUL	LOE	Loei Airport, Thailand
VNTR	TMI	Tumlingtar, Nepal	VTUQ	NAK	Nakhon Ratchasima Airport, Thailand
VNVT	BIR	Biratnagar, Nepal	VTUU	UBP	Ubon Ratchathani Airport, Thailand
VOAT	AGX	Agatti Aerodrome, India	VTUW	KOP	Nakhon Phanom Airport, Thailand
VOBG	BLR	Bangalore Intl, India	VVBM	BMV	Buon Ma Thuot Airport, Vietnam

Appendix E

ICAO	IATA	Airport & Country	ICAO	IATA	Airport & Country
VVCI	HPH	Cat Bi Airport, Vietnam	WAPP	AMQ	Pattimura Airport, Indonesia
VVCM	CAH	Ca Mau Airport, Vietnam	WBGB	BTU	Bintulu, Malaysia
VVCR	CXR	Cam Ranh Airport, Vietnam	WBGG	KCH	Kuching Intl, Malaysia
VVCS	VCS	Co Ong Airport, Vietnam	WBGK	MKM	Mukah, Malaysia
VVCT	VCA	Tra Noc Airport, Vietnam	WBGM	MUR	Marudi, Malaysia
VVDB	DIN	Dien Bien Phu Airport, Vietnam	WBGR	MYY	Miri, Malaysia
VVDL	DLI	Lien Khuong Airport, Vietnam	WBGW	LWY	Lawas, Malaysia
VVDN	DAD	Da Nang Intl, Vietnam	WBGZ	BBN	Bario, Malaysia
VVNB	HAN	Noi Bai Intl, Vietnam	WBKD	LDU	Lahad Datu, Malaysia
VVNS	SQH	Na San Airport, Vietnam	WBKK	BKI	Kota Kinabalu, Malaysia
VVPB	HUI	Phu Bai Airport, Vietnam	WBKL	LBU	Labuan I, Malaysia
VVPC	UIH	Phu Cat Airport, Vietnam	WBKS	SDK	Sandakan, Malaysia
VVPK	PXU	Pleiku Airport, Vietnam	WBKW	TWU	Tawau, Malaysia
VVPQ	PQC	Duong Dong Airport, Vietnam	WBSB	BWN	Brunei Intl, Brunei Darussalam
VVRG	VKG	Rach Gia Airport, Vietnam	WIBB	PKU	Sultan Syarif Kasim II, Indonesia
VVTH	TBB	Dong Tac Airport, Vietnam	WIIB	BDO	Husein Sastranegara, Indonesia
VVTS	SGN	Tan Son Nhat Intl, Vietnam	WIIH	HLP	Halim Perdanakusuma Intl, Indonesia
VVVH	VII	Vinh Airport, Vietnam	WIII	CGK	Soekarno, Indonesia
VYBG	NYU	Bagan, Myanmar	WIIJ	JOG	Adisucipto Intl, Indonesia
VYCZ	MDL	Chanmyathazi, Myanmar	WIIS	SRG	Achmad Yani Airport, Indonesia
VYDW	TVY	Dawei, Myanmar	WIIT	TKG	Radin Inten II Airport, Indonesia
VYGG	GAW	Gantgaw, Myanmar	WIKB	BTH	Hang Nadim Airport, Indonesia
VYGW	GWA	Gwa, Myanmar	WIKN	TNJ	Kijang Airport, Indonesia
VYHH	HEH	Heho, Myanmar	WIMM	MES	Polonia Intl, Indonesia
VYHL	HOX	Hommalin, Myanmar	WIOO	PNK	Supadio Airport, Indonesia
VYKG	KET	Kengtung, Myanmar	WIPP	PLM	Sultan Mahmud Badaruddin Indonesia
VYKI	KHM	Kanti, Myanmar	WIPT	PDG	Minangkabau Intl, Indonesia
VYKP	KYP	Kyaukpyu, Myanmar	WITT	BTJ	Sultan Iskandarmuda, Indonesia
VYKT	KAW	Kawthoung, Myanmar	WMBI	TPG	Taiping, Malaysia
VYLK	LIW	Loikaw, Myanmar	WMKA	AOR	Sultan Abdul Halim, Malaysia
VYLS	LSH	Lashio, Myanmar	WMKB	BWH	Butterworth, Malaysia
VYME	MGZ	Myeik, Myanmar	WMKC	KBR	Sultan Ismail Petra, Malaysia
VYMK	MYT	Pamti, Myanmar	WMKD	KUA	Kuantan, Malaysia
VYMM	MNU	Mawlamyine, Myanmar	WMKE	KTE	Kerteh, Malaysia
VYMN	MGU	Manaung, Myanmar	WMKI	IPH	Sultan Azlan Shah, Malaysia
VYMO	MOE	Momeik, Myanmar	WMKJ	JHB	Sultan Ismail, Malaysia
VYMS	MOG	Mong, Myanmar	WMKK	KUL	Kuala Lumpur Intl, Malaysia
VYMT	MGK	Mong Tong, Myanmar	WMKL	LGK	Langkawi Intl, Malaysia
VYMW	MWQ	Magway, Myanmar	WMKM	MKZ	Malacca, Malaysia
VYNS	NMS	Namsang, Myanmar	WMKN	TGG	Sultan Mahmud, Malaysia
VYPA	PAA	Hpa, Myanmar	WMKP	PEN	Penang Intl, Malaysia
VYPK	PAU	Pauk, Myanmar	WMSA	SZB	Sultan Abdul Aziz Shah, Malaysia
VYPN	BSX	Pathein, Myanmar	WPDL	DIL	Comoro Airport, East Timor
VYPP	PPU	Hpapun, Myanmar	WPMN	MPT	Maliana Airport, East Timor
VYPT	PBU	Putao, Myanmar	WRBB	BDJ	Syamsudin Noor Airport, Indonesia
VYPU	PKK	Pakhokku, Myanmar	WRBP	PKY	Tjilik Riwut Airport, Indonesia
VYPY	PRU	Pyay, Myanmar	WRKK	KOE	El Tari Airport, Indonesia
VYSW	AKY	Sittwe, Myanmar	WRLR	TRK	Juwata Airport, Indonesia
VYTD	SNW	Mazin, Myanmar	WRLS	SRI	Temindung Airport, Indonesia
VYTL	THL	Tachilek, Myanmar	WRRA	AMI	Selaparang Airport, Indonesia
VYYE	XYE	Ye, Myanmar	WRSJ	SUB	Juanda Airport, Indonesia
VYYY	RGN	Yangon Intl, Myanmar	WRSQ	SOC	Adisumarmo Intl, Indonesia
WAAA	UPG	Hasanuddin Airport, Indonesia	WSSS	SIN	Singapore Changi Intl, Singapore
WAAU	KDI	Wolter Monginsidi Airport, Indonesia	YABA	ALH	Albany Airport, Australia
WABB	BIK	Frans Kaisiepo Airport, Indonesia	YAPH	ABH	Alpha Airport, Australia
WADD	DPS	Bali Intl, Indonesia	YARA	ARY	Ararat Airport, Australia
WAJJ	DJJ	Sentani Airport, Indonesia	YARM	ARM	Armidale Airport, Australia
WALL	BPN	Sepinggan Intl, Indonesia	YAVV	AVV	Avalon Airport, Australia
WAML	PLW	Mutiara Airport, Indonesia	YAYE	AYQ	Ayers Rock Airport, Australia
WAMM	MDC	Sam Ratulangi Airport, Indonesia	YBAS	ASP	Alice Springs Airport, Australia

PlanePlotter User Guide

ICAO	IATA	Airport & Country	ICAO	IATA	Airport & Country
YBBN	BNE	Brisbane Airport, Australia	YFTZ	FIZ	Fitzroy Crossing Airport, Australia
YBCG	OOL	Gold Coast Airport, Australia	YGDH	GUH	Gunnedah Airport, Australia
YBCK	BKQ	Blackall Airport, Australia	YGEL	GET	Geraldton Airport, Australia
YBCS	CNS	Cairns Intl, Australia	YGFN	GFN	Grafton Airport, Australia
YBCV	CTL	Charleville Airport, Australia	YGLA	GLT	Gladstone Airport, Australia
YBHI	BHQ	Broken Hill Airport, Australia	YGLB	GUL	Goulburn Airport, Australia
YBHM	HTI	Hamilton Island Airport, Australia	YGLI	GLI	Glen Innes Airport, Australia
YBIE	BEU	Bedourie Airport, Australia	YGPT	GPN	Garden Point Airport, Australia
YBKE	BRK	Bourke Airport, Australia	YGTE	GTE	Groote Eylandt Airport, Australia
YBLA	BLN	Benalla Airport, Australia	YGTH	GFF	Griffith Airport, Australia
YBLN	BQB	Busselton Regional Airport, Australia	YHAY	HXX	Hay Airport, Australia
YBMA	ISA	Mount Isa Airport, Australia	YHBA	HVB	Hervey Bay Airport, Australia
YBMC	MCY	Sunshine Coast Airport, Australia	YHID	HID	Horn Island Airport, Australia
YBMK	MKY	Mackay Airport, Australia	YHLC	HCQ	Halls Creek Airport, Australia
YBNA	BNK	Ballina, Australia	YHML	HLT	Hamilton Airport, Australia
YBNS	BSJ	Bairnsdale Airport, Australia	YHOT	MHU	Mount Hotham Airport, Australia
YBOU	BQL	Boulia Airport, Australia	YHPN	HTU	Hopetoun Airport, Australia
YBPN	PPP	Proserpine, Australia	YHSM	HSM	Horsham Airport, Australia
YBRK	ROK	Rockhampton Airport, Australia	YIVL	IVR	Inverell Airport, Australia
YBRM	BME	Broome Intl, Australia	YKER	KRA	Kerang Airport, Australia
YBRN	BZD	Balranald Airport, Australia	YKII	KNS	King Island Airport, Australia
YBRW	BWQ	Brewarrina Airport, Australia	YKMB	KRB	Karumba Airport, Australia
YBTH	BHS	Bathurst Airport, Australia	YKMP	KPS	Kempsey Airport, Australia
YBTI	BRT	Bathurst Island Airport, Australia	YKOW	KWM	Kowanyama Airport, Australia
YBTL	TSV	Townsville Intl, Australia	YKRY	KGY	Kingaroy Airport, Australia
YBTR	BLT	Blackwater Airport, Australia	YKSC	KGC	Kingscote Airport, Australia
YBUD	BDB	Bundaberg Airport, Australia	YLEC	LGH	Leigh Creek Airport, Australia
YBWP	WEI	Weipa Airport, Australia	YLEO	LNO	Leonora Airport, Australia
YCAR	CVQ	Carnarvon Airport, Australia	YLHI	LDH	Lord Howe Island Airport, Australia
YCBA	CAZ	Cobar Airport, Australia	YLHR	IRG	Lockhart River Airport, Australia
YCBB	COJ	Coonabarabran Airport, Australia	YLIS	LSY	Lismore Airport, Australia
YCBP	CPD	Coober Pedy Airport, Australia	YLRD	LHG	Lightning Ridge Airport, Australia
YCBR	CRB	Collarenebri Airport, Australia	YLRE	LRE	Longreach Airport, Australia
YCCA	CCL	Chinchilla Airport, Australia	YLST	LER	Leinster Airport, Australia
YCCY	CNJ	Cloncurry Airport, Australia	YLTV	LTB	Latrobe Valley Airport, Australia
YCDU	CED	Ceduna Airport, Australia	YMAY	ABX	Albury Airport, Australia
YCEE	CVC	Cleve Airport, Australia	YMBA	MRG	Mareeba Airport, Australia
YCKI	CKI	Croker Island Airport, Australia	YMCO	XMC	Mallacoota Airport, Australia
YCMT	CMQ	Clermont Airport, Australia	YMDG	DGE	Mudgee Airport, Australia
YCNM	CNB	Coonamble Airport, Australia	YMEK	MKR	Meekatharra Airport, Australia
YCOE	CUQ	Coen Airport, Australia	YMEN	MEB	Essendon Airport, Australia
YCOM	OOM	Cooma, Australia	YMER	MIM	Merimbula Airport, Australia
YCOR	CWW	Corowa Airport, Australia	YMGD	MNG	Maningrida Airport, Australia
YCRG	CYG	Corryong Airport, Australia	YMHB	HBA	Hobart Intl, Australia
YCTM	CMD	Cootamundra Airport, Australia	YMIA	MQL	Mildura Airport, Australia
YCUE	CUY	Cue Airport, Australia	YMLT	LST	Launceston Airport, Australia
YCWR	CWT	Cowra Airport, Australia	YMMB	MBW	Moorabbin Airport, Australia
YDBI	DRN	Dirranbandi Airport, Australia	YMML	MEL	Melbourne Intl, Australia
YDBY	DRB	Derby Airport, Australia	YMNE	WME	Mount Keith Airport, Australia
YDLQ	DNO	Deniliquin Airport, Australia	YMOG	MMG	Mount Magnet Airport, Australia
YDPO	DPO	Devonport Airport, Australia	YMOR	MRZ	Moree Airport, Australia
YDYS	DYA	Dysart Airport, Australia	YMRB	MOV	Moranbah Airport, Australia
YECH	ECH	Echuca Airport, Australia	YMRY	MYA	Moruya Airport, Australia
YELD	ELC	Elcho Island Airport, Australia	YMTG	MGB	Mount Gambier Airport, Australia
YEML	EMD	Emerald Airport, Australia	YNAR	NRA	Narrandera Airport, Australia
YESP	EPR	Esperance Airport, Australia	YNBR	NAA	Narrabri Airport, Australia
YFBS	FRB	Forbes Airport, Australia	YNGU	RPM	Ngukurr Airport, Australia
YFLI	FLS	Flinders Island Airport, Australia	YNRM	QRM	Narromine Airport, Australia
YFRT	FOS	Forrest Airport, Australia	YNTN	NTN	Normanton Airport, Australia
YFTZ	FIZ	Fitzroy Crossing Airport, Australia	YNWN	ZNE	Newman Airport, Australia

Appendix E

ICAO	IATA	Airport & Country	ICAO	IATA	Airport & Country
YORB	RBS	Orbost Airport, Australia	YWLM	NTL	Newcastle, Australia
YORG	OAG	Orange Airport, Australia	YWLU	WUN	Wiluna Airport, Australia
YPAD	ADL	Adelaide Intl, Australia	YWOL	WOL	Wollongong Airport, Australia
YPAG	PUG	Port Augusta Airport, Australia	YWRN	QRR	Warren Airport, Australia
YPBO	PBO	Paraburdoo Airport, Australia	YWSL	SXE	West Sale Airport, Australia
YPCC	CCK	Cocos Island Intl, Australia	YWTN	WIN	Winton Airport, Australia
YPDN	DRW	Darwin Intl, Australia	YWWL	WWA	West Wyalong Airport, Australia
YPGV	GOV	Gove Airport, Australia	YWYY	BWT	Burnie Airport, Australia
YPIR	PPI	Port Pirie Airport, Australia	YYNG	NGA	Young Airport, Australia
YPJT	JAD	Jandakot Airport, Australia	ZBAA	PEK	Beijing Capital Intl, China
YPKA	KTA	Karratha Airport, Australia	ZBBB	NAY	Beijing Nanyuan Airport, China
YPKG	KGI	Kalgoorlie, Australia	ZBCF	CIF	Chifeng Airport, China
YPKS	PKE	Parkes Airport, Australia	ZBCZ	CIH	Changzhi Airport, China
YPKT	PKT	Port Keats Airport, Australia	ZBDT	DAT	Datong Airport, China
YPKU	KNX	Kununurra Airport, Australia	ZBHH	HET	Hohhot Baita Intl, China
YPLC	PLO	Port Lincoln Airport, Australia	ZBLA	HLD	Hailar Dongshan Airport, China
YPLM	LEA	Learmonth Airport, Australia	ZBOW	BAV	Baotou Airport, China
YPMQ	PQQ	Port Macquarie Airport, Australia	ZBSH	SHP	Qinhuangdao Shanhaiguan, China
YPOD	PTJ	Portland Airport, Australia	ZBSJ	SJW	Shijiazhuang Daguocun Intl, China
YPPD	PHE	Port Hedland Intl, Australia	ZBTJ	TSN	Tianjin Binhai Intl, China
YPPH	PER	Perth Intl, Australia	ZBTL	TGO	Tongliao Airport, China
YPXM	XCH	Christmas Island Airport, Australia	ZBUL	HLH	Ulanhot Airport, China
YQLP	ULP	Quilpie Airport, Australia	ZBXT	XNT	Xingtai Airport, China
YRED	RCL	Redcliffe Airport, Australia	ZBYN	TYN	Taiyuan Wusu Airport, China
YREN	RMK	Renmark Airport, Australia	ZGBH	BHY	Beihai Airport, China
YROI	RBC	Robinvale Airport, Australia	ZGGG	CAN	Guangzhou Baiyun Intl, China
YROM	RMA	Roma Airport, Australia	ZGHA	CSX	Changsha Huanghua Intl, China
YRTI	RTS	Rottnest Island Airport, Australia	ZGHY	HNY	Hengyang Airport, China
YSBK	BWU	Bankstown Airport, Australia	ZGKL	KWL	Guilin Liangjiang Intl, China
YSCB	CBR	Canberra Intl, Australia	ZGNN	NNG	Nanning Wuxu Intl, China
YSCH	CFS	Coffs Harbour Airport, Australia	ZGOW	SWA	Shantou Airport, China
YSCN	CDU	Camden Airport, Australia	ZGSD	ZUH	Zhuhai Intl, China
YSDU	DBO	Dubbo Airport, Australia	ZGSY	SYX	Sanya Fenghuang Intl, China
YSGE	SGO	Saint George Airport, Australia	ZGSZ	SZX	Shenzhen Bao'an Intl, China
YSHT	SHT	Shepparton Airport, Australia	ZGWZ	WUZ	Wuzhou Changzhoudao Airport, China
YSMI	SIO	Smithton Airport, Australia	ZGZH	LZH	Liuzhou Airport, China
YSNB	SNB	Snake Bay Airport, Australia	ZGZJ	ZHA	Zhanjiang Airport, China
YSNF	NLK	Norfolk Island Airport, Australia	ZHAY	AYN	Anyang Airport, China
YSPT	SHQ	Southport Airport, Australia	ZHCC	CGO	Zhengzhou Xinzheng Intl, China
YSRN	SRN	Strahan Airport, Australia	ZHHH	WUH	Wuhan Tianhe Airport, China
YSSY	SYD	Sydney Intl, Australia	ZHLY	LYA	Luoyang Airport, China
YSTH	HLS	St Helens Airport, Australia	ZHNY	NNY	Nanyang Airport, China
YSTW	TMW	Tamworth Airport, Australia	ZHSS	SHS	Shashi Airport, China
YSWG	WGA	Wagga Wagga Airport, Australia	ZHXF	XFN	Xiangfan Airport, China
YSWH	SWH	Swan Hill Airport, Australia	ZHYC	YIH	Yichang Airport, China
YSWL	SWC	Stawell Airport, Australia	ZJHK	HAK	Haikou Meilan Intl, China
YTAM	XTO	Taroom Airport, Australia	ZKPY	FNJ	Sunan, North Korea
YTEM	TEM	Temora Airport, Australia	ZKPY	FNJ	Sunan, South Korea
YTGM	XTG	Thargomindah Airport, Australia	ZLAN	LHW	Lanzhou Airport, China
YTIB	TYB	Tibooburra Airport, Australia	ZLDH	DNH	Dunhuang Airport, China
YTMU	TUM	Tumut Airport, Australia	ZLGM	GOQ	Golmud Airport, China
YTNG	THG	Thangool Airport, Australia	ZLHZ	HZG	Hanzhong Airport, China
YTNK	TCA	Tennant Creek Airport, Australia	ZLIC	INC	Yinchuan Helanshan Airport, China
YTRE	TRO	Taree Airport, Australia	ZLJN	JNG	Jining Airport, China
YTWB	TWB	Toowoomba Airport, Australia	ZLJQ	CHW	Jiuquan Airport, China
YWDH	WNR	Windorah Airport, Australia	ZLLL	ZGC	Lanzhou Zhongchuan Airport, China
YWGT	WGT	Wangaratta Airport, Australia	ZLQY	IQN	Qingyang Airport, China
YWHA	WYA	Whyalla Airport, Australia	ZLXN	XNN	Xining Airport, China
YWKB	WKB	Warracknabeal Airport, Australia	ZLXY	XIY	Xi'an Xianyang Intl, China
YWLG	WGE	Walgett Airport, Australia	ZLYA	ENY	Yan'an Airport, China

ICAO	IATA	Airport & Country	ICAO	IATA	Airport & Country
ZLYL	UYN	Yulin Airport, China	ZSWY	WUS	Nanping Wuyishan Airport, China
ZMAH	AVK	Arvaikheer, Mongolia	ZSWZ	WNZ	Wenzhou Intl, China
ZMAT	LTI	Altai, Mongolia	ZSXZ	XUZ	Xuzhou Airport, China
ZMBH	BYN	Bayankhongor, Mongolia	ZSYT	YNT	Yantai Laishan Airport, China
ZMBN	UGA	Bulgan, Mongolia	ZSYW	YIW	Yiwu Airport, China
ZMBU	UUN	Baruun, Mongolia	ZSZS	HSN	Zhoushan Airport, China
ZMCD	COQ	Choibalsan, Mongolia	ZUCK	CKG	Chongqing Jiangbei Intl, China
ZMUB	ULN	Buyant, Mongolia	ZUGY	KWE	Guiyang Longdongbao Airport, China
ZPBS	BSD	Baoshan Airport, China	ZULS	LXA	Lhasa Gonggar Airport, China
ZPJH	JHG	Jinghong Gasa Airport, China	ZUNC	NAO	Nanchong Airport, China
ZPLJ	LJG	Lijiang Airport, China	ZUUU	CTU	Chengdu Shuangliu Intl, China
ZPLX	YUM	Luxi Mangshi Airport, China	ZUXC	XIC	Xichang Qingshan Airport, China
ZPPP	KMG	Kunming Wujiaba Intl, China	ZUYB	YBP	Yibin Airport, China
ZPSM	SYM	Simao Airport, China	ZUZY	ZYI	Zunyi Airport, China
ZPZT	ZAT	Zhaotong Airport, China	ZWAK	AKU	Aksu Airport, China
ZSAM	XMN	Xiamen Gaoqi Intl, China	ZWAT	AAT	Altay Airport, China
ZSAQ	AQG	Anqing Airport, China	ZWFY	FYN	Fuyun Airport, China
ZSBB	BFU	Bengbu Airport, China	ZWHM	HMI	Hami Airport, China
ZSCG	CZX	Changzhou Airport, China	ZWKC	KCA	Kuqa Airport, China
ZSCN	KHN	Nanchang Intl, China	ZWKL	KRL	Korla Airport, China
ZSFY	FUG	Fuyang Airport, China	ZWKM	KRY	Karamay Airport, China
ZSFZ	FOC	Fuzhou Chengle Intl, China	ZWSH	KHG	Kashgar Airport, China
ZSGZ	KOW	Ganzhou Airport, China	ZWTN	HTN	Hotan Airport, China
ZSHC	HGH	Hangzhou Xiaoshan Intl, China	ZWWW	URC	Diwopu Airport, China
ZSJD	JDZ	Jingdezhen Airport, China	ZWYN	YIN	Yining Airport, China
ZSJJ	JIU	Jiujiang Lushan Airport, China	ZYAS	AOG	Anshan Airport, China
ZSJN	TNA	Jinan Yaoqiang Airport, China	ZYCC	CGQ	Changchun Longjia Intl, China
ZSLG	LYG	Lianyungang Airport, China	ZYDD	DDG	Dandong Airport, China
ZSLQ	HYN	Huangyan Luqiao Airport, China	ZYHB	HRB	Harbin Taiping Intl, China
ZSNB	NGB	Ningbo Lishe Intl, China	ZYHE	HEK	Heihe Airport, China
ZSNJ	NKG	Nanjing Lukou Intl, China	ZYJL	JIL	Jilin Airport, China
ZSOF	HFE	Hefei Luogang Intl, China	ZYJM	JMU	Jiamusi Airport, China
ZSPD	PVG	Shanghai Pudong Intl, China	ZYJZ	JNZ	Jinzhou Airport, China
ZSQD	TAO	Qingdao Liuting Intl, China	ZYMD	MDG	Mudanjiang Airport, China
ZSQZ	JJN	Quanzhou Jinjiang Airport, China	ZYQQ	NDG	Qiqihar Airport, China
ZSSS	SHA	Shanghai Hongqiao Airport, China	ZYTL	DLC	Dalian Zhoushuizi Intl, China
ZSSZ	SZV	Suzhou Guangfu Airport, China	ZYTN	TNH	Tonghua Liuhe Airport, China
ZSTX	TXN	Huangshan Tunxi Intl, China	ZYTX	SHE	Shenyang Taoxian Intl, China
ZSWF	WEF	Weifang Airport, China	ZYYJ	YNJ	Yanji Airport, China
ZSWX	WUX	Wuxi Airport, China			

Bibliography

Christian Wolff. Radartutorial (2010). Available: http://www.radartutorial.eu. Last accessed July 2010.

Daryl Phillips. (1990). Mode A and Mode C. Available: http://www.airsport-corp.com/modec.htm. Last accessed July 2010.

Dave. (2009). Shipplotter Chart Generator. Available: http://emit.demon.co.uk/map2.php. Last accessed September 2009.

Defense Electronics. (December 2005). Understanding mode S technology. Last accessed July 2010.

ERA Corporation. (2009). Multilateration Executive Reference Guide.

G. E. Rogers Sr. (2009). Design your own J-Pole Antenna. Available: http://www.dxzone.com/cgi-bin/dir/jump2.cgi?ID=9284. Last accessed September 2009.

Greg Goebel. (2009). Secondary Radars. Available: http://www.vectorsite.net/ttradar_6.html#m3. Last accessed July 2010.

Kirt Blattenberger. (1989). IFF - Identification - Friend or Foe. Available: http://www.rfcafe.com/references/electrical/ew-radar-handbook/iff-identification-friend-or-foe.htm. Last accessed July 2010.

MIT Lincoln Laboratory. (2000). ATCRBS Overview. Available: http://web.mit.edu/6.933/www/Fall2000/mode-s/atcrbs.html. Last accessed July 2010.

René Harte. (2009). Icom IC-M80. Available: http://discriminator.nl/ic-m80/index-en.html. Last accessed September 2009.

Satellite Signals. (2009). Latitude and Longitude. Available: http://www.satsig.net/maps/lat-long-finder.htm. Last accessed September 2009.

V.A. Orlando. The Lincoln Laboratory Journal, Volume 2, Number 3 (1989). The Mode S Beacon Radar System. Last accessed July 2010.

Index

Symbols
> 73, 81

A
ABCD buttons 45, 81, 149
About 81
ACARS 1, 2, 3, 11, 12, 14, 15, 168
ACARS Decoder 52
Airborne Collision Avoidance System 11
Aircraft list display options 48
Aircraft symbol 56, 58, 75, 77
Aircraft view options 60
Airmaster 62
AirNav Systems Radar Box 21
AirRep 49, 56
Air Traffic Control 11, 18
Alerts 141
Alert Shell 68
Alert Zone 68, 141
altitude 41, 44, 46, 56, 57, 58, 65, 67
Antenna switching 67
Audio 54
AVI output 64
AVR Receiver 54
Azimuth offset 67

B
Bibliography 207
Blue Spanner 142

C
Calibration 60, 61, 90
Centre chart 58, 77
Chart 41, 44, 47, 48, 49, 50, 51, 55, 59, 66
Chart display options 50
Chart Files folder 23
Chart Generators 87
Chart Options 41, 44, 47, 48, 49, 50, 51, 55
Checking Comms Port 165, 166
Circle sharers 60
CLB files 89
COAA Server 107
Contrails 57, 58, 77
Cosecant Squared 9
course 44, 45, 46, 56, 58
Customize button 74

D
database 47, 48, 50, 51, 52, 53, 66, 111, 113

DDE server 63
Decca Navigator 119
Default Configuration 44, 45
Delete after 56
Denial of Service (DoS) 126
DF 49, 54, 55, 58, 61, 67
DF17 14, 16, 17
DF vector 77
Direction finding 2, 61, 67
Directories 66
DME 16
DPSK 12, 13

E
Electronic Warfare 155
Emitter Category 105
Exit 45

F
Factory default settings 45
FAQs 153
File button 73
File Menu Commands 41
FindFlight 136
Flags 48, 69
Flight 81
FlightDisplay 138
Forum 153
FTP 135

G
Google Earth 2, 133, 134
Google Earth server 63
Google Maps 2, 133, 134
Google Photo 150
GPS 119
GPX overlay 50, 58, 59, 77, 96
Graphical output 64
Ground Stations 122, 123, 124, 130

H
Heavy 105
Help 81
Help Menu 71
HF 19, v
HFDL 1, 3, v
High Vortex 105
Home Altitude 67
Home Latitude & Longitude 67

Home location 49, 67
hyperbola 119, 121
hyperboloid 119, 120, 121, 122, 130
hypersharing 66

I

I/O Settings 78
IATA Codes 173
ICAO Codes 173
IFF 10, 11, 12
Image Files 26
Input data 61
Interested 143
IP address 52, 53, 54, 63

J

JPG output 64

K

Kinetic SBS-1 19, 20, 52, 53, 70
KML files 134

L

l/O settings 61
Labels 56, 75
LAN address 125, 128
Load Restore point 45, 149
Load Settings 44
Local Photo 150
logs 111, 113
Log Alerts 57
Log desig acft 62
Log Files 111, 112, 113
Log Files folder 23
Log Mode-S 62

M

ManTMA 91
MapIt 139
MAP file 89
Master User 122, 123, 124, 127, 128, 130
Match List 146
Memory map 62
Messages 51, 74
Mixer 54, 55
Mlat 48, 50, 51, 61, 119, 122
Mode-S receiver 52
ModeSCountry 103
Mode 1 12

Mode 2 12
Mode 3, 12
Mode A 12, 13
Mode C 12
MoveItFreely 135
Multilateration 119, 122
Multimap 83
MyCircles 139
My Sky 51

N

NAT 126
New Aircraft 144
No-Reg 152
None 144
NoniMapView 87, 88, 89
nose colour 98
Not Interested 144
NTP Utility 157

O

Omit after 56
Open access 52
Open New Chart 41
Open Street Map 2
Options Menu 52
OSM 59, 80, 83, 85, 86, 93
Outline 59
Outline Colours 152
Output data 62

P

PC clock 157
Permanent trails 58, 77
PlaneGadget Radar 20, 53, 70, 163
planesymbol.txt 97
Plane Symbols 97
Plot aircraft 57, 76
port 9742 125
positionless 65
Pos rep 41
PP2GM 134, 135
PPDataGrid 139
preamble 14, 15
Predict positions 58, 77
Process Menu 51

Q

Quick Charts 59, 81, 150

R

Radar 45, 46
RadarBox 52, 66, 69, 70
Radar Bands 155
Radar Cross Section 8
Radar Range Equation 7
Range rings 59, 78
Registering PlanePlotter 71
Registration 28, 29
Replay 70
Reverse mouse wheel 58
Reverse rotation 67
Review Menu 69
Rotorcraft 105
route 46, 56, 65
Router 125, 126
RRE 5
RxControl 21, 54

S

Satellite 59, 80
Satellite button 73
Save Report 44
Save Restore point 149
SBSResources 93
Scale bar 58
Script 69, 81
Secondary sharing server 66
Secondary Surveillance Radar 11
SelCal 2, 3, 46, 55, 61
Setup Comm port 53, 54
Setup Receiver 53
Setup Report 44
Setup Wizard 31, 33
Share 80, 81
Sharers Database 109
Share button 108
Share Code 108, 109
Share identifier 65
Share ID on Labels 109
Sharing 65, 66, 107, 108
ShipPlotter Map Generator 87, 88
Ships 50
Signal 51, 74
silhouette 98
SMS 145
SMS Alerts 57, 60, 76
Source 54, 55

speed 44, 45, 46, 49, 56, 58, 64
SQB database 48, 50, 51, 66
SQL 144
SQLite 100
SQLite Editors 117
squawk code 46, 56, 76
squitter 15, 16, 17, 119, 120
Start 51, 64, 73
Start button 74
Start MLat 130
Status Bar 45
Stop 51, 64, 73, 78
Stop button 74
SymbolMaker2 97
symbols 97

T

TCP/IP 125
TCP/IP server 63
TDOA 119, 120, 121
Test Networking 125, 126, 127, 128, 129
Toolbar 45, 46, 48, 59, 73
Transparency 60
type 56

U

UDP/IP output 63
UHF v
update database 53
Update registrations 52, 53, 100
User Tag 101

V

Vector background 58, 78
Vector button 74
VHF 19, 20, v
View Menu 45
Virtual Radar 1
Virtual Server 125, 126
VOR roses 94

W

Wave button 74
Waypoint 68

Z

ZoneMe 138
Zoom 68, 74

Other Books by The Author

The author has published a number of other books most of which are available from Amazon or Lulu.com.

ShipPlotter - An Illustrated User Guide
 ISBN 978-1445215310

> ShipPlotter is a unique piece of software that enables a user to have a live radar type display of shipping in their local coastal region or other regions and waterways around the world. The software decodes radio signals, received using a VHF radio receiver or scanner, from ships transmitting digital data using the marine Automatic Identification System (AIS). The book provides an excellent description of the AIS system and messaging.

> ShipPlotter visually displays the position and identification of each ship either as radar view or on a chart created from a graphic image file, a satellite image download or a downloaded Open Street Map.

The Early Days of Digital Computing in the British Army
 ISBN 978-0955675362

> The story of the first British Army battlefield digital computer then called FADAC and later to become known as FACE. A fascinating insight into the development of digital computers in the 1960's.

PlanePlotter - An Illustrated User Guide
 ISBN 978-446130636

> PlanePlotter is a unique piece of software receives and decodes live digital position reports from aircraft and plots them on a chart. With PlanePlotter users see a radar-like display of all the aircraft in range of the users radio receiver that are transmitting digital messages in a number of formats such as ACARS, ADS-B and HFDL